Intelligent Image Analysis for Plant Phenotyping

Intelligent Image Analysis for Plant Phenotyping

Edited by
Ashok Samal and Sruti Das Choudhury

CRC Press is an imprint of the
Taylor & Francis Group, an **informa** business

First edition published 2021
by CRC Press
6000 Broken Sound Parkway NW, Suite 300, Boca Raton, FL 33487-2742

and by CRC Press
2 Park Square, Milton Park, Abingdon, Oxon, OX14 4RN

CRC Press is an imprint of Taylor & Francis Group, LLC

Library of Congress Cataloging-in-Publication Data

Names: Samal, Ashok, editor. | Choudhury, Sruti Das, editor.
Title: Intelligent image analysis for plant phenotyping / Ashok Samal, Sruti Das Choudhury.
Description: First edition. | Boca Raton, FL : CRC Press, 2021. | Includes bibliographical references and index. | Summary: "Domesticated crops are the result of artificial selection for particular phenotypes and, in some case, natural selection for an adaptive trait. Intelligent Image Analysis for Plant Phenotyping reviews information on time-saving techniques using computer vision and imaging technologies. These methodologies provide an automated, non-invasive and scalable mechanism to define and collect plant phenotypes. Beautifully illustrated with numerous color images, this book is invaluable for those working in the emerging fields at the intersection of computer vision and plant sciences"– Provided by publisher.
Identifiers: LCCN 2020020511 | ISBN 9781138038554 (hardback) | ISBN 9781315177304 (ebook)
Subjects: LCSH: Image processing–Digital techniques. | Computer vision. | Phenotype.
Classification: LCC TA1637 .I475 2021 | DDC 576.5/30285642–dc23
LC record available at https://lccn.loc.gov/2020020511

ISBN: 9781138038554 (hbk)
ISBN: 9781315177304 (ebk)

Typeset in Times
by Deanta Global Publishing Services, Chennai, India

My late father Nakul Chandra, my mother Sulochana,
my wife Jeetu, and my daughters Mala, Maya, and Mili.
- Ashok Samal

My late father Mr. Gopal Das Choudhury and
my mother Mrs. Aditi Das Choudhury.
- Sruti Das Choudhury

Contents

Preface

Researchers in the life sciences have been studying phenotypes, i.e., observable traits, for centuries. Phenotypes have played a central role in categorizing and understanding the evolutionary history of species. Until recently, however, the process of compiling the phenotypic data for plants was carried out manually, in many cases destructively, and on a relatively small scale (only a few plants per species), primarily because of the amount of labor needed. For example, if we wanted to measure the area of leaves, one had to physically remove the leaves from the plant and measure the area of each leaf manually.

Image-based technology provides an alternate pathway to deriving plant phenotypes. Instead of determining the properties of the plant after harvesting, we can image the intact plant and derive the properties from the images. Image-based plant phenotyping is preferable to many other methods of collecting plant phenotypic data since it is (a) non-invasive, i.e., phenotypic traits can be measured without damaging the plants being measured; (b) tractable to automation, i.e., with properly engineered systems, very little, or no manual intervention or physical human labor is required; and (c) scalable, i.e., large population of plants can be analyzed in a short period of time. Furthermore, systems can image plants in a wide range of modalities (red-green-blue, near-infrared, fluorescence, hyperspectral, etc.), from different viewpoints, and at different frequencies throughout the development of plants. This opens up a new frontier in plant and crop sciences, in terms of understanding plant growth, and potentially being able to select plants with a unique combination of traits suitable for new climates and management regimes.

As the cost of the instrumentation for imaging technologies has plummeted over the past two decades, and research in the field of computer vision has continued to mature, the field of image-based plant phenotyping has emerged as a serious research area. There is now an ecosystem of researchers, practitioners, and even businesses, where image-based tools and technologies for plant phenotyping are the focus.

While image-based plant phenotyping holds much promise to make significant impacts, ranging from narrow research studies to help in achieving global food security, there are significant challenges as well. First and foremost, new algorithms need to be developed to address the challenges of extracting biologically meaningful phenotyping information from images and image sequences. The nature of plants is unique in many ways and is different from the objects that have been studied in much of computer vision research. Transformations applicable to rigid bodies are not suitable for modeling plants, which are characterized by non-uniform growth and development of new organs throughout their life cycle. Researchers are already working on developing novel algorithms specifically for computing plant phenotypes from images, and there is no doubt that significant breakthroughs will take place in the years to come.

The more critical challenge is to build the community around image-based plant phenotyping, an interdisciplinary research field that needs meaningful collaborations from a disparate range of researchers, including plant science, agronomy, computer science, statistics, and others. Just as importantly, perhaps more so, is the need to develop a bridge between what computer vision can provide and what the consumers of phenotyping information derived from images need and use. Many researchers and practitioners in plant science and related fields do not have a resource to obtain an authoritative perspective on both the potential of and the challenges to computer vision-based techniques, to solve the problems of interest to them. Similarly, the researchers in computer vision do not have an accurate understanding of the research challenges associated with the problems in plant phenotyping. The motivation behind the concept of the book is to address this research gap. Accordingly, the book is organized into three core content groups.

1. *Background*: What is the basis of plant phenotyping? What are the core technologies and techniques used in image analysis?
2. *Techniques*: What are some of the state-of-the-art computer vision algorithms used for plant phenotyping?
3. *Practice*: How have the plant scientists been using information derived from image analysis-based techniques?

Therefore, the primary audience for the book includes researchers and practitioners in plant science, agronomy, computer vision, statistics, among others. This is reflected in the contributors, who come from diverse disciplines that constitute this interdisciplinary research ecosystem. Advanced graduate students in these disciplines will find the content useful for their research, and the book may be used as a supplement for courses in computer vision, where phenotyping is an important element, or as an interdisciplinary graduate seminar course. The hope is that it will immediately assist them in their research and practice.

ORGANIZATION OF THE BOOK

The focus in the book is on numeric (quantitative) phenotypes, since they are amenable to analysis using a wide array of statistical genetic tools (e.g., quantitative trait loci mapping, genome-wide association studies, and genomic prediction). Furthermore, we will focus on phenotypes measured from individual plants under controlled experimental conditions, although research related to field-based phenotyping will also be discussed in this book.

Plants grown in soils consist of two main structures. Roots grow below the ground and anchor the plant in addition to serving as the transport system for water and nutrients. Shoots grow above the ground and enable plants to convert light energy into the chemical energy of sugar. Since the imaging systems for roots are not as developed as for shoots, the development of algorithms for image-based phenotyping for shoots is more advanced. Therefore, in this book, we focus on phenotypes that can be derived from shoots, and the algorithms to compute them.

PART I: BASICS

Part I consists of an introduction to the image-based approach to plant phenotyping and general background. Chapter 1 presents an overview of the significance of phenotyping in research, breeding, and agriculture in general. A discussion on high-throughput plant phenotyping platforms is provided in Chapter 2. Chapter 3 gives an overview of the core image-processing techniques used to analyze the images in plant phenotyping applications. It also provides a brief discussion on widely used imaging modalities in plant phenotyping.

PART II: TECHNIQUES

Part II provides some of the most recent breakthrough research methods in computer vision for plant phenotyping analysis. There is, of course, a large body of work on algorithms for computing phenotypes from plant images. We include a sample of algorithms, ensuring as much diversity as possible in terms of image analysis techniques, imaging modalities, and computed phenotypes.

Segmentation is often one of the indispensable early processing steps to any computer vision task, and the choice of best segmentation technique for plant phenotyping depends on the phenotyping platform (i.e., controlled greenhouse or field), imaging modality (e.g., visible light, fluorescent, infrared, near infrared, or hyperspectral), and type of phenotyping analysis, i.e., either by considering the plant as a whole (holistic) or the individual organs of the plant (component). Chapter 4 provides a review of the segmentation algorithms that have been used in the domain of image-based plant phenotyping research. It also provides a brief explanation of the publicly available benchmark datasets to facilitate algorithm development and systematic performance evaluation.

Many different phenotypes have been proposed in the literature. Chapter 5 describes a taxonomy that provides some structure to the diverse phenotypes and algorithms to compute new holistic and component phenotypes. It also discusses two datasets disseminated to spur research in this area. Chapter 6 summarizes three different modeling approaches to the realistic 3D reconstruction of plants and discusses in depth the state-of-the-art in data-driven approaches for reconstruction of plant geometry from real data. In Chapter 7, an advanced algorithm to segment the leaves of a plant from multiview images is described. The algorithm infers the 3D branch structure, using a deep-learning approach. Graphics-based modeling has been widely used in many applications to study the properties of plants under different environmental conditions and in understanding the physiology of plants. Chapter 8 presents a data mining approach to examining holistic and component phenotypes. Data analytic techniques, with phenomic and genomic data, provide opportunities for plant breeding. In Chapter 9, an overview of hyperspectral imaging for plant phenotyping is presented. Some of the most commonly used data-driven techniques used in hyperspectral image analysis are also described.

Finally, in Part II, statistical and machine learning-based algorithms relevant to plant phenotyping are discussed. Plant phenotyping, using digital images, may be treated as feature extraction and trait prediction problems that can be solved using machine learning and statistical approaches. Chapter 10 discusses core statistical techniques that are useful in image-based plant phenotyping research and practice. Chapter 11 presents a brief overview of machine learning in general and deep learning in particular, especially as they relate to computer vision and their applications to plant phenotyping.

PART III: PRACTICE

This part focuses on how image-based phenotyping impacts researchers and practitioners in plant science and agronomy. Chapter 12 discusses the use of chlorophyll *a* fluorescence (ChlF) as a leading tool in phenotyping. ChlF is directly connected to the complex physiological traits of photosynthesis, such as the quantum efficiency of photochemistry and heat dissipation of excitation energy under stress, across a plethora of genotypes, species, and stress conditions. Chapter 13 presents three different scenarios that breeders might face when testing lines in fields. Global climate change is associated with a rise in sea levels, which will result in increased salinity of the soil. Chapter 14 examines this impact by developing a salt tolerance index for over 100 genotypes and identifying salt-tolerant genotypes. Whereas there are many benefits to automated phenotyping, its adoption as part of crop breeding faces many challenges. Chapter 15 presents an analytical framework for the complex decision-making process for its adoption.

Acknowledgments

This book would not have been possible without the contribution of many people, and we are indebted to them. Foremost, we would like to express our sincere gratitude to Randy Brehm, Senior Editor, Agriculture and Nutrition Books Program, CRC Press/Taylor and Francis Group for her encouragement and support throughout the development of the book, from its conception through to its completion. We are also extremely thankful to Julia Tanner, Editorial Assistant, Life Sciences, CRC Press/Taylor & Francis Group, for her prompt answers to many of our questions during the final draft preparation. We also thank Lillian Woodall, Project Manager at Deanta Global, for all her help during the production phase of the book. We are grateful to the reviewers for their insightful comments on the book proposal. Their suggestions were very helpful in organizing the contents of the book.

We are extremely grateful to the plant phenotyping research ecosystem at the University of Nebraska-Lincoln (UNL), USA. The environment nurtured under the leadership of Dr. Tala Awada, Associate Dean at Agricultural Research Division and Associate Director of Nebraska Agricultural Experiment Station is centered around the UNL Greenhouse Innovation Complex located at the Nebraska Innovation Campus (NIC), which houses an automated, high-throughput plant phenotyping platform (LemnaTec Scanalyzer 3D). Her encouragement and constant support have been instrumental in our research work for the past five years. She contributed to the Introduction chapter and provided valuable suggestions to improve the quality and presentation of the chapter.

We want to thank all the staff at the UNL Greenhouse Innovation Complex, who manage and provide access to the facility, have generously given of their time, and, most importantly, shared images that we have used extensively in our research. We are especially grateful to Dr. Vincent Stoerger, the Plant Phenomics Operations Manager, for his support in setting up experiments that focused on addressing computer vision problems in plant phenotyping analysis. We are grateful to Dr. Awada and Dr. Stoerger for educating us with essential background knowledge about plant science over the years.

Our research has been supported by the Agricultural Research Division and the Nebraska Research Initiative at the UNL, and by a Digital Agriculture mini-grant by the Midwest Big Data Spoke project in UAS (Plant Sciences, and Education), McIntire-Stennis Forest Research Funding from United States Department of Agriculture (USDA), and the US National Science Foundation (NSF), to whom we are grateful.

We gratefully acknowledge the contributions of all the members of the UNL Plant Vision Initiative research group (students, staff, and collaborators). Over the years, we have had the good fortune of working with many talented undergraduate and postgraduate students, including Bhushit Agarwal, Rajesh Adloori, Venkata Satya Siddhartha Gandha, Suraj Gampa, Srinidhi Bashyam, Srikanth Maturu, Rengie Gui, and Rubi Quinones, who have contributed in many different ways to our research program by collecting and managing datasets, studying various problems, and developing algorithms and tools. We especially want to express our gratitude to Srinidhi Bashyam for his untiring help in supporting our research and for his assistance in the book development process.

We have benefited significantly from our discussions with many collaborators from different disciplines. We are thankful to our collaborators for sharing datasets, writing collaborative grant proposals, hosting academic visits, and creating an environment for exchanging research ideas.

We particularly appreciate Dr. Benes Bedrich, Professor, Department of Computer Graphics Technology, Purdue University, IN, USA, and Dr. Carmela Rosaria Guadagno, Associate Research Scientist, Department of Botany, University of Wyoming at Laramie, WY, USA, for sharing their expertise. Last but not least, we would like to thank all the authors for taking the time from their own active research programs to contribute to this book. We are grateful to them for their patience in the development of the book and for managing the unexpected delays that happened in this endeavor.

Editors

Ashok Samal is a Professor in the Department of Computer Science and Engineering at the University of Nebraska-Lincoln, USA. He received the Bachelor of Technology from the Indian Institute of Technology, Kanpur, India, and his Ph.D. from the University of Utah, Salt Lake City, UT, USA. His research interests include computer vision and data mining, and he has published extensively in these areas. More recently, he has focused on plant phenotyping and co-leads the Plant Vision Initiative research group at the University of Nebraska-Lincoln.

Sruti Das Choudhury is a Research Assistant Professor in the School of Natural Resources and the Department of Computer Science and Engineering at the University of Nebraska-Lincoln, USA. Previously, she was a Postdoctoral Research Associate in the Department of Computer Science and Engineering at the University of Nebraska-Lincoln and an Early Career Research Fellow in the Institute of Advanced Study at the University of Warwick, UK. She was awarded the Bachelor of Technology in Information Technology from the West Bengal University of Technology, and the Master of Technology in Computer Science and Application from the University of Calcutta, India. She obtained her Ph.D. in Computer Science Engineering from the University of Warwick, UK. Her research focus is on biometrics, data science, and most recently, image-based plant phenotyping analysis. She co-leads the Plant Vision Initiative research group at the University of Nebraska-Lincoln.

Contributors

Lana Awada
Johnson Shoyama School of Public Policy
University of Saskatchewan
Saskatoon, SK, Canada

Tala Awada
School of Natural Resources
University of Nebraska-Lincoln
Lincoln, NE, USA

and

Agricultural Research Division
University of Nebraska-Lincoln
Lincoln, NE, USA

Sanjiv Bhatia
Department of Mathematics and Computer Science
University of Missouri
St. Louis, MO, USA

Gustavo Bonaventure
BASF SE
Innovation Center Gent
Gent, Belgium

Amlan Chakrabarti
A.K. Choudhury School of Information Technology
University of Calcutta
Kolkata, West Bengal, India

Ayan Chaudhury
INRIA Grenoble Rhône-Alpes
Team MOSAIC
Le Laboratoire Reproduction et Développement des Plantes
Université Lyon
Lyons, France

Malay Das
Department of Life Sciences
Presidency University
Kolkata, India

Sruti Das Choudhury
School of Natural Resources and Department of Computer Science and Engineering
University of Nebraska-Lincoln
Lincoln, NE, USA

Shayani Das Laha
Department of Life Sciences
Presidency University
Kolkata, West Bengal, India

and

Department of Microbiology
Raiganj University
Raiganj, West Bengal, India

Brent E. Ewers
Department of Botany
University of Wyoming
Laramie, WY, USA

and

Program in Ecology
University of Wyoming
Laramie, WY, USA

Suraj Gampa
Department of Computer Science and Engineering
University of Nebraska-Lincoln
Lincoln, NE, USA

Christophe Godin
INRIA Grenoble Rhône-Alpes
Team MOSAIC
Le Laboratoire Reproduction et Développement des Plantes
Université Lyon
Lyons, France

Saptarsi Goswami
A.K. Choudhury School of Information Technology
University of Calcutta
Kolkata, West Bengal, India

Carmela Rosaria Guadagno
Department of Botany
University of Wyoming
Laramie, WY, USA

Suman Guha
Department of Statistics
Presidency University
Kolkata, West Bengal, India

Reka Howard
Department of Statistics
University of Nebraska-Lincoln
Lincoln, NE, USA

Ayaka Ide
The Institute of Scientific and Industrial Research
Osaka University
Osaka, Japan

Takahiro Isokane
The Institute of Scientific and Industrial Research
Osaka University
Osaka, Japan

Marcus Jansen
LemnaTec GmbH
Aachen, Germany

Diego Jarquin
Department of Agronomy and Horticulture
University of Nebraska-Lincoln
Lincoln, NE, USA

Yasuyuki Matsushita
Graduate School of Information Science and Technology
Osaka University
Osaka, Japan

Hossain Ali Mondal
School of Crop Improvement
College of Post Graduate Studies in Agricultural Sciences
Central Agricultural University
Umiam, Meghalaya, India

Amlan Jyoti Naskar
Department of Life Sciences
Presidency University
Kolkata, West Bengal, India

Fumio Okura
Osaka University
Osaka, Japan

and

PRESTO
Japan Science and Technology Agency
Osaka, Japan

Stefan Paulus
Institute of Sugar Beet Research
Göttingen, Germany

Peter W. B. Phillips
Johnson Shoyama School of Public Policy
University of Saskatchewan
Saskatoon, SK, Canada

Rubi Quiñones
Department of Computer Science and Engineering
University of Nebraska-Lincoln
Lincoln, NE, USA

Eleanor Quint
Department of Computer Science and Engineering
University of Nebraska-Lincoln
Lincoln, NE, USA

Ashok Samal
Department of Computer Science and Engineering
University of Nebraska-Lincoln
Lincoln, NE, USA

Tanmay Sarkar
Department of Life Sciences
Presidency University
Kolkata, West Bengal, India

Stephen Scott
Department of Computer Science and Engineering
University of Nebraska-Lincoln
Lincoln, NE, USA

Stuart J. Smyth
Department of Agricultural and Resource Economics
University of Saskatchewan
Saskatoon, SK, Canada

Bashyam Srinidhi
Department of Computer Science and Engineering
University of Nebraska-Lincoln
Lincoln, NE, USA

Cong Wu
Department of Computer Science and Engineering
University of Nebraska-Lincoln
Lincoln, NE, USA

Alencar Xavier
Department of Agronomy
Purdue University
West Lafayette, IN, USA

Zheng Xu
Department of Mathematics and Statistics
Wright State University
Dayton, OH, USA

Yasushi Yagi
The Institute of Scientific and Industrial Research
Osaka University
Osaka, Japan

Part I

Basics

1 Image-Based Plant Phenotyping

Opportunities and Challenges

Ashok Samal, Sruti Das Choudhury, and Tala Awada

CONTENTS

1.1 INTRODUCTION

A plant phenotype is defined as the quantifiable and observable morphological, physiological, and biochemical properties of a plant, resulting from the interaction of its genotype with the environment. The concept of "genotype to phenotype" was first introduced by Wilhelm Johannsen in 1909, while working on germination studies in barley and common beans (Roll-Hansen 2009). Since a plant's pattern of growth progression is influenced by climatic variations, phenotypes may record the ontogenetic properties of a plant, which refers to the developmental history of the plant during its lifetime, from germination to senescence. Phenotypes can be measured either by considering the whole plant (holistic phenotypes), or its individual organs (component phenotypes). Examples of holistic phenotypes include plant height, total leaf area, and biomass. Examples of component phenotypes include the area of individual leaves, the average cross-sectional area of the stem, and the volume of a fruit (Das Choudhury et al. 2019).

Phenotypes are regulated by the plant's genetics and by its interaction with the environment, so that different genotypes will probably produce different phenotypes, and plants of the same genotype might have different phenotypes under different environmental conditions. It is similarly possible for plants of different genotypes to have similar phenotypic traits. Thus, the relationship between phenotypes to genotypes is many-to-many in nature. Some traits are influenced more by genetic factors, while the plant's growing environment is more determinative for others. Ultimately, phenotypes are significant, as they influence resource acquisition by the plant and its yield.

1.2 IMPORTANCE OF PHENOTYPING RESEARCH

We are in the midst of a significant change in global climate patterns, where environmental stresses, e.g., droughts, floods, extreme temperatures, and disease outbreaks, are expected to increase in frequency and intensity in many regions and are predicted to reduce crop yields in the impacted areas. This is expected to be exasperated by the need to ensure food security for a growing world population in the presence of dwindling natural resources (FAO 2017). These complex challenges make it necessary to accelerate research for boosting plant yield and adaptation to local environments. Increasing and, in some cases, maintaining yields will require implementation of novel techniques in gene discovery and plant breeding, as well as advanced technologies such as precision agriculture (Chawade et al. 2019). Precise and fast assessment of phenotypes for crop varieties or populations is needed to accelerate breeding programs (Lobos et al. 2017), and to aid in the development of advanced management practices for precision agriculture. Plant phenotyping tools have the potential to accomplish these goals (Camargo and Lobos 2016).

Phenotyping can also play an essential role in the understanding of critical plant physiological processes. Traits, especially complex ones, reflect the biological processes that govern growth, phenology, and productivity of plants (Lynch and Walsh 1998). Understanding the underlying processes can lead to the "elucidation of the mechanisms impacting important ecophysiological traits" (Pauli et al. 2016).

1.3 PLANT PHENOTYPING ANALYSIS FRAMEWORK

Figure 1.1 shows an overall schematic of a typical high-throughput plant phenotyping framework. The images of the plants are captured by multiple cameras of different modalities in an automated high-throughput plant phenotyping facility at regular intervals for a specific period of study. The images thus captured are first stored in a centralized database, and then transferred to the centralized server for analysis and integration with other datasets and/or for distribution to the different end-users for scientific discovery. The computing and data can be hosted locally or delivered through a cloud computing platform over the internet. The plant phenotyping analysis is a multidisciplinary research field at the intersection of computer vision, statistics, engineering, machine learning, plant science, and genetics. To fully exploit the benefits of image-based plant phenotyping, raw images, metadata, and computed data, as well as genomic data, must be stored in a searchable database so that they can be efficiently queried and analyzed for data mining and knowledge discovery. Significant insights can be obtained using these techniques, both within experiments and across multiple experiments in plant phenotyping, particularly in a large phenotyping facility that supports diverse experiments generating heterogeneous, high-velocity and high-volume data.

1.4 PLANT PHENOTYPING NETWORKS

Due to the growing significance of phenotyping analysis and the need to develop long-term plans, a number of international, national, and regional plant phenotyping networks (summarized in Table 1.1) have been organized over the past decade (Carroll et al. 2019), with the following common missions:

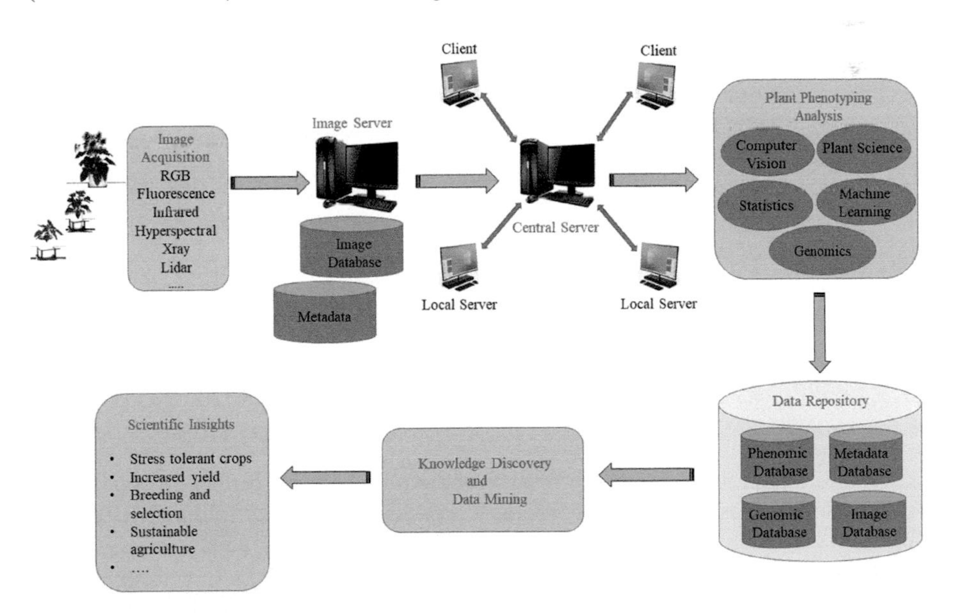

FIGURE 1.1 A typical framework for high-throughput plant phenotyping analysis.

TABLE 1.1
Plant Phenotyping Networks around the World

Network	Name	Web Location
APPF	Australian Plant Phenotyping Facility	http://www.plantphenomics.org.au/
APPN	Austrian Plant Phenotyping Network	http://www.appn.at/
DPPN	German Plant Phenotyping Network	https://dppn.plant-phenotyping-network.de/
EPPN	European Plant Phenotyping Network	https://www.plant-phenotyping-network.eu/
FPPN	French Plant Phenotyping Network	https://www.phenome-fppn.fr/
IPPN	International Plant Phenotyping Network	https://www.plant-phenotyping.org/
LatPPN	Latin American Plant Phenotyping Network	https://www.frontiersin.org/articles/10.3389/fpls.2016.01729/full
NaPPI	Finland National Plant Phenotyping Infrastructure	https://www.helsinki.fi/en/infrastructures/national-plant-phenotyping
NPPN	Nordic Plant Phenotyping Network	https://www.forageselect.com/nppn
Phen-Italy	Italian Plant Phenotyping Network	http://www.phen-italy.it/
UKPPN	UK Plant Phenotyping Network	http://www.ukppn.org.uk/

1. Accelerating plant phenotyping research by representing a multidisciplinary community, comprising plant biologists, ecologists, engineers, agronomists, and computational scientists, to foster collaboration, innovation, and the initiation of multi-investigator and multi-institution projects.
2. Promoting a framework for data standards to facilitate data sharing, accessibility, interoperability, and reusability worldwide.
3. Incentivizing mutually beneficial research between public and private sectors.
4. Facilitating the interdisciplinary training needed for effective translational plant phenotyping research.

As Table 1.1 shows, most of the efforts are concentrated around Western Europe. However, there are some networks located in other parts of the world, including Asia, Australia, and North America. These networks have considerably advanced the research in plant phenotyping, by managing extensive involvement of individuals in transdisciplinary and transinstitutional research collaborations globally.

1.5 OPPORTUNITIES AND CHALLENGES ASSOCIATED WITH HIGH-THROUGHPUT IMAGE-BASED PHENOTYPING

Low-throughput plant phenotyping has traditionally been performed manually, with the help of various sensors or instruments, e.g., leaf area meters, chlorophyll meters, infrared gas analyzers, etc. Whereas some measurements, e.g., plant height, can be performed non-destructively and repeatedly over time, many other traits necessitate destructive measurements, e.g., weight, plant architecture, nutrient content, and water relations. Additionally, low-throughput phenotyping is labor-intensive and time-consuming, particularly for large experiments with hundreds or thousands of plants that require measurement of multiple traits (Reynolds et al. 2019). Furthermore, manual measurements, even by trained individuals, naturally entail some degree of inconsistency, which poses some challenges for plant scientists, where repeatability and accuracy are critical. Also, there are many complex phenotypes, e.g., bi-angular convex-hull area ratio, stem angle, and early indication of fruit decay, which are either difficult or not possible to determine manually, even by skilled individuals. However, they are crucial because they "govern key biological processes that influence overall plant productivity and adaptability" (Pauli et al. 2016), and hence are essential for crop breeding and management. Whereas harvesting the plant to measure phenotypes is useful to determine some key traits, e.g., biomass, it prevents the recording of ontogenetic properties.

High-throughput image-based plant phenotyping has the potential to overcome many of the challenges described above. Deriving the phenotypes from images requires limited manual involvement. Thus, multiple phenotypes of many plants can be obtained repeatedly, efficiently, and in a relatively short time, making this approach easier to extrapolate. The plant can be imaged in a variety of modalities, using different cameras (e.g., visible, fluorescent, infrared (IR), near infrared (NIR), and hyperspectral), to capture the desirable traits. Computing the phenotypes from images is non-invasive, so it is possible to compute ontogenetic phenotypes by imaging the same plant at multiple points in its lifecycle. Since the multimodal image sequences store abundant information about the plant, the phenotypes can be computed in the present, allowing the possibility of applying novel tools for trait discovery in the future.

Whereas image-based plant phenotyping has many inherent advantages, there are also challenges associated with it. First, it is expensive to set up and maintain a high-throughput plant phenotyping infrastructure and imaging facilities. Alternative, low-cost approaches and set-ups have been proposed for small-scale experiments (e.g., Armonienė et al. 2018). Cameras in the visible spectrum are relatively inexpensive, but some imaging systems, especially hyperspectral cameras, are significantly more costly. Whereas there is minimal human involvement in image acquisition, domain experts are needed for operating and maintaining the infrastructure, and for data management and analysis. Furthermore, whereas high-throughput plant phenotyping systems are quite efficient at acquiring and managing abundant and high-quality images of the target plants, the ability to translate these images into biologically informative quantitative traits remains a significant bottleneck. Therefore, algorithms that accurately compute the phenotypes need to be developed before the true potential can be realized.

Image-based plant phenotyping research is still in its infancy, with vast opportunities for exploration. Since plants are living organisms with increasing complexity in architecture over time, skilled researchers with domain and computer vision and machine-learning expertise need to be employed to develop novel algorithms for investigating meaningful traits. In addition, trained personnel dedicated to day-to-day image analysis for computing basic phenotypes, e.g., height and plant volume, using established image-analysis pipelines, are also required for genomic prediction.

1.6 IMAGE-BASED PLANT PHENOTYPING ANALYSIS

Image analysis approaches for plant phenotyping can be broadly divided into two categories: traditional computer-vision approach and deep learning. Figure 1.2 provides an overview of the two approaches. Traditional image-processing pipelines typically consist of a sequence of processing steps, starting with one or more preprocessing algorithms to clean up the image. It is then followed by a segmentation algorithm to isolate the plant (or a specific part of the plant) from the background. Then, a set of feature detectors is used to identify particular features in the plant image. Finally, either generic or custom algorithms are used to compute diverse phenotypes, e.g., stem angle as a potential measure of a plant's susceptibility to lodging (Das Choudhury et al. 2017), detection of emergence timing for each leaf and counting of total number of leaves present at any time point for growth stage determination in maize (Das Choudhury et al. 2018), and tracking of the growth of maize ears and silks (Brichet et al. 2017). A comprehensive summary of the state-of-the-art methods for computing physiological plant phenotypes, based on the traditional image-processing pipeline, was presented by Das Choudhury et al. (2019). The tasks include early detection of drought stress, using hyperspectral image sequences of barley plants (Römer et al. 2012), determination of salt stress in wheat (Moghimi et al. 2018), food quality assessment (Huang et al. 2014), and early prediction of yield (González-Sanchez et al. 2014).

More recently, research in computer vision has been greatly influenced by deep-learning algorithms, and various tasks in plant phenotyping can benefit from these algorithms. A deep neural network is trained to generate either a phenotype directly or to identify the whole plant and its specific objects (e.g., flowers), using prelabeled data. Alternatively, the network is used to locate the objects of interest in an image, which are then used to compute the necessary phenotypes, using traditional methods. Researchers have already used these approaches in performing various tasks, e.g., computing plant stalk count and stalk width (Baweja et al. 2017), leaf counting in rosette plants (Giuffrida et al. 2018; Ubbens et al. 2018), maize tassel counting (Lu et al. 2017), cotton flower detection (Xu et al. 2018),

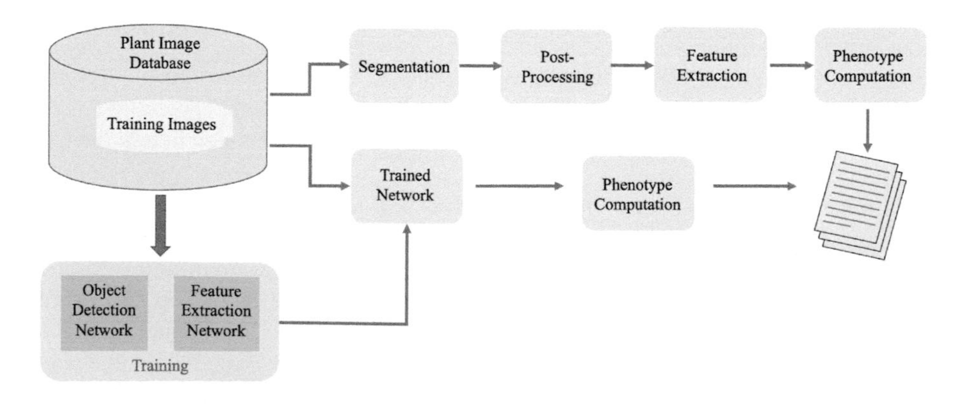

FIGURE 1.2 Two different approaches for image-based plant phenotype computation.

wheat spike detection (Hasan et al. 2018), and rice panicle segmentation (Xiong et al. 2017). A comprehensive summary of deep-learning algorithms for identification, classification, quantification, and prediction of plant stress phenotypes is presented by Singh et al. (2018).

1.7 DATA MANAGEMENT FOR PLANT PHENOTYPING

Most of the phenotypic traits are highly dependent on genotype-by-environment interactions, which decreases the likelihood of a phenotyping experiment being exactly reproducible, because it is challenging to completely control the environmental conditions during the plant's life cycle. Hence, the reusability of phenotypic data is essential not only for data accessibility and discovery, but also to allow integration across experiments, to decipher the genetic architecture of traits across environments, and to allow the prediction of genotype performances in the context of the climate change adaptations.

As the body of researchers and practitioners in plant phenotyping grows, the amount of image data generated for plant phenotyping will increase. More plant genotypes will be imaged from different parts of the world, and more plants will be studied under stress conditions, both biotic and abiotic. A dataset consisting of multimodal and multiview images of a number of plants with a wide variety of genotypes for the plants' entire life cycle contains enormous information in the domain of large-scale phenotypic search space, i.e., structural, physiological, and ontogenetic, ranging from vegetative to reproductive stages, would require multiple research groups to continue investigations over time to make the best use of the dataset. It is a "Big Data" challenge, and data centers need to be able to accommodate the high-volume, high-velocity, heterogeneous, and complex datasets, to ensure the integration and reusability of the data, at present and into the future. Standardized solutions for storage, analysis, query, visualization, and appropriate web interfaces remain a bottleneck, and need to be addressed for large-scale research.

Long-term management of image datasets for modern phenomic research, following the FAIR principles (findability, accessibility, interoperability, and reusability), can serve as a solution. The findability of the datasets may be ensured through fast querying mechanisms; accessibility is guaranteed by archiving the dataset under long-term storage. The reusability and interoperability rely on data curation, which is defined as the long-term management of data throughout its life cycle, from the time of its creation to the time when it becomes obsolete, to ensure datasets are reliably retrievable for reuse in a variety of phenotyping research efforts (Pommier et al. 2019). Several initiatives have been undertaken to develop tools for standardizing phenotypic datasets, e.g., the Minimal Information About Plant Phenotyping Experiment (MIAPPE) defined the set of information necessary to enable data reuse (Ćwiek-Kupczyńska et al. 2016). Whereas data standardization defines a set of rules for reusability, data integration relies on interoperability by linking different datasets together to avoid data silos (Lenzerini 2002; Pommier et al. 2019). It may be achieved by following the linked-data principle, which refers to the set of best practices for establishing links between data from different sources and publishing them on the internet (Bizer et al. 2009).

1.8 COMPUTATIONAL CHALLENGES IN IMAGE-BASED PLANT PHENOTYPING

To successfully leverage image-based phenotyping, research in computer vision must utilize the knowledge of plant processes and decades of advances in remote sensing research to identify biologically meaningful phenotypes from plant images, and to develop efficient algorithms to compute them. In addition, scientists and researchers, who use phenotypic data to answer scientific questions, need to exploit state-of-the-art computer vision algorithms to compute advanced phenotypes. In this section, we focus on the computational barriers to a successful deployment of image-based plant phenotyping.

1.8.1 Computational Resources

Computing traits from images in a high-throughput phenotyping system requires the development of an infrastructure for data architecture and configuration. After the multimodal images are captured and stored in a local or central server, they are transferred to a local client computer and/or institution, or to cloud-based servers for management and computation of phenotypes. The image transfer process can be time-consuming as well; even with a high-speed network, it can take hours or even days, depending on the volume and resolution of images, and the regularity of data transfer. The process of computing the phenotypes requires significant computational resources. Images of multiple plants, for multiple days in multiple modalities, are analyzed, and typically many different phenotypic traits are computed, from simple (e.g., plant height) to complex (e.g., stress index). This is currently accomplished by using a powerful local computer, a shared node in a local data center, or through a cloud-based server.

There has been a tendency among institutions to overlook the efforts, expertise, and resources needed to manage and share the data as part of the research process (Nowogrodzki 2020). To be successful, emerging data centers need to offer the capabilities for data architecture, configuration and analytic technologies that allow the researchers to store (i.e., for short- and long-term storage), visualize, show workflow, use version control (i.e., keep previous versions), integrate heterogeneous data, capture metadata that can be expandable, guarantee compatibility between systems, offer onsite analysis to avoid downloading of data, and finally recognize and manage issues related to data rights and privacy (Nowogrodzki 2020).

1.8.2 Algorithm Robustness

Robustness of the computer vision-based algorithms is a challenge in plant phenotyping. Four primary factors contribute to this problem: (a) scale change, (b) illumination variation, (c) plant rotation, and (b) occlusion. Figure 1.3 illustrates the effects of all these factors, using sample images captured at the High-Throughput Plant Phenotyping Core Facility (Scanalyzer 3D and HTS Scanalyzer, LemnaTec Gmbh, Aachen, Germany), at the University of Nebraska-Lincoln, USA.

Scale change: Accurate representation of ontogenetic trajectories of phenotypes can provide important information about the growth patterns of plants. However, to

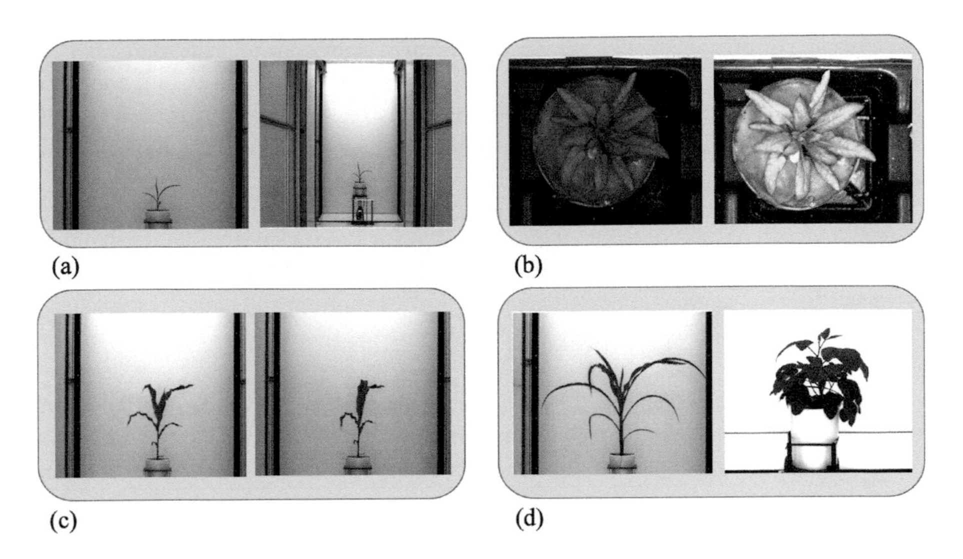

FIGURE 1.3 Illustration of challenges for image-based plant phenotyping algorithms.

capture the full extent of the growing plant, periodic adjustment of the zoom level of the camera is required. The structural and physiological trajectory-based phenotypes, computed from a sequence of images, e.g., leaf elongation rate, plant growth rate, and rate of change of stem cross-section area (Das Choudhury et al. 2019), should account for the change of scale. Figure 1.3 (top left) shows two images of a maize plant captured on two consecutive days with different camera zoom levels. This problem can be addressed by applying an image registration technique prior to segmentation (Das Choudhury et al. 2018).

Illumination variation: The robustness of the segmentation of the plant, even in high-throughput systems with controlled imaging conditions, is not solved in the general case. Color-based segmentation is a commonly used technique for segmenting plants in high-throughput phenotyping systems (Hamuda et al. 2016; Zheng et al. 2009). However, the threshold used to extract the plant pixels is often determined manually and is not consistent across the entire image sequence of the plant. Different plants have different values of hue, whereas lighting conditions and shadows impact the color as well, and the shades of different parts of the plant change from the plant's emergence to its senescence. Figure 1.3 (top right) shows two sample images of an Arabidopsis plant captured at different times, with significant differences in illumination. The deep-learning-based approaches hold significant promise in addressing this challenge and must be developed further.

Plant rotation: Plants can alter their foliage architecture in response to the environment. Maize plants, for instance, can change the orientation of the whole shoot, as well as those of individual leaves, to optimize light interception, minimizing shade from neighboring plants (Maddonni et al. 2002). This significantly changes the plant's topology in a 2D image, and hence, poses a challenge to determine some trajectory-based phenotypes accurately. For example, the height of the plant, as measured by the height of the bounding rectangle enclosing the plant, is not affected by the change in phyllotaxy, i.e., the arrangement of leaves around the stem, but the count of the total number of plant pixels or the volume of the convex-hull enclosing the plant (as a measure of the plant's biomass) may be significantly affected (Figure 1.3 (bottom left)). Computing the phenotypes from a 3D reconstruction of the plant can address this challenge.

Occlusion: Occlusion, involving overlap between individual plant parts, poses a significant challenge in many image-based plant phenotyping tasks, particularly in later stages in a plant's life, e.g., leaf detection, counting and alignment, leaf tracking, and measurement of leaf size. Figure 1.3 (bottom right) shows the sample images of two plants with different architectures (maize and soybean), both with occluded leaves.

The segmentation of individual leaves in the presence of crossovers is challenging to address when using traditional segmentation techniques. Hence, a more sophisticated approach, based on globally consistent correspondence detection of multiple feature sets, obtained from multiple side views using proximal Gauss-Seidel relaxation, has been introduced by Yu et al. (2016) to address this problem. Furthermore, the examination of the image sequence that exploits the information about a plant's growth history may be more effective than analyzing the individual images.

1.8.3 Inference from Incomplete Information

A fundamental challenge in computer vision is that images do not carry all the information about the scene which they represent. The imaging process is a projection that maps the 3D scene (a plant in our case) onto a 2D image. This process necessarily entails a loss of information, and hence, retrieving the 3D information from the projection is an ill-posed problem in general. Even if the plant is imaged from multiple side views and top views, as is often done in many applications, an accurate reconstruction is not always feasible due to the thin ribbon-like shape of leaves, the presence of occlusions, or the absence of textural surfaces.

As mentioned before, there are several different types of phenotypes, and they cannot all be determined using images from any single sensor, as each sensor captures only one or a few, but not all, aspects of the plant. For example, a red-green-blue (RGB) image can serve as a proxy for greenness but does not measure the chlorophyll content of the plant, nor does it capture its water content, both of which are relevant phenotypes that relate to essential processes in the plant. Therefore, many different imaging modalities are often employed to capture the electromagnetic spectrum over a broad range of frequencies. The spectral resolution of the imaging sensor also poses a challenge. Signals for some phenomena are only available over very narrow wavelengths, and, if a broadband sensor is used, the signal becomes diffused and is not useful. Therefore, hyperspectral imaging has become widespread in plant phenotyping. However, identification of the wavelengths that encode important biological processes related to phenotyping is an open problem, that can be addressed by using non-guided and/or guided process-based computer vision, artificial intelligence, and deep learning, as well as working closely with the remote sensing research community to investigate the possibility of downscaling validated vegetation indices for images acquired at close range.

1.8.4 Large Phenotype Search Space

Interaction of the genotype with the environment has a differential impact on the phenotypes. Identification of the affected phenotype, however, depends on the phenotype in question. Whereas some phenotypes, e.g., physical measurements, are easier to observe and determine, more complex phenotypes, e.g., branching structure, may not be as obvious or as easy to define. Attempting to compute them from images that do not always preserve all the information about the plant only adds to the challenge. One approach taken by researchers to address this problem is to compute as many different phenotypes (structural, physiological and ontogenetic) as can be derived from images captured by cameras using different modalities, e.g., visible light, fluorescent, NIR, IR, and hyperspectral, and analyze which phenotype(s) can show the variations needed to study the effects of the genotype-by-environment ($G \times E$) interaction.

Although many phenotypes have been proposed in the literature, the space of possible phenotypes is large, and many interesting and useful phenotypes have likely not been determined yet. Most of the phenotypes have either been direct measurements of the plant (e.g., height, aspect ratio, etc.) or properties that have been proposed for general shapes in computer vision (e.g., convex-hull volume, solidity, etc.). Higher-order phenotypes, that describe the geometry of the different elements of the plant, have not been widely explored. Also, meaningful combinations of two or more phenotypes, referred to as a derived phenotype (Das Choudhury et al. 2016), may interpret important information about the plant's behavior. For example, bi-angular convex-hull area ratio and plant aspect ratio can help explain plant rotation due to light interception and canopy architecture, respectively (Das Choudhury et al. 2016). However, the number of possible derived phenotypes is vast, and identifying those with significance in plant science is challenging and depends on the research objectives. Hence, extensive cross-disciplinary research efforts are essential in this relatively unexplored domain.

1.8.5 Analysis of Image Sequences

Since the plant is a dynamic entity, the growth patterns record important aspects of the genotype-by-environment interaction, and identifying phenotypes that capture them can be of significant value. In many cases, the images of the plants are captured at regular intervals over a period of time during the plant life cycle to facilitate the computation of phenotypes that reflect the ontogenetic characteristics of the plant. Hence, researchers have attempted to compute phenotypes from individual images of the sequence and then used graphical demonstrations to represent the rate of change of these phenotypes as a function of time, referred to as "trajectory-based phenotypes" (Das Choudhury et al. 2019).

Image-sequence analysis has greater potential for solving many complex unsolved computer vision tasks in phenotyping, especially at later growth stages, when plants assume complex architecture. The ambiguity in detecting and counting leaves, resulting from leaf crossovers, can be potentially resolved by examining the imaging of plants early in the life cycle, when the leaves did not overlap, and tracking their growth over the time sequence. Similarly, the emergence of plants or organs can be more accurately determined by analyzing image sequences well before a more concrete indication of the new growth is visible in an image, and, depending on the interval between imaging times, various optical flow approaches may be used to address this problem.

Undoubtedly, there is great potential in analyzing image sequences to compute phenotypes. However, researchers need to address the challenges associated with image sequence analysis. Whereas plants are solid objects, a plant and its parts do not strictly follow the affine transformation models. For example, a plant under stress may have some leaves that fold, curl and droop irregularly, and thus do not conform to regular rigid body transformation. Furthermore, even if a plant is growing, its area in images in a sequence may not be monotonic. A plant may sometimes rotate to increase its exposure to light, and hence its projections with the same camera parameters on a subsequent day may be smaller, even though the plant may actually have grown.

1.8.6 Lack of Benchmark Datasets

Access to large datasets with ground-truth information has proved to be very beneficial to research progress. ImageNet (Deng et al. 2009), which has over three million labeled images in over 5000 categories, is credited with the boom in deep-learning-based research in image analysis, and artificial intelligence, in general. Benchmark datasets are useful for two essential purposes in image-based plant phenotyping: (a) *algorithm development* and (b) *performance evaluation*. Well-developed datasets engender the development of novel algorithms to compute phenotypes. Also, they are critical for the systematic evaluation of any computer-vision-based algorithm. A database of images, with known values of results from different processing steps and phenotypes, is essential to compare the performance of competing algorithms. Furthermore, deep-learning approaches, increasingly useful going forward, require many labeled images.

While there have been efforts to publicly release benchmark datasets with ground-truths for specific tasks, this approach remains a challenge for several reasons, including the nature of siloed research, the lack of data management and sharing policies, digital rights, and a shortage of domain experts. Datasets for only a few plant species, typically some model plants, e.g., Arabidopsis, or a cereal crop, e.g., maize, are available for specific computer vision tasks. Table 1.2 summarizes the benchmark datasets available for image-based plant phenotyping.

A key challenge in this regard is not only developing image-based ground-truths, but also obtaining accurate measurements of phenotypic traits of plants, using hand-held devices. For example, exact values of plant height, total plant volume, or area of individual leaves are essential to validate the results obtained by the computer vision algorithms that compute these phenotypes. While some of the manual measurements would require the destruction of the plant, such an effort could prove to be worthwhile in the long run in stimulating research into image-based plant phenotyping.

TABLE 1.2
Public Benchmark Datasets for Image-Based Plant Phenotyping Analysis

Dataset Name	Plant(s)	Phenotyping Tasks	Reference
Leaf Segmentation Challenge Dataset	Arabidopsis and tobacco	• Leaf segmentation • Tracking and counting	http://www.plant-phenotyping.org/CVPPP2014-challenge.
Michigan State University Plant Imagery Dataset	Arabidopsis and bean	• Leaf segmentation • Counting, alignment, and tracking • 3D leaf reconstruction	Cruz et al. 2016
Panicoid Phenomap-1	Maize	• Evaluation of holistic phenotypes	Das Choudhury et al. 2016
UNL Component Plant Phenotyping Dataset	Maize	• Leaf detection and tracking • Leaf segmentation • Evaluation of phenotypes	Das Choudhury et al. 2018
UNL 3D Plant Phenotyping Dataset	Maize	• 3D reconstruction • 3D holistic and component phenotypes	https://plantvision.unl.edu/dataset
Komatsuna Dataset	Komatsuna	• 3D reconstruction	Uchiyama et al. 2017
Deep Phenotyping Dataset	Arabidopsis	• Classification	Namin et al. 2018

1.9 LOOKING INTO THE FUTURE

Whereas many advances have been made in image-based plant phenotyping analysis, progress in many important directions is still expected as we look forward. In this section, we summarize some future trends in image-based plant phenotyping analysis.

1.9.1 Imaging Platforms

The cost of digital-imaging products continues to decrease, and, probably, cheaper alternatives, with customized sensors, with which to monitor a plant's growth and to identify its phenological stages, will become more common and accessible in the future. The imaging devices are becoming smaller and cheaper, making them easier to use in smaller laboratory environments and to deploy on drone-based platforms. Cameras of different modalities are becoming widespread, due to economies of scale. Fluorescence and hyperspectral cameras are expected to become more common, with potentially significant breakthroughs in plant phenotyping, with a focus on plant functions. In particular, identification of crucial wavelengths for monitoring essential plant processes by hyperspectral imaging will lead to more compact (and cheaper and hence more easily deployable) multispectral imaging systems.

The less widely used modalities have the potential to become more widespread. In particular, sensors which monitor root growth (e.g., X-ray computed tomography (CT)) will increasingly become more available to researchers (Atkinson et al. 2019).

Other imaging, like light detection and ranging (LiDAR), positron emission tomography (PET), and magnetic resonance imaging (MRI), also hold promise in this regard. Technologies to directly generate 3D structures of plants will also become increasingly available and accessible. While some tools are available to take advantage of the new technologies, a great deal of research will need to be performed in this field.

1.9.2 Integrated Phenotypes

Not surprisingly, most of the image-based phenotyping research has focused on the part of the plant that is above ground. Imaging roots is neither easy nor accurate, using the current technologies (Atkinson et al. 2019). Therefore, algorithm development for automated root phenotyping analysis is also somewhat limited, although image-processing techniques for the characterization of roots, based on destructive methods, have been extensively explored (Slota et al. 2017). As the root-imaging technologies become more accessible, there will be significant progress in image-based algorithms to compute root-based traits, including ontogenetic phenotypes.

Once research in root phenotyping becomes more mature, the opportunity to examine the plant as a whole (above and below ground) will become increasingly feasible, and integrated phenotypes, relating the properties of the shoot with those of the root system, will be developed and used in the phenotyping practice. In particular, the growth patterns of the root, and its relation to the growth patterns in the shoot, will be developed. They will provide the basis for understanding the dynamics at a finer spatial and temporal resolution. Efforts have been made for the early detection of stress in plants, e.g., drought and mineral deficiency (Chen and Wang 2014; Kim et al. 2011; van Maarschalkerweerd and Husted 2015) and the temporal spread of drought stress based on image sequence analysis (Gampa 2019). However, the categorization of different stress levels, e.g., mild, moderate, and extreme, as well as the determination of the speed of recovery from various stress levels, is still an unsolved problem. Uga et al. (2013) demonstrated that controlling root growth angle contributes to drought tolerance. Thus, an integrated phenotype to quantify the speed of recovery from different stress levels in the event of de-stressing, in relation to controlled root angle, may be of importance.

In a high-throughput plant phenotyping platform, multiple cameras with different modalities are becoming more prevalent for imaging the plants for the extraction of complementary phenotypic traits. For example, visible light encodes morphological characteristics of the plants or their organs (Das Choudhury et al. 2017; Gage et al. 2017). In contrast, IR images can serve as a proxy for a plant's temperature, which, in turn, can be used to detect differences in stomatal conductance, a measure of the plant's response to water status, and transpiration rate under abiotic stress responses (Li et al. 2014). Hyperspectral cameras are uniquely suited to providing insights into plant functions and processes, e.g., early detection and temporal spread of drought stress (Gampa 2019). In addition to providing diverse phenotypes, the use of multimodal imagery can assist in image analysis tasks. For example, the contrast between a plant and its background in IR or NIR images becomes significantly reduced, making the segmentation more challenging. Use of the RGB images, with appropriate geometric transformations, to guide the segmentation of the IR or NIR images will make the results more accurate.

Finally, ontogenetic phenotypes will become more common as algorithms that analyze the whole-image sequence of the plant are developed. It will lead to greater exploration of phenotypes that capture the growth of the plant as a whole and in part (organs) at different stages of growth. Second-order phenotypes, that relate the ontogeny of different organs (e.g., leaves, flowers, and fruits) also need to be developed to provide a greater understanding of the underlying processes.

1.9.3 Learning-Based Approaches

As the number of images captured for phenotyping increases, the recent machine-learning algorithms, especially deep-learning models, will have a significant impact on image-based plant phenotyping. They have significantly altered the computer-vision landscape in the past decade in tasks ranging from segmentation, classification, feature detection, etc. Deep-learning-based approaches have improved the performance of these tasks significantly in comparison with traditional computer-vision methods. They hold similar promise in image-based plant phenotyping as well.

One of the advantages of deep-learning-based methods is that, whereas training the network can be expensive, in terms of both resources (e.g., the need for graphic-processing unit (GPU)-based processors) and time (it can take days, or even weeks to train the network), the actual task is relatively quick, and hence they have the potential to be deployed in real-time applications, especially field phenotyping. Another advantage is that, as the size of the datasets grows, the accuracy of the deep-learning methods is likely to increase, as they learn from new data and become more robust.

Deep architectures, particularly convolutional neural networks (CNNs), have been shown to capture properties of images at different levels of hierarchy, from edges to lines and blobs to low-level features to higher-order features. This strategy fits naturally to plant imaging since the plant is naturally organized as a modular composite of its parts, and it is possible that the intermediate representations in deep architectures correspond to plant structures, such as leaf junctions, leaf tips, stem segments, etc. A particular challenge in the use of deep learning in plant phenotyping is whether to train the networks for only one species or to make them universal. For example, it may be possible to develop an accurate CNN-based architecture that can identify flowers of a particular type (e.g., sunflower); however, a network that can recognize all types of flowers may be quite challenging. On the other hand, strategies such as transfer learning, i.e., improving a learner in one domain by transferring the learner characteristics from a related domain (Wen et al. 2013), may make it easier to first train a network with flower images of a particular species and then use that to train the network to either build a universal flower detection network or to recognize flowers of related species.

A fundamental problem still remains concerning the lack of large datasets with publicly available metadata, a key requirement for successfully training a deep neural network. Currently, however, there is a dearth of such data in most phenotyping applications. It is expected that, as more datasets are collected and labeled, the training data will increase over time. Manual image labeling will be critical to improving both the quality and the quantity of the data needed for deep learning. Furthermore, data augmentation strategies can increase the size of the training data by using various geometric and photometric transformations to improve network accuracy.

Two different types of deep networks will be common in the context of plant phenotyping, based on their scope. Feature extraction networks are trained to compute individual features directly from images as phenotypes. On the other hand, object detection networks (which often include a feature extraction network, followed by a detection network) are trained to detect parts of a plant, e.g., leaves, flowers, and fruits, for specific phenotyping tasks, such as emergence detection of a new leaf, counting the total number of flowers in an image, and tracking the individual fruits in the image sequence for automated growth estimation.

1.9.4 Shape Modeling and Simulation for Phenotyping

Modeling the complex geometry of plants has received considerable attention in computer graphics (Rozenberg and Salomaa 1980). There is a large body of research involved in interactively modeling trees, that are represented as a collection of articulated rigid bodies, and in simulating their motion under the influence of wind and obstacles (Quigley et al. 2017). The enormous progress in this research field over the years has been successful in developing very high quality and extremely precise models of complex geometrical structures, and even plants with hairs (Fuhrer et al. 2004). The 3D modeling of plants, using multiview image sequences, has paved a new avenue of applications in plant phenotyping. The increasing popularity of high-throughput plant phenotyping systems has made the multiview image sequences easier to acquire, making the reconstruction and subsequent modeling and phenotyping analysis more feasible. Accurate modeling of the plant geometry can be used to simulate the behavior of the actual plants exposed to different environmental conditions and then to compute useful phenotypes, e.g., branching structure, internode distance, and the cross-section area of the stem, using a fully automated approach.

For example, Pirk et al. (2012) have developed a dynamic tree modeling and representation technique, that allows the 3D models of a tree to react interactively to changes in environmental conditions, as a result of light interception and proximity of obstacles, that cause mechanical stress to the plants. The method may be adopted to study differential growth patterns of plants of different genotypes in response to a variety of environmental stresses, e.g., drought and salinity. The method of Wang et al. (2005), for real-time rendering of plant leaves in response to variation in illumination, has the potential to quantify plant-light interaction, including subsurface scattering at the level of individual photons. Whereas there has been some recent work in using shape modeling to segmenting leaves for application in phenotyping (Chen et al. 2019), vast opportunities remain for the exploration of shape modeling techniques for plant phenotyping analysis. Another useful avenue of investigation of shape analysis is in the generation of augmented datasets, that may prove critical in deep-learning applications.

1.9.5 Event-Based Phenotypes

One of the unique aspects of plants is that they continue to develop throughout their life cycle, producing new tissues and organs. Furthermore, different parts of a plant have different growth habits. Plants not only develop new organs and bifurcate to different components during their life cycle, but they also show symptoms of senescence. Studying the important biological events in a plant's life, i.e., the phenology, may provide insightful phenotypes. Possible events may include germination, leafing, flowering, the onset of senescence, etc. The timing of important events is crucial in the understanding of the vigor of the overall plant, which is likely to vary with the climate, and these are referred to as event-based phenotypes.

Event detection in plants is a challenge since the change patterns are unlike the visual tracking of rigid bodies, e.g., vehicles and pedestrians, and hence it requires a different problem formulation. Whereas the movement of rigid bodies is characterized by a change in location, the shapes of different parts of a plant are continuously, but not uniformly, changing in both geometry and topology. Furthermore, the rate of change (both growth and senescence) is typically more gradual than the rigid body motion. Phyllotaxy, the plant's mechanism to optimize light interception by repositioning the leaves, adds another layer of complexity, that leads to self-occlusions and leaf crossovers. Some notable research in this domain includes detection of budding and bifurcation events from 4D point clouds, using a forward-backward analysis framework (Li et al. 2013) and plant emergence detection and tracking of the coleoptile based on adaptive hierarchical segmentation and optical flow, using spatiotemporal image sequence analysis (Agarwal 2017).

Recent developments in hyperspectral image analysis provide another pathway for accurate detection of the timing of important events, because the images of the plant can be captured at very narrow wavelengths, using a hyperspectral camera. Thus, events that generate signals at specific wavelengths can be captured in this imaging modality but not by cameras (e.g., visible light) that image only at low chromatic resolutions, i.e., over a wide band of wavelengths. Hence, hyperspectral imaging has the potential to accurately compute various event-based phenotypes, including the transition from vegetative to reproductive stages, detection of buds occluded by leaves, quantification of mineral content in a stalk, and internal decay of a fruit. However, hyperspectral imaging, with its many narrow bands, causes a massive increase in the amount of data that needs to be stored, and also requires data analysis techniques to find the best combination of bands uniquely suited for a particular phenotyping task.

1.10 SUMMARY

Phenotypes refer to various traits of plants and have been studied by researchers since the origin of plant science. Phenotypes are central to understanding the relationship between the genotype of the plant and the environment, both biotic and abiotic. Recent developments in image-based phenotyping have overcome many challenges associated with traditional approaches, and provide an opportunity to systematically study the relationship that can help address food security in this time of global climate change and growing population. This chapter provided a discussion on the national and regional plant phenotyping networks that have been organized, with the common mission of advancing research in the field of plant phenotyping analysis. It also emphasized that the long-term management of imaging datasets must adhere to FAIR principles, i.e., findability, accessibility, interoperability, and reusability, which can be achieved by data standardization and data integration. There is already a solid body of research in image-based plant phenotyping. However, several important challenges must be addressed to extract phenotyping information from image sequences as we move forward. The most salient challenges are summarized in this chapter. Furthermore, the emerging trends in image-based plant phenotyping are discussed. This research is in its infancy but holds significant promise to address some of the most important problems in plant and crop science.

REFERENCES

Agarwal, B. 2017. Detection of plant emergence based on spatio temporal image sequence analysis. MS Thesis, University of Nebraska-Lincoln, Lincoln, NE.

Armonienė, R., F. Odilbekov, V. Vivekanand, and A. Chawade. 2018. Affordable imaging lab for noninvasive analysis of biomass and early vigour in cereal crops. *BioMed Research International* 2018:1–9. doi: 10.1155/2018/5713158.

Atkinson, J., A. M. P. Pound, M. J. Bennett, and D. M. Wells. 2019. Uncovering the hidden half of plants using new advances in root phenotyping. *Current Opinion in Biotechnology* 55:1–8.

Baweja, H. S., T. Parhar, O. Mirbod, and S. Nuske. 2017. Stalknet: A deep learning pipeline for high-throughput measurement of plant stalk count and stalk width. In *Field and Service Robotics*, eds. M. Hutter and R. Siegwart. 271–284. Springer, Cham.

Bizer, C., T. Heath, and T. Berners-Lee. 2009. Linked data-the story so far. *International Journal on Semantic Web and Information Systems* 5(3):1–22.

Brichet, N., C. Fournier, O. Turc, O. Strauss, S. Artzet, C. Pradal, C. Welcker, F. Tardieu, and L. Cabrera-Bosquet. 2017. A robot-assisted imaging pipeline for tracking the growths of maize ear and silks in a high-throughput phenotyping platform. *Plant Methods* 13:96. doi: 10.1186/s13007-017-0246-7.

Camargo, A. V., and G. A. Lobos. 2016. Latin America: A development pole for phenomics. *Frontiers in Plant Science* 7:1729.

Carroll, A. A., J. Clarke, N. Fahlgren, M. A. Gehan, C. J. Lawrence-Dill, and A. Lorence. 2019. NAPPN: Who we are, where we are going, and why you should join us! *The Plant Phenome Journal* 2(1):1. doi: 10.2135/tppj2018.08.0006.

Chawade, A., J. van Ham, H. Blomquist, O. Bagge, E. Alexandersson, and R. Ortiz. 2019. High-throughput field-phenotyping tools for plant breeding and precision agriculture. *Agronomy* 9(5):258.

Chen, L., and K. Wang. 2014. Diagnosing of rice nitrogen stress based on static scanning technology and image information extraction. *Journal of Soil Science and Plant Nutrition* 14:382–393. doi: 10.4067/S0718-95162014005000030.

Chen, Y., S. Baireddy, E. Cai, C. Yang, and E. J. Delp. 2019. Leaf segmentation by functional modelling. In *Proceedings of the IEEE Conference on Computer Vision and Pattern Recognition Workshops.* Long Beach, CA.

Cruz, J. A., X. Yin, X. Liu, S. M. Imran, D. D. Morris, D. M. Kramer, and J. Chen. 2016. Multimodality imagery database for plant phenotyping. *Machine Vision and Applications* 27(5):735–749. doi: 10.1007/s00138-015-0734-6.

Ćwiek-Kupczyńska, H., T. Altmann, D. Arend et al. 2016. Measures for interoperability of phenotypic data: Minimum information requirements and formatting. *Plant Methods* 12:44. doi: 10.1186/s13007-016-0144-4.

Das Choudhury, S., V. Stoerger, A. Samal, J. C. Schnable, Z. Liang, and J.-G. Yu. 2016. Automated vegetative stage phenotyping analysis of maize plants using visible light images. In *KDD Workshop on Data Science for Food, Energy and Water*. San Francisco, CA.

Das Choudhury, S., S. Goswami, S. Bashyam, A. Samal, and T. Awada. 2017. Automated stem angle determination for temporal plant phenotyping analysis. In *ICCV Workshop on Computer Vision Problems in Plant Phenotyping* 41–50. Venice, Italy.

Das Choudhury, S., S. Bashyam, Y. Qiu, A. Samal, and T. Awada. 2018. Holistic and component plant phenotyping using temporal image sequence. *Plant Methods* 14:35.

Das Choudhury, S., A. Samal, and T. Awada. 2019. Leveraging image analysis for high-throughput plant phenotyping. *Frontiers in Plant Science* 10:508.

Deng, J., W. Dong, R. Socher, L.-J. Li, K. Li, and F.-F. Li. 2009. ImageNet: A large-scale hierarchical image database. In *IEEE Conference on Computer Vision and Pattern Recognition* 248–255. Miami Beach, FL.

FAO. 2017. The future of food and agriculture–Trends and challenges. *Annual Report*. http://www.fao.org/3/a-i6583e.pdf.

Fuhrer, M., H. W. Jensen, and P. Prusinkiewicz. 2004. Modeling hairy plants. In *Proceedings of the 12th Pacific Conference on Computer Graphics and Applications* 217–226. Seoul, South Korea.

Gage, J. L., N. D. Miller, E. P. Spalding, S. M. Kaeppler, and N. de Leon. 2017. TIPS: A system for automated image-based phenotyping of maize tassels. *Plant Methods* 13:21.

Gampa, S. 2019. A data-driven approach for detecting stress in plants using hyperspectral imagery. MS Thesis, University of Nebraska-Lincoln, Lincoln, NE.

Giuffrida, M. V., P. Doerner, and S. A. Tsaftaris. 2018. Pheno-Deep Counter: A unified and versatile deep learning architecture for leaf counting. *The Plant Journal* 96(4):880–890. doi: 10.1111/tpj.14064.

González-Sanchez, A., J. Frausto-Solis, and W. Ojeda. 2014. Predictive ability of machine learning methods for massive crop yield prediction. *Spanish Journal of Agricultural Research* 12(2):313–328. doi: 10.5424/sjar/2014122-4439.

Hamuda, E., M. Glavin, and E. Jones. 2016. A survey of image processing techniques for plant extraction and segmentation in the field. *Computers and Electronics in Agriculture* 125:184–199.

Hasan, M. M., J. P. Chopin, H. Laga, and S. J. Miklavcic. 2018. Detection and analysis of wheat spikes using convolutional neural networks. *Plant Methods* 14:100. doi: 10.1186/s13007-018-0366-8.

Huang, H., L. Liu, and M. O. Ngadi. 2014. Recent developments in hyperspectral imaging for assessment of food quality and safety. *Sensors* 14(4):7248–7276. doi: 10.3390/s140407248.

Kim, Y., D. M. Glenn, J. Park, H. K. Ngugi, and B. L. Lehman. 2011. Hyperspectral image analysis for water stress detection of apple trees. *Computers and Electronics in Agriculture* 77(2):155–160. doi: 10.1016/j.compag.2011.04.008.

Lenzerini, M. 2002. Data integration: A theoretical perspective. In *Proceedings of the twenty-first ACM SIGMOD-SIGACT-SIGART Symposium on Principles of Database Systems* 233–246.

Li, Y., X. Fan, N. J. Mitra, D. Chamovitz, D. Cohen-Or, and B. Chen. 2013. Analyzing growing plants from 4D point cloud data. *ACM Transactions on Graphics* 32(6):1–10.

Li, L., Q. Zhang, and D. Huang. 2014. A review of imaging techniques for plant phenotyping. *Sensors* 14(11):20078–20111. doi: 10.3390/s141120078.

Lobos, G. A., A. V. Camargo, A. del Pozo, J. L. Araus, R. Ortiz, and J. H. Doonan. 2017. Plant phenotyping and phenomics for plant breeding. *Frontiers in Plant Science* 8:2181.

Lu, Y., Y. Huang, and R. Lu. 2017. Innovative hyperspectral imaging-based techniques for quality evaluation of fruits and vegetables: A review. *Applied Sciences* 7(2):189. doi: 10.3390/app7020189.

Lynch, M., and B. Walsh. 1998. *Genetics and Analysis of Quantitative Traits*. Sinauer, Sunderland, MA.

Maddonni, G. A., M. E. Otegui, B. Andrieu, M. Chelle, and J. J. Casal. 2002. Maize leaves turn away from neighbors. *Plant Physiology* 130(3):1181–1189.

Moghimi, A., C. Yang, M. E. Miller, S. F. Kianian, and P. M. Marchetto. 2018. A novel approach to assess salt stress tolerance in wheat using hyperspectral imaging. *Frontiers in Plant Science* 9:1182. doi: 10.3389/fpls.2018.01182.

Namin, T., S. M. Esmaeilzadeh, M. Najafi, T. B. Brown, and J. O. Borevitz. 2018. Deep phenotyping: Deep learning for temporal phenotype/genotype classification. *Plant Methods* 14:66. doi: 10.1186/s13007-018-0333-4.

Nowogrodzki, A. 2020. Eleven tips to work with large data sets. *Nature* 577(7790):439–440.

Pauli, D., S. C. Chapman, R. Bart, C. N. Topp, C. J. Lawrence-Dill, J. Poland, and M. A. Gore. 2016. The quest for understanding phenotypic variation via integrated approaches in the field environment. *Plant Physiology* 172(2):622–634.

Pirk, S., O. Stava, J. Kratt, M. A. M. Said, B. Neubert, R. Měch, B. Benes, and O. Deussen. 2012. Plastic Trees: Interactive self-adapting botanical tree models. *ACM Transactions on Graphics* 31(4):1–10. doi: 10.1145/2185520.2185546.

Pommier, C., C. Michotey, G. Cornut, et al. 2019. Applying FAIR principles to plant phenotypic data management in GnpIS. *Plant Phenomics* 2019:1671403. doi: 10.34133/2019/1671403.

Quigley, E., Y. Yu, J. Huang, W. Lin, and R. Fedkiw. 2017. Real-time interactive tree animation. *IEEE Transactions on Visualization and Computer Graphics* 24(5):1717–1727.

Reynolds, D., F. Baret, C. Welcker et al. 2019. What is cost-efficient phenotyping? Optimizing costs for different scenarios. *Plant Science* 282:14–22.

Roll-Hansen, N. 2009. Sources of Wilhelm Johannsen's genotype theory. *Journal of the History of Biology* 42(3):457–493.

Römer, C., M. Wahabzada, A. Ballvora et al. 2012. Early drought stress detection in cereals: Simplex volume maximization for hyperspectral image analysis. *Functional Plant Biology* 39(11):878–890. doi: 10.1071/FP12060.

Rozenberg, G., and A. Salomaa. 1980. *Mathematical Theory of L Systems*. Academic Press, Inc., Orlando, FL.

Singh, A. K., B. Ganapathysubramanian, S. Sarkar, and A. Singh. 2018. Deep learning for plant stress phenotyping: Trends and future perspectives. *Trends in Plant Science* 23(10):883–898. doi: 10.1016/j.tplants.2018.07.004.

Slota, M., M. Maluszynski, and I. Szarejko. 2017. Root phenotyping pipeline for cereal plants. In *Biotechnologies for Plant Mutation Breeding*, eds. J. Jankowicz-Cieslak, T. H. Tai, J. Kumlehn, and B. J. Till. Springer, Cham.

Ubbens, J., M. Cieslak, P. Prusinkiewicz, and I. Stavness. 2018. The use of plant models in deep learning: An application to leaf counting in rosette plants. *Plant Methods* 14:6. doi: 10.1186/s13007-018-0273-z.

Uchiyama, H., S. Sakurai, M. Mishima, D. Arita, T. Okayasu, A. Shimada, and R. Taniguchi. 2017. An easy-to-setup 3D phenotyping platform for Komatsuna dataset. In *Proceedings of the IEEE International Conference on Computer Vision Workshops* 2038–2045. Venice, Italy.

Uga, Y., K. Sugimoto, S. Ogawa et al. 2013. Control of root system architecture by DEEPER ROOTING 1 increases rice yield under drought conditions. *Nature Genetics* 45(9):9. doi: 10.1038/ng.2725.

van Maarschalkerweerd, M., and S. Husted. 2015. Recent developments in fast spectroscopy for plant mineral analysis. *Frontiers in Plant Science* 6:169. doi: 10.3389/fpls.2015.00169.

Wang, L., W. Wang, J. Dorsey, X. Yang, B. Guo, and H. Shum. 2005. Real-time rendering of plant leaves. *ACM Transactions on Graphics* 24(3):712–719.

Wen, L., L. Duan, D. Xu, and I. W. Tsang. 2013. Learning with augmented features for supervised and semi-supervised heterogeneous domain adaptation. *IEEE Transactions on Pattern Analysis and Machine Intelligence* 36:1134–1148.

Xiong, X., L. Duan, L. Liu, H. Tu, P. Yang, D. Wu, G. Chen, L. Xiong, W. Yang, and Q. Liu. 2017. Panicle-SEG: A robust image segmentation method for rice panicles in the field based on deep learning and superpixel optimization. *Plant Methods* 13:104. doi: 10.1186/s13007-017-0254-7.

Xu, R., C. Li, A. H. Paterson, Y. Jiang, S. Sun, and J. S. Robertson. 2018. Aerial images and convolutional neural network for cotton bloom detection. *Frontiers in Plant Science* 8:2235. doi: 10.3389/fpls.2017.02235.

Yu, J.-G., G.-S. Xia, A. Samal, and J. Tian. 2016. Globally consistent correspondence of multiple feature sets using Proximal Gauss-Seidel relaxation. *Pattern Recognition* 51:255–267.

Zheng, L., J. Zhang, and Q. Wang. 2009. Mean-shift based color segmentation of images containing green vegetation. *Computers and Electronics in Agriculture* 65(1):93–98.

2 Multisensor Phenotyping for Crop Physiology

Stefan Paulus, Gustavo Bonaventure, and Marcus Jansen

CONTENTS

2.1 CROP PHENOTYPING

2.1.1 Breeding for Crop Performance and Yield

Crop plants deliver various materials for human use (Franke 1997). They are not only the prime source of human food and feed for domestic animals, but they also provide raw materials like fibers, chemicals, medical substances, and material for heating and energy. Depending on the intended use, different parts of the plants represent the desired economic yield (Brown et al. 2015). The harvested material can be leaves in the case of many vegetables, seeds of grain crops, tubers, fruits from other crops, fibers from stems or leaves, pigments and compounds from any plant organ, or the overall biomass to use as fuel. Moreover, the correct phenotypic appearance of leaves and flowers is essential for ornamental plants. Thus, it is important that the crops deliver the appropriate phenotypic properties in terms of quantity and quality

(Tardieu et al. 2017). As the yield is determined by phenotypic traits, measuring phenotypes is essential for many purposes in crop science and breeding (Barabaschi et al. 2016). A second important factor beyond the case-specific yield is the performance of crop growth and development that is necessary to achieve the yield. For this performance, high requirements are imposed in terms of efficiency. This means that crops need to reach the yield with limited use of resources, mainly water and nutrients, which are limited in most cases, with resource availability possibly changing with time during the growth period. Crop efficiency can comprise tolerance to high or low temperatures, to excess light, to the wind or other environmental constraints. A further prominent factor in crop performance is resistance to pathogens and pests, that, in turn, would mitigate the demand for crop protection chemicals in agriculture. Therefore, determining crop growth and development at the phenotypic level in time-resolved studies is an important tool with which to address plant responses to the environment, particularly to stress factors.

Crop breeding drives the transition from wild plants to plants better suited to human use, with continuous improvement (Brown et al. 2015). Thereby, particularly in the early stages of breeding, the phenotypic appearance of the plants is the key selection factor that determines the success of the breeding effort. Before knowledge on genetics was available, phenotypic analyses were the only way to assess plant quality in breeding efforts. In the genomic era, phenotyping is still highly important as it documents and quantifies the success of all breeding, conventional or molecular (Cabrera-Bosquet et al. 2012). Newly bred plant material must deliver desirable phenotypes not only under the controlled conditions of the breeder's field but particularly under the variable conditions under which the farmers grow the new varieties.

Modern crop physiology and breeding rely strongly on genomic work and use molecular markers to elucidate physiological functions and to accelerate breeding (Barabaschi et al. 2016). All efforts in genomic improvement of crops are valuable only if the genomic potential can be turned into improved phenotypic performance under the relevant cultivation conditions. This means that, despite the environment's influence on the phenotypic development of the plants, crops require robust response ranges to ensure adequate performance, even under suboptimal environments. Therefore, the essential phenotypic traits that deliver the superior yield must be resilient to adverse environmental factors, so that yield penalties stay within acceptable ranges under stresses that occur under usual agricultural practice. Ideally, yield-determining phenotypic properties should exhibit low amplitudes of responses, so that yield is not completely lost, for instance, when extreme weather events occur during cultivation. To achieve this, crop plants need strong overall plasticity to cope with various environmental situations during their growth.

2.1.2 Purpose of Phenotypic Image Analysis

When analyzing phenotypic traits of crops, researchers and breeders increasingly make use of image-based technologies that allow measurement of properties of the plants without touching, damaging, or destroying the current plant sample (Jansen et al. 2014; Walter et al. 2015). Thus, the use of various optical instruments to record such data is followed by sophisticated procedures to derive information out of the recordings, that relate to the developmental, morphological, and physiological status of the plant. Thereby, the non-invasive recording serves as a direct or indirect measure of the plant's properties. Direct measures, such as height or width of a plant, mainly refer to structural and morphological traits (Walter et al. 2007), whereas the indirect measures address physiological properties, e.g. chlorophyll fluorescence as an indicator of stress response (Hairmansis et al. 2014). Indirect measures usually require mathematical or algorithmic approaches to translate the measure into biologically interpretable information. In current phenotyping, there are two main trends of purposes for which sensor recordings are used: (a) comprehensive characterization and (b) measures targeting decision making.

Firstly, scientists record broad ranges of information, frequently with multi-sensor platforms, with high temporal and spatial resolution, and use the data for the comprehensive characterization of the samples. This frequently occurs in basic research and is driven by scientific interest. Such cases frequently focus on the phenome as such, and ultimately intend to further develop and improve phenotyping algorithms and tools as well.

Secondly, industrial users or scientists who employ phenotyping methods as a tool intend to make decisions based on certain well-defined phenotypic properties of the samples. For instance, users may demand information about whether seeds have germinated, or whether a certain substance is toxic. For example, the phenotypic appearance of duckweed changes upon contact with toxic substances in the growth medium, and these changes serve as an indicator in ecotoxicology, when testing agrochemicals, food additives, technical chemicals, or drugs.

2.1.3 Intelligent Image Analysis

After recording, the images and related sensor records need further processing to transform the raw data into information (Jansen et al. 2018; Paulus et al. 2016). This process is purpose-driven, and thus the term "intelligent" is used, depending on what type of information needs to be derived from the recordings (Chen et al. 2014; Rousseau et al. 2015).

As described in the previous section, many applications require decision making, based on phenotypic information. In such cases, processing intelligence implies that the sensor records are analyzed in a way that the system automatically derives the decision from the data. For instance, with seed germination detection, as long as the seed shape stays as it was in the beginning, seeds have not yet germinated, but, as soon as a cylindrical outgrowth, i.e. a root, appears, germination has taken place. The analysis then changes the seed status from "not germinated" to "germinated." Detecting root outgrowth, using image processing, then allows a decision maker to determine whether germination had taken place. Moreover, intelligent image processing allows taking the plant performance as an indicator of proper growth conditions, when cultivating plants under conditions of a controlled environment. Phenotypic traits can be used to monitor proper irrigation in greenhouse cultivation using mathematical models (Guo et al. 2017). Thereby, the phenotypic performance of the plants allows decisions to be taken on the irrigation procedure during cultivation. Thus, the intelligence of the image processing lies in enabling a decision to be made, that facilitates the next steps in a defined process, thereby ensuring a certain level of quality.

In research-driven approaches, where comprehensive characterizations are required, intelligent analysis of the recorded images aims not at immediate decision making, but to provide a broad range of plant-related phenotypic data that can be studied in relation to molecular and biochemical data of the same set of plants. In this case, the phenotypic data help explain gene functions, facilitate the screening of populations, or provide insight into responses to the environment.

Overall, intelligent image analysis combines advanced technology for processing images, with interpretation techniques to connect measured parameters with plant traits. Therefore, intelligent image processing requires a clear understanding of the critical traits of the plants, so that the image-based algorithms can compute the trait-related information.

2.1.4 Critical Traits for Seed Crop Improvement

Image analysis methods have typically employed fixed sequences of image processing and measurement processes, crafted by their respective designers to suit specific purposes, that originally might be unrelated to plant analysis. As a result, applying a specific tool to a slightly different problem or environment – which commonly occurs in plant science research – often required a substantial modification of the software, if not a near-complete rewrite (Tardieu et al. 2017). Moreover, present analysis methods are very limited in their capacity to detect and quantify those plant traits which are critical to improving crop performance. In other words, most critical crop performance-related traits are usually either not detected or are detected very inefficiently by current image-processing methods.

Seed yield, for example, is one of the most important and complex traits in crops and reflects the interaction of the environment with all the growth and development processes that occur throughout the life cycle of the plant. Seed yield is usually defined as the weight of seeds per unit of land area at a specific percentage of moisture content. Although extensive research into the genetic basis of yield and yield-associated traits has been reported, in each such experiment the genetic architecture and determinants of seed yield have remained ambiguous and largely dependent on environmental conditions. Being a very complex trait, seed yield is usually studied as the result of a combination of multiple yield-component traits, such as seed weight and seed number per plant, and yield-related traits such as growth rate, canopy architecture, root functionality, resistance to biotic stresses and tolerance to abiotic ones, photosynthetic capacity, etc. To be useful for breeders and scientists studying yield-related problems, phenotype analysis tools need to provide the biological information which is critical to improving crop performance.

In rice (*Oryza sativa*), for example, slow grain filling is a yield-limiting factor in cultivars with large panicles and many spikelets. Rice cultivars with these characteristics often fail to achieve their high yield potential due to poor grain filling of inferior spikelets (located on secondary branches of the panicle) in contrast to superior spikelets, located on primary branches of the panicle (Zhu et al. 2011). Therefore, information on reproductive traits such as the rate of seed growth, number of empty and filled seeds, distribution of empty and filled seeds in the panicle, number of panicle branches, and vegetative traits such as photosynthetic capacity, stalk functionality, and senescence rate are critical for rice breeders and plant scientists. Additional important reproductive traits for rice include number of seeds per plant, seed mass, panicle size, number of panicles per plant, number of primary, secondary, and higher order panicle branches, length of panicle branches, number of spikelets per unit of panicle branch length, panicle growth rate, and grain growth during filling stage. Among vegetative traits for rice are leaf area duration, leaf area index (green leaf area per unit ground surface area), photosynthetic capacity, and rate of senescence.

Similarly, for maize (*Zea mays*), critical reproductive traits for yield are the number of kernels per plant, ear size, kernel abortion, grain-filling period, and flowering date. Among important vegetative traits for corn are growth rate between specific growth and reproductive stages, root growth and functionality, photosynthetic capacity, and stalk functionality (e.g., transport). Finally, for soybean (*Glycine max*), some critical reproductive traits for seed yield are the number of seeds per plant, seed mass, and the number of viable pods per plant. Some critical vegetative traits include crop growth duration and rate, number of root branches, root length, and photosynthetic capacity.

In summary, most high-precision/high-throughput phenotyping platforms are still not taking full advantage of all the information contained in the captured images and hence there are opportunities for significant developments and improvements in the future. Advances in the areas of image analysis methods and sensor data fusion are fundamental to extracting biologically meaningful information (e.g., yield-related traits mentioned above for rice, maize, and soybean), with which researchers and breeders can exploit the full potential of the acquired data. Moreover, the development of methods to perform meta-analyses of phenotypic data, including environmental conditions, and the incorporation of data collected by farmers will be critical to the pathway from sensor to knowledge (Tardieu et al. 2017) and hence to the improvement of crop performance. The fusion of images obtained with diverse sensors in conjunction with new advances in deep-learning methods, including approaches such as convolutional neural networks (CNNs), are starting to show promising and generic solutions for the identification and quantification of important yield-related traits (Hawkesford and Lorence 2017; Pound et al. 2017; Sadeghi-Tehran et al. 2017a; Tardieu et al. 2017).

2.2 CAMERAS AND SENSORS FOR CROP MEASUREMENTS

2.2.1 RGB Imaging

Consumer-grade digital cameras have been in the market for more than 20 years and are widely used in many applications. As their imaging mode corresponds to human vision, such red-green-blue (RGB) images are most intuitive for plant rating, and information derived from RGB images is used in a broad range of applications (Berger et al. 2012; Fahlgren et al. 2015). Digital RGB imaging uses lens configurations and CCD (charge-coupled device) chips to acquire brightness information. Filters on the CCD array, sensitive to the red, green, and blue wavelengths, enable the capturing of color information. The combination of pixels of different colors is named de-mosaicking. There are many de-mosaicking algorithms, each resulting in a slightly different end image. Based on the sensor and lens configuration, the imaged area can be adapted for microscopic, near-field, wide-field, or even telecentric imaging.

In the field of plant phenotyping, the demands for image processing, using digital RGB images, started at the turn of the millennium. Beginning with automated rating of duckweed (*Lemna minor*) for ecotoxicological tests, high-throughput screening and automated analysis of larger plants for laboratory, greenhouse, and complete field screening systems have been developed (Arvidsson et al. 2011; Cleuvers and Ratte 2002; Rajendran et al. 2009; Virlet et al. 2017).

2.2.1.1 Greenhouse Parameters Recorded with RGB Cameras

Today, high-throughput phenotyping greenhouses include high levels of automation. In plant-to- sensor systems like the Greenhouse Scanalyzer, the plants are located on a conveyor system that moves the plants to measuring and handling units, that perform automated weighing, watering, spraying, or imaging. Images are recorded from top or side view to capture comprehensive sets of parameters. Typical parameters are plant height, width, diameter, and convex-hull, based on geometry or the ratio of green–yellow–brown pixels to detect the influence of biotic or abiotic stresses.

An experiment with tomato plants was conducted on a Greenhouse Scanalyzer plant-to-sensor system. The plant measurements were taken for 33 days, using RGB cameras for top and side images, illuminated by halogen lights. For the side views, each plant was imaged at 90° intervals for a total of four images. The size of the plant was determined to be the average number of pixels in the top view and the average (of the four) side views. Ten plants in two treatment groups were measured. Five plants were watered with 400 ml irrigation, whereas the other five plants were watered with 600 ml irrigation. The number of pixels was extracted for each image, using a sequence of image-processing steps (Figure 2.1) and averaged to derive one measurement for each treatment per day. Furthermore, the number of green and yellow pixels was measured; yellow pixels served as an indicator for senescent plant

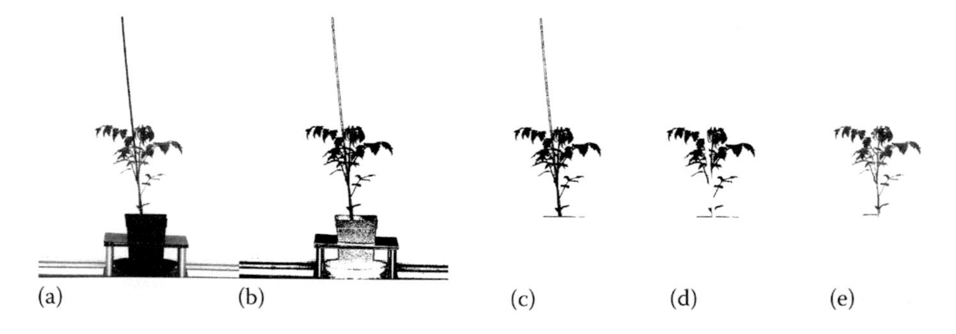

FIGURE 2.1 Image processing steps to derive the size of the plants. (a) shows the original RGB image, (b) shows the image after foreground/background separation, (c) and (d) show the removal from the stick and the plant pot and (e) shows the final cut-out of the plant and its analysis.

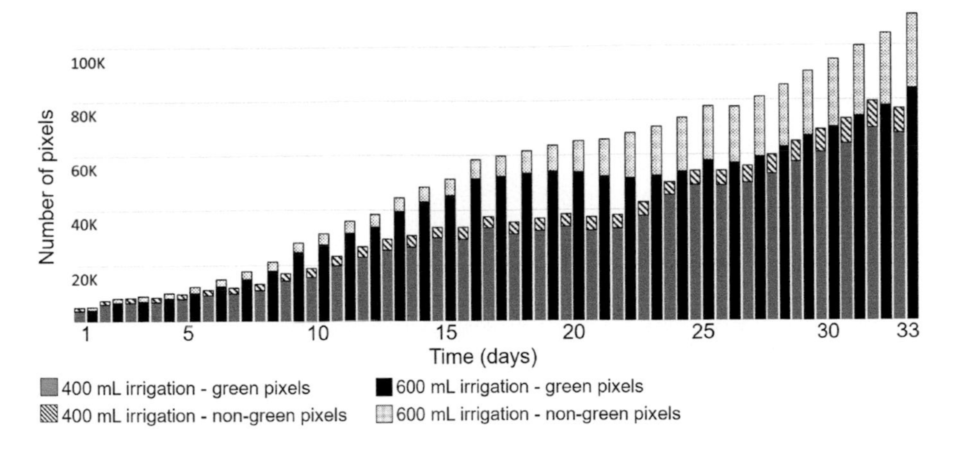

FIGURE 2.2 The growth curve, including the number of green/yellow pixels for the plants. This indicates that an improvement in growth is associated with a larger number of yellow or non-green pixels.

tissue. The resulting growth curves shown in Figure 2.2 showed that greater growth in the group with the higher irrigation levels was associated with a significantly higher fraction of yellow or non-green pixels, indicating that the higher irrigation promoted senescence progression.

The complete workflow was fully automated and could be repeated without any human intervention. Experiments with other treatments, a larger number of plants per treatment, or even with different plant species could easily be adapted.

2.2.1.2 Field Parameters

In general, plant analysis on a field scale is significantly different from greenhouse and laboratory-scale experiments (Cendrero-Mateo et al. 2017). Common problems, such as illumination, reflections, movement due to the wind, differences in solar radiation, shadowing, and changes due to a moving illumination from the sun are either absent or are minor under controlled conditions in the greenhouse/laboratory where imaging is typically performed under clear conditions.

Field measurements can be recorded by hand-held cameras, mobile devices with a camera- illumination setup, by using gantry systems, or by using unmanned aerial vehicles (UAVs). Imaging is typically a trade-off between image quality and the size/cost of the imaging setup. UAVs carry small, light-weight cameras, while handheld devices have low positioning accuracy and tend to move during data acquisition, which results in lower-quality images. Gantry systems can record images with the highest technical diversity, as they have a huge payload and thus can carry a full range of imaging and sensing equipment. They also have the greatest accuracy in positioning the measuring head. The trade-offs between cost, size, and quality have to be adapted, based on the goals of an experiment.

The LemnaTec Field Scanalyzer, a multi-sensor system on a gantry, carries cameras with different modalities, including visible light, fluorescence, hyperspectral, thermal, and laser, along with a set of sensors to measure environmental variables

FIGURE 2.3 LemnaTec Field Scanalyzer gantry system in the construction phase on an experimental field at the University of Arizona. The multi-sensor box (the white box in the middle) can be moved in three directions over an area of 20 m × 200 m. The measuring distance between the camera and plants can be adjusted.

(Virlet et al. 2017). Such systems enable the use of stereo setups for the generation of 3D point clouds from RGB images. Furthermore, very high-resolution images can be obtained, as the hard drives can be mounted on the system for storage. A gantry-based system, that covers a field of the size of 20 m × 200 m, carrying three stereo-RGB camera pairs with a resolution of eight megapixels each, is shown in Figure 2.3. The gantry can record in a stop-move-stop mode or continuously at lower speeds. In addition to the RGB cameras, the system is also equipped with a chlorophyll fluorescence imaging system, laser scanners for 3D imaging, and thermal cameras.

RGB images from the field can be used to derive different parameters, based on morphology and surface color. By identifying plants by their color, the images can be used to estimate ground cover, thus deriving information on the developmental status of the canopy (Sadeghi-Tehran et al. 2017b). Whereas, in laboratory and greenhouse studies, data are frequently recorded for single plants, field studies usually work at the canopy level. A piece of important information on canopy development involves determining the plant growth stages, which indicate the progress of vegetative and reproductive development. By implementing advanced image-processing tools, growth stages can be determined using a Field Scanalyzer (Sadeghi-Tehran et al. 2017a).

Furthermore, additional details about the plants, such as the number of ears on cereal plants, can be determined, which helps in computing the ear density per crop area (Sadeghi-Tehran et al. 2017a; Virlet et al. 2017).

2.2.2 3D Laser Imaging

Laser scanning in plant measurements typically uses the laser triangulation technique to measure the plant's geometry. The method uses a laser emitter with an adapted laser wavelength and a 3D camera, that uses a pre-calibrated matrix to record 3D points in space. This results in a very accurate point cloud. A point cloud can be depicted as a set of pointwise coordinate measurements of the plant's surface with an accuracy between microns to millimeters. A bottleneck in the laser triangulation principle requires a trade-off between measurable volume and resolution. Either a whole plant can be measured, or the details of plant organs, such as leaves, flowers, or fruit can be resolved, but not both.

In addition to a single laser scanner, a combination of laser scanners to cover different points of view are often used to capture a plant's geometry. The latter can be designed by a moving sensor adapted to a linear axis, a rotary table, or a robotic arm. Additional views are combined to reduce the effects of occlusion in the resulting point cloud. Although different techniques exist to measure the 3D surface of plants, including structured light (Bellasio et al. 2012) approaches, time of flight techniques (Alenya et al. 2011), or even structure from motion, using a set of RGB images (Rose et al. 2015), the laser triangulation method is considered to be state-of-the-art for accuracy of representation, and is widely used.

A variety of representations can be used for the analysis of the 3D data. In addition to the standard type of XYZ data, that describes a point in the 3D space, more informative datatypes, like the use of XYZN datatypes, that also store the point normal, or XYZRGB, that can also store the RGB color, can be used. The native output of the laser scanners is XYZI, which includes an intensity value (I) that encodes the reflectance along with the location coordinates. Reflectance and associated laser color enable the derivation of plant tissue information, like chlorophyll content.

2.2.2.1 Greenhouse Parameters Recorded with Laser Scanners

A direct measurement within the point cloud is possible by measuring distances between single points to obtain a measure of plant height and width (Paulus et al. 2014b). Furthermore, it is possible to derive the surface normal at the points by taking account of its neighborhood to obtain a measure of the leaf angle. The 3D scanning devices are the only sensors that enable to achieve the differentiation between leaf movement/rotation and leaf growth (Dornbusch et al. 2012; Dornbusch et al. 2014). When recording time courses on leaf movements, it is essential to make sure that the position of the plant is not changed between single scans, because this would result in substantial discrepancies when merging the scans.

For plant phenotyping, common industrial laser scanners can be used (Paulus et al. 2014b). Also, scanners with adapted wavelengths in a spectrum that have no or only minor absorption by the plant material may be used (Dornbusch et al. 2015; Virlet et al. 2017). Adaption of the wavelength is essential as a red laser color is highly absorbed by the plant tissue, whereas blue, or better, near-infrared (NIR) laser wavelengths are not (Dupuis et al. 2015).

By using manual or machine-learning-based techniques, it is possible to distinguish the individual organs of the plant (Paulus et al. 2013). This enables monitoring of the growth of single organs, like leaves and stems, over time (Paulus et al. 2014a). The occlusion of the plant by its own organs (e.g. leaf or stem) becomes worse when monitoring for a longer time. In a case study, we recorded point clouds of a wheat plant and analyzed it to derive plant-related information (Figure 2.4). For instance, a height map indicates the overall height together with heights of the plant's leaves.

FIGURE 2.4 A point cloud measured from a barley plant with a high-accuracy 3D laser scanner (a). The Z-axis visualizes a color-coded height map (b). Visualization of the orientation of the point normal where similar normal directions result in similar point colors (c). Using machine- learning algorithms for classification, the different plant organs can be distinguished, as shown here for leaf and stem (d).

Common parameters that can be derived are plant height and width as measurements of plant size (Busemeyer et al. 2013). Furthermore, leaf area and inclination, as well as the volume, using a convex-hull algorithm, can be derived from scanned data to address the dimensions and the spatial extent of the plant and its parts (Azzari et al. 2013; Bellasio et al. 2012). New parameters that have been developed recently due, to the convergence of high-resolution imaging and machine-learning algorithms, to identify specific organs, include single leaf growth, stem size, and growth (Paulus et al. 2014b). An essential step for agronomic use of such data is predicting traits from measured parameters, such as measuring the ear volume as a predictor of grain yield (Paulus et al. 2013).

2.2.2.2 Field Parameters Recorded with Laser Scanners

In the field, the recognition of single plant organs is more challenging than in indoor settings. Depending on the scanning resolution, organs are either not recognizable or just very fuzzy. Simple parameters, that can be collected by state-of-the-art field phenotyping platforms (Dornbusch et al. 2015; Virlet et al. 2017), include height maps, showing the height of the plant encoded in color, or inclination maps, showing the inclination of the surface encoded in color (Figure 2.5). This enables the derivation of growth information for single field plots and can be directly linked to soil, nutrition, and fertilizer information, as it is used in precision agriculture.

Additional algorithms to detect and measure the properties of single organs will be developed in the coming years, using machine- and deep-learning techniques. These methods will use 3D features like point cloud feature histograms (Rusu et al. 2009) to encode the neighborhood geometry of a source point, based on the surface normal at each point. As points do not have a normal, it has to be calculated, using the points in the neighborhood, and by applying a plane- fitting algorithm or by principal component analysis. Machine-learning routines, such as support vector machines (Paulus et al. 2013) or boosted trees, can use this to separate the geometry of ears, stems, and leaves. Thus, an organ-specific parameterization, using triangle meshes or cylinder approximation, is computable. Newer approaches overcome the

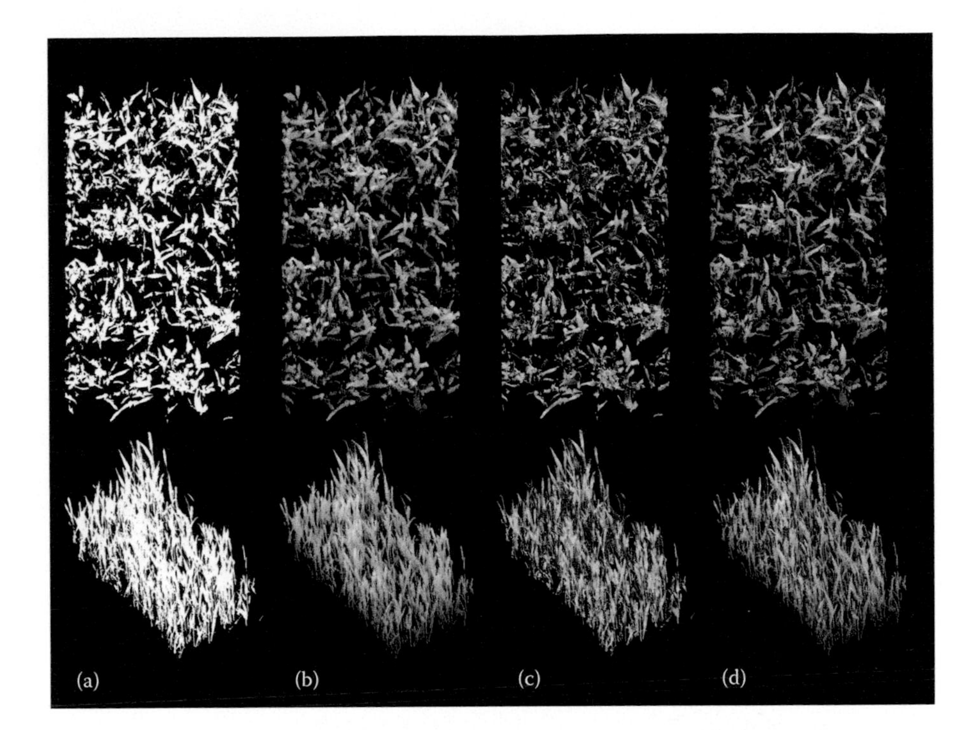

FIGURE 2.5 A wheat field plot measured with a 3D laser scanner. Different views are shown from the top (first row) and the side (second row). The result is a highly accurate 3D point cloud (a). Color-coded height map of the canopy. Bottom points are shown in blue, top points are shown in red (b). Visualization of the orientation of the point normal. Similar normal directions result in a similar point color (c). A state-of-the-art problem in plant phenotyping is the detection of the ears (red) within the point cloud of the plot consisting of ground, stem, leaf, and ear points (d).

use of supervised learning methods and switch to unsupervised methods. Using clustering in the feature space enables the separation of non-trained classes (Wahabzada et al. 2015). One advantage of this approach is the separation of classes that may not have been the focus of the researcher/practitioner. The users typically focus on the separation of leaf and stem, but the unsupervised approach may identify new classes that are, for example, leaf border points, leaf stem transition points, or leaf veins.

2.2.3 Fluorescence Imaging

Distinct from methods using reflected light, fluorescence imaging detects light that is emitted from specific fluorescent compounds, i.e., chemicals that convert a fraction of the incident light into re-emitted fluorescent light. The most prominent example in plants is chlorophyll, which exhibits dark red fluorescence upon illumination. In particular, red and blue wavelength ranges, from the solar spectrum or artificial illumination, cause excitation of fluorescence in chlorophyll. In the presence of reflected green light, the dark red chlorophyll fluorescence is invisible to humans, but, when using optical filters and appropriate excitation light sources, it can be imaged and visualized. In addition to chlorophyll, other pigments in plant tissue can show fluorescence, such as carotenoids.

Sensing and quantifying properties of fluorescence can provide insight into the physiological processes of the plant, and various measuring techniques have been developed to measure this information (Humplik et al. 2015). Predominantly, such measures focus on the presence or activity of chlorophyll, as it is the most significant fluorescent pigment in plants. The measures are based on either single-level static images or a series of images on fluorescence kinetics. Only such kinetic measurements can provide data on photosystem II status and related activities (Baker 2008; Krause and Weis 1991; Schreiber 2004; Strasser et al. 2000). Several parameters, derived from kinetic chlorophyll fluorescence analysis, are used as proxies for photosynthetic functions. Kinetic chlorophyll fluorescence imaging, in combination with visible light imaging, is used in plant phenotyping to indicate stress responses of plants (Jansen et al. 2009; Tschiersch et al. 2017).

In contrast, single-level measures of fluorescence can serve to assess the presence of fluorescent materials and to approximate their prevalence, using signal intensity. Color analysis of fluorescence signals enables differentiation between fluorescent pigments at a qualitative level. Thus, such assessments can serve to determine plant stress in studies on environmental effects.

2.2.3.1 Greenhouse Parameters

Fluorescence imaging under controlled environments is less challenging than under outdoor conditions. In closed rooms, one can use dedicated background material, avoid external light, and keep imaging conditions constant between image acquisitions. Thus, the range of applications is broader than outdoors, where measuring protocols are impacted by sunlight.

In the case of plant stress, both single-level and kinetic fluorescence imaging deliver informative parameters that can qualify and quantify the stress effects and responses. For single-level imaging, stress indicators include the decrease of red, chlorophyll-originating fluorescence and the appearance of fluorescence signals caused by chlorophyll degradation or degradation of other pigments that are usually overlaid by the chlorophyll fluorescence.

In senescence, chlorophyll degrades in multiple steps, resulting in intermediate metabolites. Some of those metabolites are fluorescent (Hörtensteiner 2006), so that a shift in fluorescence signals occurs during senescence progression. Fluorescence imaging was used as an indicator of senescence in rice plants after exposing them to salinity (Hairmansis et al. 2014).

In an example, we exposed a leaf from a tobacco plant grown under low light to one hour of full sunlight, and subsequently imaged the fluorescence (Figure 2.6); a second such tobacco leaf was kept in low light as a reference. While the leaf surface was still looking green, and no damage was visible to the human eye, nor was detectable with an RGB image (Figure 2.6, left), fluorescence imaging – single-level fluorescence and pulse-amplitude-modulation (PAM) fluorescence – both revealed damage after excess light treatment. In the single-level fluorescence image, the red signal originating from chlorophyll was much lower in the sun-exposed leaf compared to the leaf kept at low light (Figure 2.6, center). The PAM fluorescence image

FIGURE 2.6 Tobacco leaves with (left-hand leaf) and without light stress (right-hand leaf). RGB image: no damage or differences visible; single-level fluorescence image: lower intensity of red fluorescence in light-stressed leaf; PAM fluorescence image: markedly lower Fv/Fm values in the stressed leaf.

revealed that the sudden exposure to full sunlight caused damage to photosystem II, so that the variable fluorescence/maximal fluorescence (Fv/Fm) ratio, that represents the maximum potential quantum efficiency of photosystem II, if all capable reaction centers were open, was markedly decreased (Figure 2.6, right). Fluorescence imaging can visualize stress effects – here, caused by high light exposure – that are not visible to RGB imaging. Thus, such imaging technology is ideal for stress detection in plants, particularly for pre-symptomatic detection of the impacts of environmental stresses or microbe-triggered diseases.

2.2.3.2 Field Parameters

At the field level, it is much more challenging than in indoor environments, to record fluorescence images. The main reason is that fluorescence light emitted from a pigment in the leaf is much weaker than light that is reflected from a leaf surface. Therefore, the fluorescence detectors either need to be close to the leaf surface or have very high sensitivity to weak fluorescence light. For any outdoor imaging solution, high sensitivity of the sensor and proper filtering of the non-target wavelengths are essential to properly capture the fluorescence light.

In a trial reported by Virlet et al. (2017), wheat cultivars were subjected to different levels of fertilizer. Measurements of the quantum yield of photosystem II for dark-adapted plants, using the Field Scanalyzer in-built chlorophyll fluorescence imager, indicated that greater fertilization promoted higher quantum yield in all cultivars (Figure 2.7). Thus, higher fertilizer use increased the capacity to perform photosynthesis. Whereas cultivar Widgeon had low responsiveness to fertilizer increase and generally exhibited low levels of quantum yield, cultivar Soissons was highly responsive, with the lowest quantum yield occurring at the low fertilizer treatment, and being saturated at high levels in response to intermediate or high fertilizer treatments. Cultivar Avalon had a unique response, in that its quantum yield decreased from low to intermediate fertilizer levels but increased with the high-fertilizer treatment. 'Cadenza', 'Crusoe', and 'Gatsby' had quantum yields which increased with fertilizer levels. The data generated with the phenotyping technology identified cultivars which improved their photosynthetic capacity in response to fertilizer and to what extent. Thus, the study supported the concept of targeted fertilization. In turn, the study proved that fluorescence imaging is suitable for detecting fertilizer deficiency in field crops, and indicates that these are cultivar-specific responses.

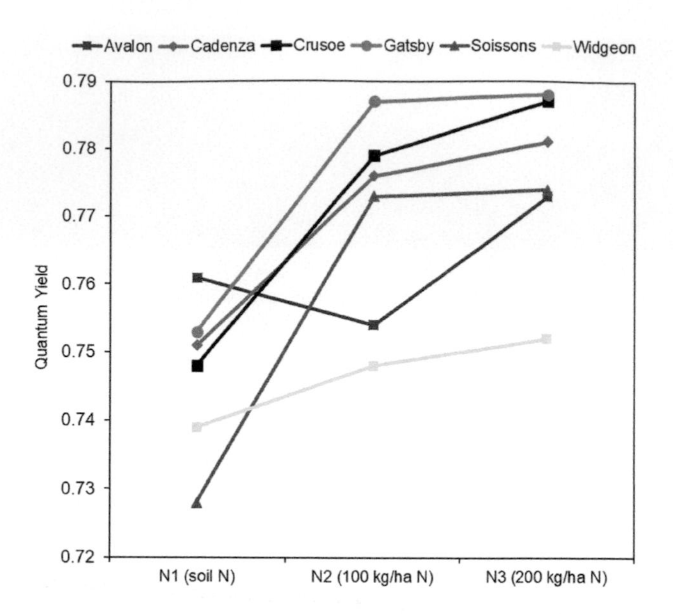

FIGURE 2.7 Quantum yield of photosystem II in wheat cultivars subjected to different nitrogen regimes. Measurements were carried out with a chlorophyll fluorescence imager mounted in the Field Scanalyzer on the experimental field at Rothamsted Research, UK.

2.3 CONCLUSIONS

Intelligent processing extracts parameters from optical recordings of biological samples that are linked to the structural and physiological properties of the organisms. Such links are usually established *via* calibration experiments that associate digital parameters with traditionally measured biological traits. For instance, connecting the projected area of plants in images with their biomass in a pre-experiment allows tracking of the temporal development of plant biomass in time-series growth experiments.

Image processing for plant phenotyping is evolving from measuring structural traits, such as dimensions and areas for growth studies, toward measuring physiological traits, such as color or fluorescence intensities, as indicators of stress responses or disease interactions. Increasingly, spectral imaging is gaining importance and the actual image processing is supported by complex data analysis and machine learning. Such efforts aim to identify spectral signatures of physiological responses and status, ranging from water and fertilizer supply over developmental processes to plant health. In developmental processes and ripening stages, the quality of the plant material, or the accumulation of biochemical substances, can be the target of the measurement. For instance, in medicinal plants, the production of health-promoting substances may be monitored with spectral imaging and corresponding data analysis. Monitoring plant status non-invasively and continuously will allow the establishment of control loops between measurement and plant management. Based on indicators that measure the state of the plants, the growers can adjust fertilizer, light, water, or climatic factors to manage the plants toward the desired phenotypic target.

ACKNOWLEDGMENT

The authors are grateful to Pouria Sadeghi-Tehran for sharing data on chlorophyll fluorescence, recorded at Rothamsted Research, UK.

REFERENCES

Alenyà, G., B. Dellen, and C. Torras. 2011. 3D modelling of leaves from color and ToF data for robotized plant measuring. In *2011 IEEE International Conference on Robotics and Automation* 3408–3414. Shanghai, China.

Arvidsson, S., P. Pérez-Rodríguez, and B. Mueller-Roeber. 2011. A growth phenotyping pipeline for Arabidopsis thaliana integrating image analysis and rosette area modeling for robust quantification of genotype effects. *The New Phytologist* 191(3):895–907.

Azzari, G., M. L. Goulden, and R. B. Rusu. 2013. Rapid characterization of vegetation structure with a Microsoft Kinect sensor. *Sensors* 13(2):2384–2398.

Baker, N. R. 2008. Chlorophyll fluorescence: A probe of photosynthesis in vivo. *Annual Review of Plant Biology* 59:89–113.

Barabaschi, D., A. Tondelli, F. Desiderio, A. Volante, P. Vaccino, G. Vale, and L. Cattivelli. 2016. Next generation breeding. *Plant Science* 242:3–13.

Bellasio, C., J. Olejníčková, R. Tesař, D. Sebela, and L. Nedbal. 2012. Computer reconstruction of plant growth and chlorophyll fluorescence emission in three spatial dimensions. *Sensors* 12(1):1052–1071.

Berger, B., B. de Regt, and M. Tester. 2012. High-throughput phenotyping of plant shoots. In *High-throughput Phenotyping in Plants*, ed. J. Normanly, 9–20. Humana Press, Totowa, NJ.

Brown, J., P. Caligari, and H. Campos. 2015. *Plant Breeding*, 2nd ed. Wiley-Blackwell, Oxford, UK.

Busemeyer, L., D. Mentrup, K. Möller et al. 2013. BreedVision—A multi-sensor platform for non-destructive field-based phenotyping in plant breeding. *Sensors* 13(3):2830–2847.

Cabrera-Bosquet, L., J. Crossa, J. von Zitzewitz, M. Serret, and J. L. Araus. 2012. High-throughput phenotyping and genomic selection: The frontiers of crop breeding converge. *Journal of Integrative Plant Biology* 54(5):312–320.

Cendrero-Mateo, M. P., O. Muller, H. Albrecht, et al. 2017. Field phenotyping: Concepts and examples to quantify dynamic plant traits across scales in the field. In *Terrestrial Ecosystem Research Infrastructures*, ed. A. Chabbi and H. W. Loescher, 77–104. CRC Press, Boca Raton, FL.

Chen, D., K. Neumann, S. Friedel, B. Kilian, M. Chen, T. Altmann, and C. Klukas. 2014. Dissecting the phenotypic components of crop plant growth and drought responses based on high-throughput image analysis. *The Plant Cell* 26(12):4636–4655.

Cleuvers, M., and H.-T. Ratte. 2002. Phytotoxicity of coloured substances: Is Lemna duckweed an alternative to the algal growth inhibition test? *Chemosphere* 49(1):9–15.

Dornbusch, T., S. Lorrain, D. Kuznetsov, A. Fortier, R. Liechti, I. Xenarios, and C. Fankhauser. 2012. Measuring the diurnal pattern of leaf hyponasty and growth in Arabidopsis – A novel phenotyping approach using laser scanning. *Functional Plant Biology* 39(11):860.

Dornbusch, T., O. Michaud, I. Xenarios, and C. Fankhauser. 2014. Differentially phased leaf growth and movements in Arabidopsis depend on coordinated circadian and light regulation. *The Plant Cell* 26(10):3911–3921.

Dornbusch, T., M. Hawkesford, M. Jansen, et al. 2015. *Digital Field Phenotyping by LemnaTec*. Lemnatec, Aachen.

Dupuis, J., S. Paulus, A.-K. Mahlein, and T. Eichert. 2015. The impact of different leaf surface tissues on active 3D laser triangulation measurements. *Photogrammetrie-Fernerkundung-Geoinformation* 6:437–447.

Fahlgren, N., M. A. Gehan, and I. Baxter. 2015. Lights, camera, action: High-throughput plant phenotyping is ready for a close-up. *Current Opinion in Plant Biology* 24:93–99.

Franke, W. 1997. *Nutzpflanzenkunde: Nutzbare Gewächse der gemäßigten Breiten, Subtropen und Tropen; 89 Tabellen, 6., neubearb. und erw. Aufl. Flexible Taschenbücher BIO.* Thieme, Stuttgart.

Guo, D., J. Juan, L. Chang, J. Zhang, and D. Huang. 2017. Discrimination of plant root zone water status in greenhouse production based on phenotyping and machine learning techniques. *Scientific Reports* 7(1):8303.

Hairmansis, A., B. Berger, M. Tester, and S. J. Roy. 2014. Image-based phenotyping for non-destructive screening of different salinity tolerance traits in rice. *Rice* 7(1):16.

Hawkesford, M. J., and A. Lorence. 2017. Plant phenotyping: Increasing throughput and precision at multiple scales. *Functional Plant Biology* 44(1):v–vii.

Hörtensteiner, S. 2006. Chlorophyll degradation during senescence. *Annual Review of Plant Biology* 57:55–77.

Humplik, J. F., D. Lazar, A. Husickova, and L. Spichal. 2015. Automated phenotyping of plant shoots using imaging methods for analysis of plant stress responses – A review. *Plant Methods* 11:29.

Jansen, M., F. Gilmer, B. Biskup et al. 2009. Simultaneous phenotyping of leaf growth and chlorophyll fluorescence via GROWSCREEN FLUORO allows detection of stress tolerance in *Arabidopsis thaliana* and other rosette plants. *Functional Plant Biology* 36(11):902–914.

Jansen, M., F. Pinto, K. A. Nagel, D. van Dusschoten, F. Fiorani, U. Rascher, H. U. Schneider, A. Walter, and U. Schurr. 2014. Non-invasive phenotyping methodologies enable the accurate characterization of growth and performance of shoots and roots. In *Genomics of Plant Genetic Resources*, ed. R. Tuberosa, A. Graner, and E. Frison, 173–206. Springer, Dordrecht.

Jansen, M., S. Paulus, K. Nagel, and T. Dornbusch. 2018. Image processing for bioassays. In *Bioassays: Advanced Methods and Applications* 263–287. Elsevier, Amsterdam, Netherlands.

Krause, G. H., and E. Weis. 1991. Chlorophyll fluorescence and photosynthesis: The basics. *Annual Review of Plant Biology* 42(1):313–349.

Paulus, S., J. Dupuis, A.-K. Mahlein, and H. Kuhlmann. 2013. Surface feature based classification of plant organs from 3D laserscanned point clouds for plant phenotyping. *BMC Bioinformatics* 14:238.

Paulus, S., J. Dupuis, S. Riedel, and H. Kuhlmann. 2014a. Automated analysis of barley organs using 3D laser scanning: An approach for high throughput phenotyping. *Sensors* 14(7):12670–12686.

Paulus, S., H. Schumann, H. Kuhlmann, and J. Léon. 2014b. High-precision laser scanning system for capturing 3D plant architecture and analysing growth of cereal plants. *Biosystems Engineering* 121:1–11.

Paulus, S., T. Dornbusch, and M. Jansen. 2016. Intuitive image analyzing on plant data processing: High throughput plant analysis with LemnaTec Image. In *Proceedings of the 13th International Conference on Precision Agriculture*. St. Louis, MO.

Pound, M. P., J. A. Atkinson, A. J. Townsend et al. 2017. Deep machine learning provides state-of-the-art performance in image-based plant phenotyping. *GigaScience* 6(10):10.

Rajendran, K., M. Tester, and S. J. Roy. 2009. Quantifying the three main components of salinity tolerance in cereals. *Plant, Cell & Environment* 32(3):237–249.

Rose, J. C., S. Paulus, and H. Kuhlmann. 2015. Accuracy analysis of a multi-view stereo approach for phenotyping of tomato plants at the organ level. *Sensors* 15(5):9651–9665.

Rousseau, D., Y. Chene, E. Belin, G. Semaan, G. Trigui, K. Boudehri, F. Franconi, and F. Chapeau-Blondeau. 2015. Multiscale imaging of plants: Current approaches and challenges. *Plant Methods* 11:6.

Rusu, R. B., N. Blodow, and M. Beetz. 2009. Fast point feature histograms (FPFH) for 3D registration. In *IEEE International Conference on Robotics and Automation* 3212–3217. Kobe, Japan.

Sadeghi-Tehran, P., K. Sabermanesh, N. Virlet, and M. J. Hawkesford. 2017a. Automated method to determine two critical growth stages of wheat: Heading and flowering. *Frontiers in Plant Science* 8:252.

Sadeghi-Tehran, P., N. Virlet, K. Sabermanesh, and M. J. Hawkesford. 2017b. Multi-feature machine learning model for automatic segmentation of green fractional vegetation cover for high-throughput field phenotyping. *Plant Methods* 13:422.

Schreiber, U. 2004. Pulse-amplitude-modulation (PAM) fluorometry and saturation pulse method: An overview. In *Chlorophyll A Fluorescence* 279–319. Springer, Dordrecht.

Strasser, R. J., A. Srivastava, and M. Tsimilli-Michael. 2000. The fluorescence transient as a tool to characterize and screen photosynthetic samples. In *Probing Photosynthesis: Mechanisms, Regulation & Adaptation* 445–483. CRC Press.

Tardieu, F., L. Cabrera-Bosquet, T. Pridmore, and M. Bennett. 2017. Plant phenomics, from sensors to knowledge. *Current Biology* 27(15):R770–R783.

Tschiersch, H., A. Junker, R. C. Meyer, and T. Altmann. 2017. Establishment of integrated protocols for automated high throughput kinetic chlorophyll fluorescence analyses. *Plant Methods* 13:895.

Virlet, N., K. Sabermanesh, P. Sadeghi-Tehran, and M. J. Hawkesford. 2017. Field Scanalyzer: An automated robotic field phenotyping platform for detailed crop monitoring. *Functional Plant Biology* 44(1):143.

Wahabzada, M., S. Paulus, K. Kersting, and A.-K. Mahlein. 2015. Automated interpretation of 3D laserscanned point clouds for plant organ segmentation. *BMC Bioinformatics* 16:248.

Walter, A., H. Scharr, F. Gilmer et al. 2007. Dynamics of seedling growth acclimation towards altered light conditions can be quantified via GROWSCREEN: A setup and procedure designed for rapid optical phenotyping of different plant species. *The New Phytologist* 174(2):447–455.

Walter, A., F. Liebisch, and A. Hund. 2015. Plant phenotyping: From bean weighing to image analysis. *Plant Methods* 11:14.

Zhu, G., N. Ye, J. Yang, X. Peng, and J. Zhang. 2011. Regulation of expression of starch synthesis genes by ethylene and ABA in relation to the development of rice inferior and superior spikelets. *Journal of Experimental Botany* 62(11):3907–3916.

3 Image Processing Techniques for Plant Phenotyping

Bashyam Srinidhi and Sanjiv Bhatia

CONTENTS

3.1 INTRODUCTION

A phenotype describes an observable property of an organism, which can include its morphology, physiology, and behavior. A plant's phenotype may also be used to reflect its biochemical properties, such as its health and maturity. Phenotypes are essential to the description of a plant's health, disease status, and evolutionary fitness (Houle et al. 2010; Mahner and Kary 1997). The set of all phenotypes expressed by an organism is known as its *phenome*.

Current advances in image processing and computer vision research have made it possible to efficiently study plant phenotypes and to draw conclusions to assist in plant breeding and selection. This has been the motivation for the US National Science Foundation recognizing the need for foundational infrastructure in plant biology research by funding the iPlant Collaborative (Goff et al. 2011). In this chapter, we explore the basic image processing techniques applicable to plant phenotyping analysis. They include techniques to isolate (segment) a plant from its background, to recover the structure of the plant, and to derive properties from the segmented plant image.

3.2 GOALS OF PLANT PHENOTYPING

Plant phenotyping plays a major role in sustainable food security (Tester and Langridge 2010), addressing the fundamental challenges in agriculture, such as achieving a higher yield on limited land area or an acceptable yield under adverse environmental conditions, through better crop selection, hybridization, and management. The analysis of phenotypic data could reveal the gene(s) responsible for a trait of the plant. For example, a crop simulation model could be used effectively for selective breeding of wheat or sorghum by analyzing the relationship between the phenome and its genome (Chapman 2007). There have also been studies on the interaction between aphid genotypes and host plant phenotypes (Service and Lenski 1982).

The architecture of a plant determines the type of phenotypes that may be derived, and the image processing steps which can be used to compute these phenotypes. In this chapter, we discuss processes to compute the phenotypes of two common architectures found in model plants, namely the rosette-like architecture in Arabidopsis or tobacco, and the distichous phyllotactic architecture in maize or sorghum (Meinke et al. 1998; van der Linden 1996). Algorithms to compute phenotypes for most other plants can be adapted using image processing operations described in this chapter.

In the next section, we briefly review the literature on the image processing techniques used in phenotyping. This is followed by sections that describe the image processing pipeline, with separate sections on image acquisition and data structures for images, some basic operations on images that are relevant in order to extract further information, image segmentation to focus on the relevant portions of an image, and morphological operations to further analyze the segmented image.

3.3 BACKGROUND AND LITERATURE SURVEY

There is a significant body of literature on the application of computer vision and image processing techniques to agriculture. Bhargava and Bansal have published a survey of computer vision techniques to sort fruits and vegetables in terms of their quality (Bhargava and Bansal 2018). Grillo et al. (2017) have reported on the use of image processing techniques, combined with statistical analysis, to identify wheat landraces, using glume size, shape, color, and texture. The work of Grillo et al. (2017) is focused on achieving classification based on the geographical origin of the landrace for product traceability. Ali et al. (2017) used image segmentation and color histograms, combined with local binary patterns, to detect and classify diseases in citrus crops. Similarly, Sharif et al. (2018) used a fusion of color, texture, and geometric features to detect lesion spots on citrus fruits and leaves, and to classify the affected crops according to disease. Liu and Chahl (2018) reported on the use of a multispectral machine vision system to detect pests on crops.

Guerrero et al. (2017) used computer vision techniques to detect weed density and overlapping in maize crops, using a mobile agricultural vehicle. They addressed the issue of the effects of vibration and other undesired movements during image acquisition and its resolution, using perspective geometry.

Jiang et al. (2018) relied on color, texture, and plant morphology, using computer vision techniques to detect drought stress in maize crops. Their work can be applied to organize irrigation to mitigate the harmful effects of drought, as a result of timely intervention. Kiani et al. (2017) combined computer vision with olfactory sensors to detect adulteration in saffron samples. They used principal component analysis, hierarchical cluster analysis, and support vector machines to improve classification into pure or adulterated samples.

3.4 IMAGE-PROCESSING METHODOLOGY

The computation of plant phenotypes, using computer vision-based techniques, enables automatic extraction of plant traits from the images. Figure 3.1 shows the overall sequence of steps in a typical image-based plant phenotyping process. The pipeline illustrates the procedure using terminologies from both the computer vision and plant science domains.

The input to the processing pipeline is typically an image of the plant. The input image may contain other objects in the scene, e.g., the plant container, or exposed soil. A set of image enhancement operations are then performed to remove imaging artifacts (noise) to improve the quality of the signal (plant) in the image. This is achieved using techniques, such as thresholding and morphological operations,

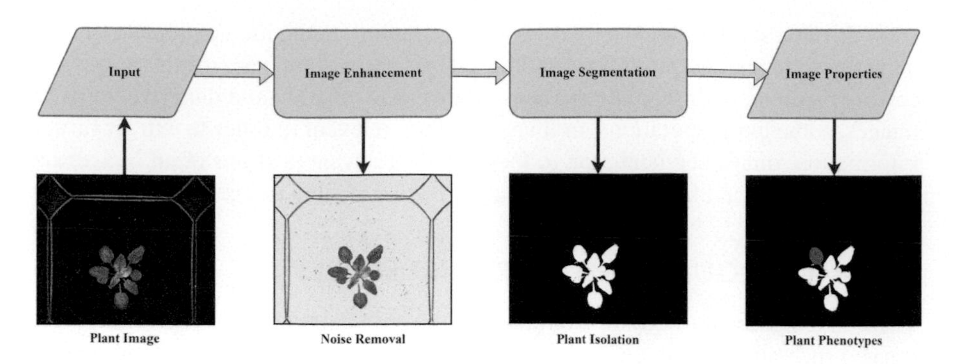

FIGURE 3.1 A typical image processing pipeline for plant phenotyping.

which are described later in this chapter. The enhanced image is then segmented to extract a specific portion of the image to generate a mask, representing only the plant or its components. Finally, the segmented image of the plant is used to compute various properties to obtain the desired phenotypic information.

3.5 IMAGE ACQUISITION/IMAGING BASICS

The first step in computer vision-based plant phenotyping is to acquire the images of the plants in a manner that preserves the relevant information. Images are typically acquired using an imaging sensor that can detect a specified range of wavelengths (band) of the electromagnetic (EM) spectrum, focused on a specific area. Sensor characteristics determine the spectral and spatial resolutions of the acquired image. The spectral resolution for an image is represented by the number and range of wavelengths that the sensor can detect. A sensor with a narrow spectral bandwidth can target more specific wavelengths. A low spectral resolution means a few large bands and *vice versa*. A full range of EM spectra, distributed across various frequencies and wavelengths, is shown in Figure 3.2.

Different types of cameras capture different ranges of wavelengths in the EM spectrum and at different spectral resolutions. If an EM band, captured by a camera, is reported as 750–760 nm, then the spectral resolution of the images captured is 10 nm. Some of the common types of cameras used for plant phenotyping are described later in this section. The most commonly used digital cameras capture only a narrow part of the spectrum, i.e., only the visible spectrum in the range of 390–700 nm. Certain bands, like gamma rays, ultraviolet, microwaves, radio waves, TV waves, and long waves, are rarely used for plant phenotyping. The spatial resolution for an image represents the minimum area (the smallest observable unit) that can be represented by each picture element, or pixel. The image appears more *detailed* with a higher spatial resolution as there are more pixels per unit area.

3.5.1 Image Data Structures

After acquisition, an image is represented as a rectangular array of pixels. Figure 3.3 shows a greatly enlarged region in an image of size 2454 × 2056 pixels. The rectangular

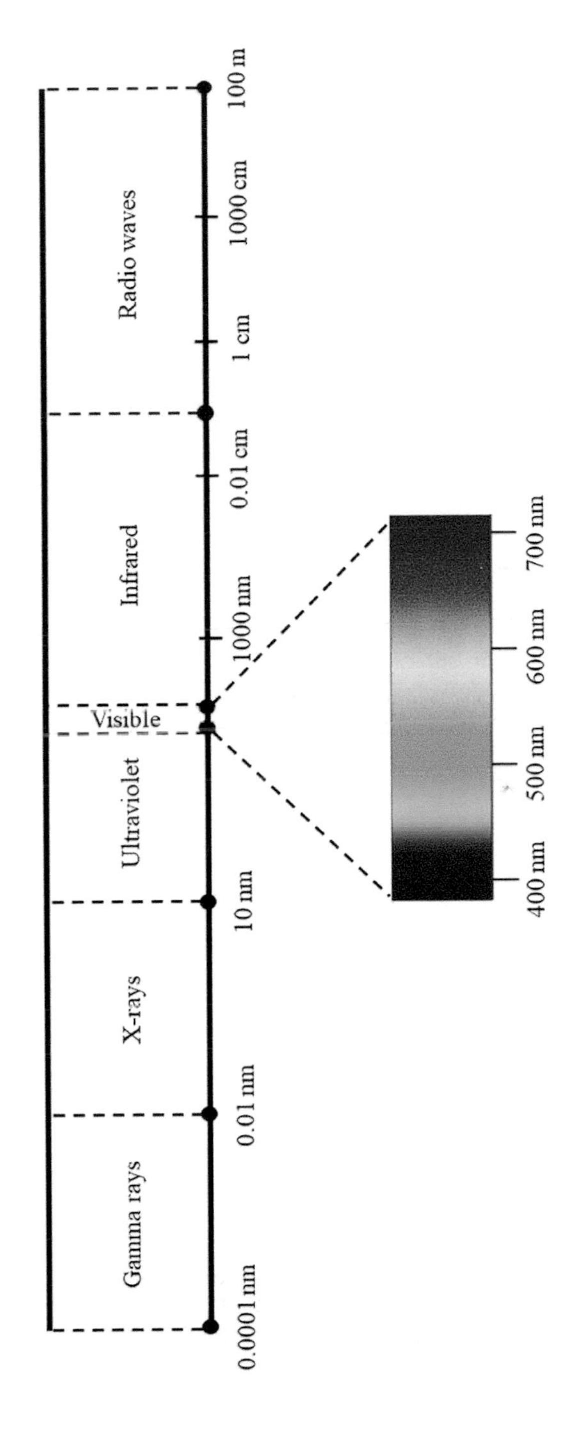

FIGURE 3.2 The electromagnetic spectrum.

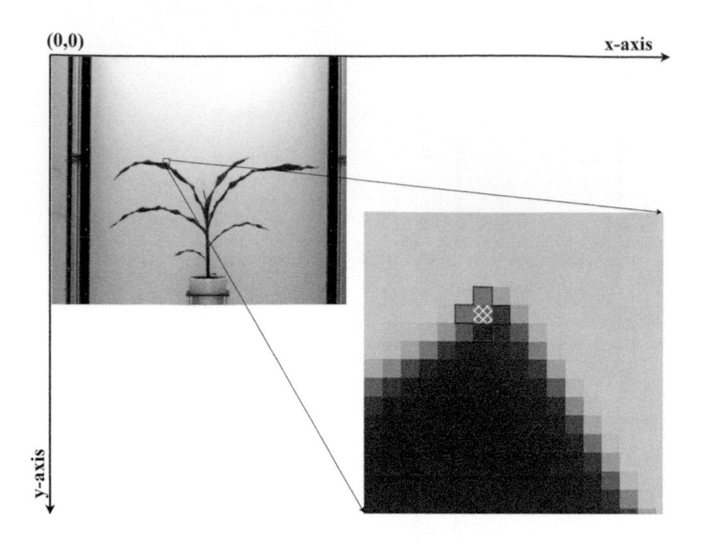

FIGURE 3.3 A pixel and its neighborhood in the image.

array or 2-dimensional pixel coordinate plane, represented here, has its origin at the top-left corner of the image with the *x*-axis along the rows (increasing left to right) and *y*-axis along the columns (increasing top to bottom) in the image. The cross-hatched area represents a randomly chosen pixel with four of its neighborhood pixels. The neighborhood pixels are useful in many image processing tasks, such as morphological operations, edge detection, filtering, and feature matching. The number and selection of neighbors depend on the operation or the algorithm being used. In Figure 3.3, we show 4-connected pixels (top, bottom, left, right) as neighbors, but some algorithms use 8-connected pixels (4-connected pixels plus pixels at corners) as neighbors.

Every pixel in an image stores a value that is proportional to the brightness (intensity) of the corresponding point in the scene that is imaged. The intensity value may vary, depending on the imaging sensor or the image format. Image formats refer to the internal layout of the images in a computer. In a monochrome or grayscale image, a pixel's intensity is typically represented by an integer between 0 and 255, where 0 represents black, 255 represents white, and the intermediate numbers represent different shades of gray. For color images, a pixel contains three intensity values, corresponding to the red, green, and blue (RGB) components of the visible spectrum. Each pixel is stored as its RGB value in the image file, with each RGB component in the range [0, 255].

In some applications, the RGB representation of color may not be appropriate. For example, humans think of colors as dark and bright, with little perceptual difference between very dark values of different colors. Thus, a system based on human visual perception, called the hue, saturation, and value (brightness) (HSV) system, was developed. Another system for analysis of images is known as CIELAB color space (defined by the International Commission on Illumination (CIE)), or $L^*a^*b^*$, where L is luminance and the a and b components represent color. The different color representations are known as color spaces and there are standard algorithms to convert images from one color space to another.

3.5.2 Visible Light Images

The visible portion of the EM spectrum is sensed by the human eye and spans a narrow range of the overall range. It consists of a range of wavelengths from roughly 390 nm to 700 nm. Since this range is easily perceived by the human eye, the sensors for this part of the spectrum are the most common. The images of a plant captured in this spectrum help in studying the physical appearance of the plant, including the color and shape. The images in the visible band are described using the three component colors and are commonly known as RGB images. Most of the storage formats use RGB values to store color images.

Many properties of objects, including plants, can be computed using only the brightness information (without any color information), stored as grayscale images (instead of RGB images). The grayscale version can be directly computed from the RGB images and is often more efficient since it contains the structural information on the plant, with the algorithms working on a single channel instead of three. For example, it is easier to compute shape properties in grayscale images than in color images.

3.5.3 Infrared Images

The wavelengths in the infrared (IR) band range roughly from 700 nm to 1 mm and are further subdivided into three categories, namely near-IR, mid-IR, and far-IR, based on their proximity to the visual spectrum. The far-IR spectrum signifies longer wavelengths, from 10 µm to 1 mm, and is generally not useful as these wavelengths are difficult to capture using currently available sensors. Most of the imaging in the infrared spectrum occurs in the mid-IR range (2.5–10 µm) and near-IR (750–2,500 nm) bands (see Figure 3.2). These wavelengths are useful to detect the heat signature of the objects, which is inversely proportional to the absolute temperature of an object, according to Wien's displacement law (Riedl 2001) given by:

$$\lambda_p = \frac{b}{T} \tag{3.1}$$

where λ_p is the peak wavelength, T is the temperature (in kelvin) of the source, and b is Wien's displacement constant, with a value of 2897.8 (µm K).

Figure 3.4 shows a near-IR image of the top view of a tobacco plant taken in a greenhouse. The mid-IR signature of an object sometimes acts as its fingerprint and helps in its identification. These signatures may be used to study the internal and external anatomy of plants.

3.5.4 Hyperspectral and Multispectral Images

Hyperspectral images capture multiple images of an object, each in a narrow band of wavelengths, across a range of wavelengths. This enables us to find a unique signature of an object in a certain wavelength of the spectrum, which may not be directly evident in the visible band. Since the wavelengths of adjacent bands are similar, the

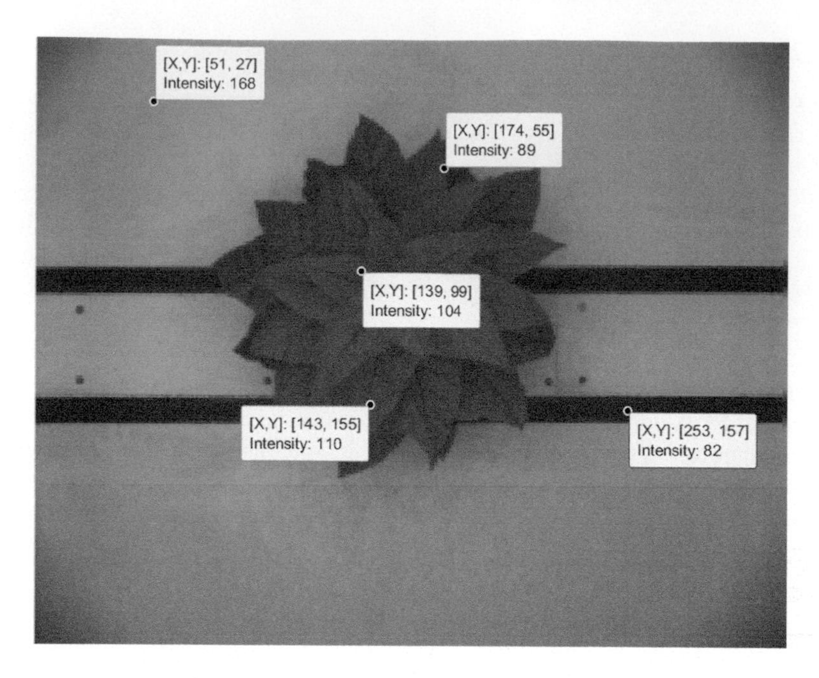

FIGURE 3.4 A near-infrared image of a tobacco plant with the intensity value of select pixels.

corresponding images are also correlated. The hyperspectral imagery is not widely exploited in the area of plant phenotyping at present, although it has been receiving attention recently. It has the potential to identify disease in crops, detect stresses, and is expected to be widely used in the future as infrastructure costs reduce and the technology becomes more widely available. Figure 3.5 shows the structure of a hyperspectral image (left) and the image for a specific band (right). When images are captured in only a few select bands, such as less than a dozen bands, they are known as *multispectral* images. Multispectral images are easier to manage and have been used for detection of pests in crops (Liu and Chahl 2018). In both hyperspectral and multispectral images, the images in different bands need to be *registered* (Brown 1992) such that the corresponding areas in different bands of images can be properly identified for subsequent analysis.

3.5.5 Fluorescent Images

In fluorescent objects, the molecules are excited to higher energy levels by exposure to radiation. As they return to a lower energy level, they emit light of a certain wavelength, depending on the energy difference between the two energy states. This process is called fluorescence and the images that are created using this light can be used to study the object's emission properties. In the case of plants, chlorophyll emits fluorescent radiation. As with multispectral images, fluorescent images are sensed at specific bands. An example of a fluorescent image is shown in Figure 3.6.

FIGURE 3.5 Hyperspectral images of a tobacco plant from 400 nm to 1600 nm.

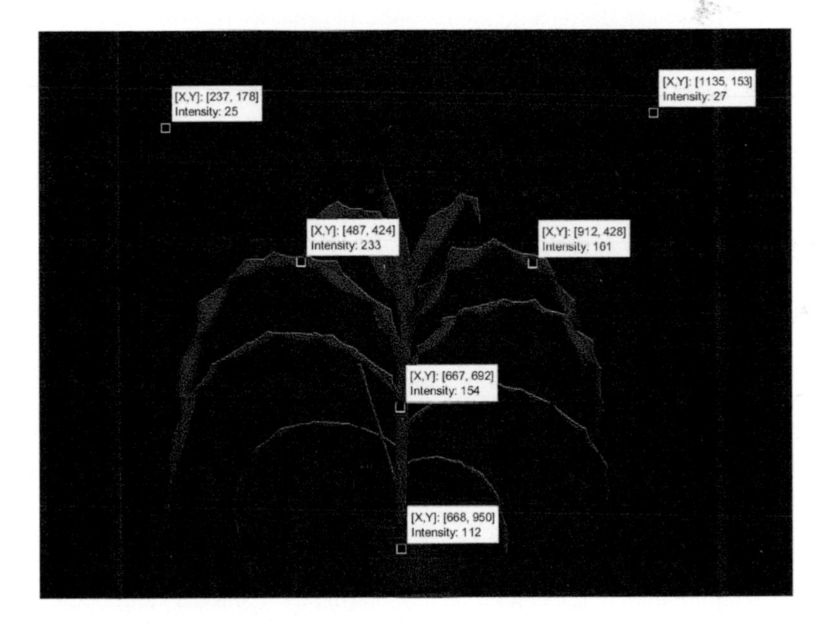

FIGURE 3.6 A fluorescent image of a sample maize plant.

3.6 BASIC IMAGE-PROCESSING OPERATIONS

Many algorithms have been developed to perform a variety of tasks to extract information from images. They range from algorithms to remove "salt and pepper" noise from images to those which can identify the expression on a human face (Szeliski 2010). This section briefly describes the image processing operations that are relevant, in the context of plant phenotyping. These operations help with the transformation of a plant image, to separate the objects of interest within the image, and

FIGURE 3.7 Sample images of two plant species used for illustration: maize (left) and Arabidopsis (right).

eventually to compute the desired phenotypes. In order to illustrate these operations, two plant species with different architectures are used, depending on the operation's suitability. These two species are maize and Arabidopsis (Figure 3.7) with rosette and distichous phyllotactic architectures, respectively.

It is also worth noting that there are many software applications available to analyze an image of a plant and to help compute the different phenotypes. For instance, many plant image analysis applications can be found online (Lobet et al. 2013). However, the choices and availability of software may vary, depending on factors such as the plant organ used (root, shoot, stem, leaf, flower), the feature type (length, width, area, color, angle, count, growth, yield), image type (RGB, IR, HS, grayscale, binary), and automation level (semi or fully automated, or manual).

3.6.1 Grayscale Conversion

Whereas the plant images are acquired in color, they are often processed as grayscale images in the computation of many phenotypes. For example, we can extract the shape of a plant by looking at its image in grayscale. There are many ways to convert a color (RGB) image to a grayscale image. The approaches generally use a weighted combination of the color information (in terms of RGB values). One widely used weighting scheme is based on the perception of the primary RGB colors by the human eye. We can convert a color image to grayscale by using the standard conversion expression given by:

$$g(x,y) = 0.21R(x,y) + 0.72G(x,y) + 0.07B(x,y) \quad (3.2)$$

where g, R, G, and B represent grayscale, red, green, and blue components of the image, respectively, and (x, y) indicates the pixel at a location in the image. The grayscale image computed in this way forms the basis of many image-processing algorithms, some of which are described next.

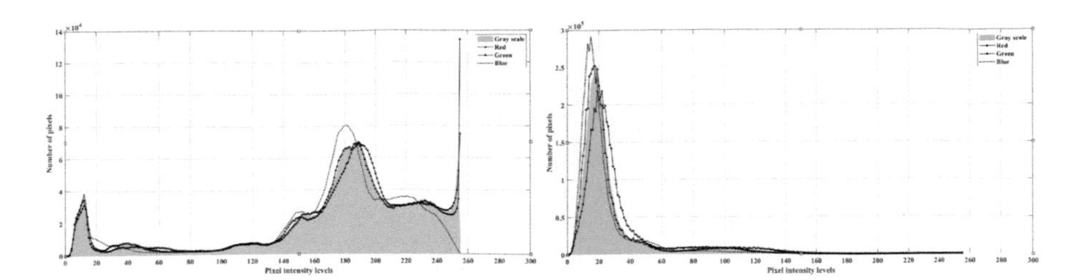

FIGURE 3.8 Histograms of the images in Figure 3.7 (*R*, *G*, and *B* channel histogram and histogram after converting the image to grayscale): maize (left), Arabidopsis (right).

3.6.2 Histogram Processing

Histogram processing involves the counting of different intensity levels in the image pixels to facilitate statistical analysis of the image based on the probabilistic distribution of those pixels. For a grayscale image, the pixel intensity levels are generally distributed in the range [0, 255]. A histogram counts the number of pixels at each of those 256 intensity levels. The computation of the histogram helps with many downstream operations, such as the computation of average intensity and contrast in the image. A visualization of the histogram of the images in Figure 3.7 is presented in Figure 3.8. The histograms of the three-color channels (R, G, B) are shown as different lines. The histogram of the grayscale image, derived using Equation 3.2, is shown as a shaded area.

3.6.3 Thresholding

In many imaging configurations, a well-contrasted (with respect to the object of interest) background is used when capturing the image of the plant (foreground). In these situations, thresholding is an easy-to-use method to separate the object from the background on the basis of the color or intensity of the image. First, an intensity threshold is chosen, either manually, based on the histogram of the image, or by automated methods (Otsu 1979). Often, the histogram of such images is bimodal (see Figure 3.8 (left)). Then, all pixels that have an intensity less than the threshold are changed to black (intensity level 0) and all the pixels with intensity greater than the threshold are changed to white (intensity level 255). Figure 3.9 shows the result of the application of thresholding on the two images shown in Figure 3.7.

In the case of plant phenotyping, the white pixels in the thresholded image correspond to the plant and may be viewed as a *mask*. Such a mask may be used to extract only the plant pixels from the original image. Figure 3.10 shows the result of applying the mask to the original image in Figure 3.7.

Thresholding, using a single threshold, can be extended to multi-level thresholding, in which a range of interested intensities is used to threshold the image to create multiple regions of interest (ROI). This is illustrated in Figure 3.11, using the Arabidopsis image shown in Figure 3.7, where some leaves have turned yellow. Using multi-level thresholding, we can generate three regions: background (intensity <65), green leaves (intensity between 65 and 150), and yellow leaves (intensity >150).

FIGURE 3.9 Result of thresholding to the two sample images shown in Figure 3.7: maize (left), Arabidopsis (right).

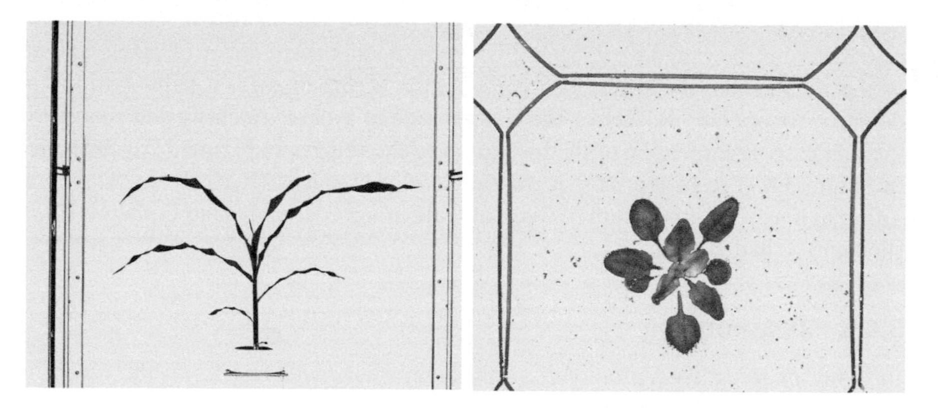

FIGURE 3.10 Extracted plant components using the masks after thresholding for the images shown in Figure 3.7 maize (left), Arabidopsis (right).

Figure 3.11 (left) shows the mask for the yellow pixels and Figure 3.11 (right) shows the yellow regions extracted from the original image.

3.6.4 Edge Detection

Edge detection is the process of finding the boundary of an object in the image. This technique is also used to distinguish objects from one another by marking their boundaries. Each enclosed boundary is treated as an individual entity in the image. The most common edge detection algorithms are Sobel's operator, Robert's cross operator, and the Canny edge detector (Gonzalez and Woods 2018). Figure 3.12 (left) shows the binarized image of the Arabidopsis plant shown in Figure 3.7, whereas Figure 3.12 (right) shows the edges detected using Sobel's operator.

3.6.5 Image Transformations

Image transformations typically take an image as input and generate another image as the output. The expectation is that the transformed image is more salient for

FIGURE 3.11 Result of multilevel thresholding for the Arabidopsis image in Figure 3.7.

FIGURE 3.12 Result of thresholding the Arabidopsis image shown in Figure 3.7 (left) and detected edges (right).

further extraction of information. Common transformations used in phenotyping are briefly described in this section.

Image cropping refers to extracting an ROI, usually a rectangular area in an image. The rectangular area is specified by the coordinates of its top left corner along with its height and width. Figure 3.13 shows an example of cropping, where a specified area of the image (Figure 3.13 [left]), is extracted into a separate, cropped image, Figure 3.13 (right).

Image scaling is a technique used to change the size of an image by increasing or decreasing the resolution of the image. Many image-processing algorithms, that may involve multiple images, require that the images be of the same scale or dimensions. For instance, the frame difference technique for segmentation (Section 3.6.6) operates on two images of the same resolution. Sometimes, when a plant is in its early stage of growth, it is imaged at a higher zoom factor than when it is larger. In order to analyze the sequence of images consistently, all the images are scaled to the same resolution.

FIGURE 3.13 Image displaying the rectangular cropping area (left); cropped section of the image (right).

FIGURE 3.14 Illustration of image scaling. Left: zoomed out image of a plant; middle: static background scene with only the pot; right: the image scaled to the same resolution as that of the static background.

Figure 3.14 (left) shows an image of a plant growing in a pot at a low zoom level. Figure 3.14 (middle) shows an empty pot at a high zoom level. Since they are imaged at different scales, it is hard to compare the two images and to extract the plant. Figure 3.14 (right) shows the scaled (and cropped) version of the plant that matches the scale of the image in Figure 3.14 (middle). The operations of image scaling and other geometric transformations, such as rotations, are commonly implemented, using affine transformations (Gonzalez and Woods 2018).

Image filtering in the spatial domain is a process of transforming the image, using a mathematical function (filter). The transformation function is applied locally to each pixel in the image. Based on the intensity of the pixel and its neighborhood, a new intensity value of the pixel is determined. The filters may be classified as linear or non-linear, depending on the function used. A filter is classified as linear when the neighborhood pixels are linearly combined to compute the output pixel. Common linear filters include mean, Gaussian, Sobel, Prewitt, and Laplace filters. Filters using mathematical operations, such as min, max, median, and percentile filters, are non-linear (Gonzalez and Woods 2018). Filters are widely used in image-processing operations, such as smoothing (mean, median, and Gaussian filters), sharpening (Laplace filters), and edge enhancement (Sobel, Prewitt, and Laplace filters).

3.6.6 Segmentation

Segmentation is the process of grouping pixels in an image that potentially describes an object. These pixels may be grouped based on their intensity, color, texture, or on several other criteria. In the context of plant phenotyping, an object may be a whole plant or any of its organs, e.g. a fruit, a flower, etc. There are different techniques for segmentation, depending on the content of the image and the type of objects to be identified. For example, consider the segmentation of the plants shown in Figure 3.7. Here, our goal is to divide the pixels into two groups or regions, namely a foreground region, that contains only the plant pixels, and a background region that has all the other pixels. In this section, we look at different segmentation techniques that are relevant to extracting the plant images for phenotyping and ignore the other objects that are part of the background (or noise).

3.6.6.1 Frame Difference Segmentation

Image difference, a simple pixel-by-pixel difference between the two images captured with a static camera, is often used to identify changes in the scene. In plant phenotyping, it is possible to ensure that the background remains constant with respect to the camera parameters. In such an environment, we can capture an image containing only the background and then introduce the plant into the scene and capture another image. The difference between the two images allows us to extract just the pixels of the plant. Figure 3.15 (left) shows the static background image and Figure 3.15 (middle) shows the plant image with the same background. Figure 3.15 (right) shows the difference between the two images, in which the plant is clearly marked. The difference image can be further processed to remove any artifacts and to derive an accurate mask for the plant.

3.6.6.2 Color-Based Segmentation

If the color of the foreground or background is known, then segmentation can be performed by retaining certain pixels corresponding to the foreground or by removing certain pixels of the background, based on their color characteristics. For example, Figure 3.16 demonstrates the retention of foreground in HSV color space using Matlab's Color Thresholder (Mathworks Inc. 2019). The HSV threshold values are shown in Figure 3.16 (left). After color-based segmentation of the maize plant (Figure 3.7), the image is shown in Figure 3.16 (right); the background is inverted from black to white for better clarity. Different color spaces can be used to achieve color-based

FIGURE 3.15 Illustration of frame difference segmentation.

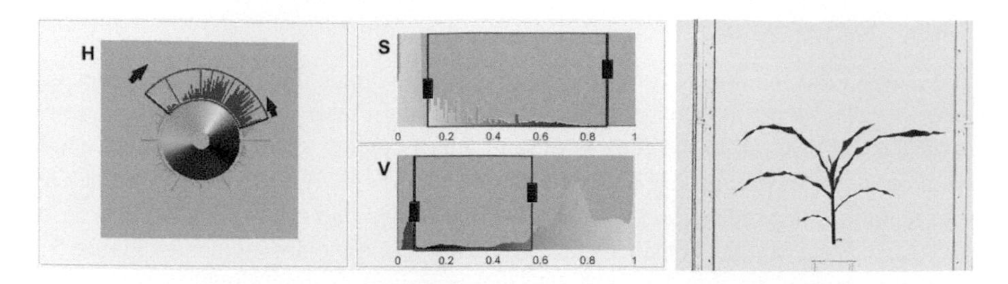

FIGURE 3.16 Sample HSV thresholds applied (left) to the maize plant (Figure 3.7) and the result of HSV thresholding (right).

segmentation. The segmented image may require additional processing to remove noise and other artifacts. It should be noted that the same value for thresholds will not apply to all images and must be adjusted from one class of images to the next.

3.6.7 Morphological Operations

Morphological operations are used to remove imperfections from the structure of the image. These operations are performed generally with binary images derived after segmentation. The black pixels are considered to be the background, with white pixels as the foreground in the binary image. The morphological operations are based on a *structuring element* that can be used to guide the change in morphology of the objects (the foreground pixels). The morphological operations described in this section can be performed in a specific area or shape in the input image. Dilation, erosion, opening, and closing are considered the core morphological operations, and advanced operations, like top-hat or watershed transforms, are also useful for plant phenotyping applications.

3.6.7.1 Dilation and Erosion

The operation of dilation is used to fill holes in an image and to bridge the gaps. The holes and gaps may occur due to noise in the input data, or inaccuracy in image acquisition. Dilation is used to increase the boundary of the foreground pixels and decrease the size of holes. The overall result is an increase in the size of the foreground or thickening of the object. The filling of new pixels depends on the size and shape of the structuring element. In plant phenotyping, if a leaf has specks of any residue or has small spots, the intensity of which is different from its surroundings, perhaps because of disease or stress, dilation can fill those holes in the segmented image. The operation of erosion is the opposite of dilation. It is used to remove bridges, spurs, and thin connections that are not real. Erosion is used to remove pixels from the boundary of the foreground pixels, resulting in a shrinking of the object. The number of pixels to be removed depends on the size and shape of the structuring element.

3.6.7.2 Opening and Closing

The operation of opening is the application of the erosion operator followed by the dilation operator, using the same structuring element. It is used to smoothen the contours of an object, breaking narrow isthmuses, and eliminating thin protrusions.

The operation of closing is the opposite of opening; it is the application of the dilation operator followed by the erosion operator, using the same structuring element. It is used to smoothen sections of contours, fusing narrow breaks and long thin gulfs, eliminating small holes, and filling gaps in the contour.

3.6.8 Thinning

The thinning operator is applied to a binary image to create a simplified image that is topologically equivalent to the original. It is mostly used to tidy up an image after the application of other operations, such as to reduce the width of lines that are determined by the edge detector operator.

3.6.9 Connected Component Analysis

A connected component in an image is a collection of pixels that have the same intensity and are connected to each other through 4-pixel or 8-pixel connectivity. The input to the connected component analysis is typically a binarized image, and the extracted component may be used for further analysis, or in operations such as the creation of masks, skeletonization, and measurement of area. In phenotyping applications, it is used to eliminate the regions that do not correspond to the plant, which has the largest area.

3.6.10 Skeletonization

Skeletonization is the process of obtaining a medial axis of a connected segment in a binary image. It is used to extract new descriptions of a shape. Figure 3.17 depicts the

FIGURE 3.17 Illustration of the skeleton (medial lines) of a plant image.

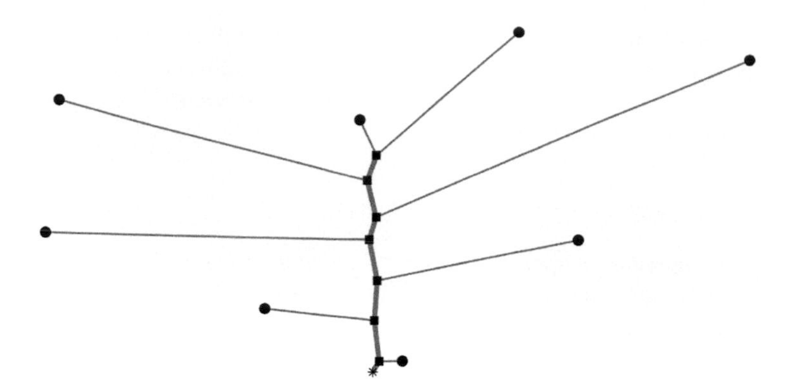

FIGURE 3.18 The graphical representation of the skeleton in Figure 3.17. The edges and nodes are numbered randomly. Key: '●'– leaf tips; '■'– collars and '*'– base.

skeleton of the binarized image of a plant with distichous phyllotactic architecture. However, skeletonization may not be suitable for plants with rosette architecture.

3.6.10.1 Graphical Representation

The skeleton of a plant may be represented as a simple mathematical graph with nodes and edges. The concept of nodes and edges may be further extended to assigning values to the edges (for example, length, position, and orientation of plant leaves). Figure 3.18 shows the graphical representation of the skeleton shown in Figure 3.17. A graphical structure facilitates higher-level analyses of a plant's structure. For example, in the graph, all leaf tips are nodes with degree 1, and collars are nodes with degree 3. The stem can be derived from this representation by starting from the base and simply connecting the edges that represent the collars.

3.7 FEATURE COMPUTATION

Once the plant or its components have been isolated in the image, the image can be used to compute its phenotypes. Many of the properties/features of objects, in general, may be used directly as phenotypes. First, simple shape-based properties are discussed. Then, properties based on color or other observations in the electromagnetic spectrum are described.

3.7.1 Basic Shape Properties

In this section, we describe some basic shape properties that can be computed from a plant object for use as phenotypes.

3.7.1.1 Length

The length of a component of a plant can be computed based on its skeleton. For example, the length of a leaf can be measured from its junction (the point where the leaf emerges from the stem) to its tip, when the leaf is kept flat. Depending on the level of accuracy required, the length of the leaf in an image can be computed by

FIGURE 3.19 A segmented plant image with the skeleton of a leaf shown as a dotted line.

simply counting the number of pixels in the medial line in the skeleton of the leaf, or with the help of complex methods, such as curve fitting. Figure 3.19 shows the skeleton of just one leaf in the plant.

3.7.1.2 Area

The total number of pixels in a connected component is used to compute its area. This can be used to compute the area of the whole plant, or that of any individual organ, e.g., a leaf as shown in Figure 3.20. Counting the total number of pixels will give an estimate of the plant canopy area.

FIGURE 3.20 A segmented plant image with the area of a leaf shown in gray.

FIGURE 3.21 A segmented plant image with its normal bounding box.

3.7.1.3 Bounding Box

A bounding box is a rectangle, the edges of which pass through the farthest point on each of the four sides of the segment. The rectangle provides general information about the width and height of the component. The bounding boxes may be classified as:

- *Normal bounding box*: It is the smallest rectangle containing the plant segment, the edges of which are aligned to the x- and y-axes.
- *Minimum bounding box*: It is the smallest rectangle containing the plant segment in any orientation. This can be more accurate when the object is not aligned with the axes, for example, if a maize plant is tilted.

Figure 3.21 shows the bounding boxes for the maize plant illustrated in Figure 3.7. In this case, the minimum and the normal bounding boxes happen to be the same.

3.7.1.4 Aspect Ratio

The aspect ratio is defined as the ratio of the height of the plant to its width. This gives the measure of a plant's shape signature, which is usually different for different plants or plants of different genotypes. It is easy to see that the aspect ratio is trivial to compute from the bounding box.

3.7.1.5 Convex Hull

A convex hull is the smallest possible polygon containing the plant segment. It is typically used in the construction of geometric structures. A convex hull is defined by a set of vertices, such that a line drawn from any point within the graph to any other point within the graph is completely enclosed inside the polygon. The convex hull can yield a measure of the perimeter of a leaf or a plant. Figure 3.22 shows the convex hull of the maize plant presented in Figure 3.7.

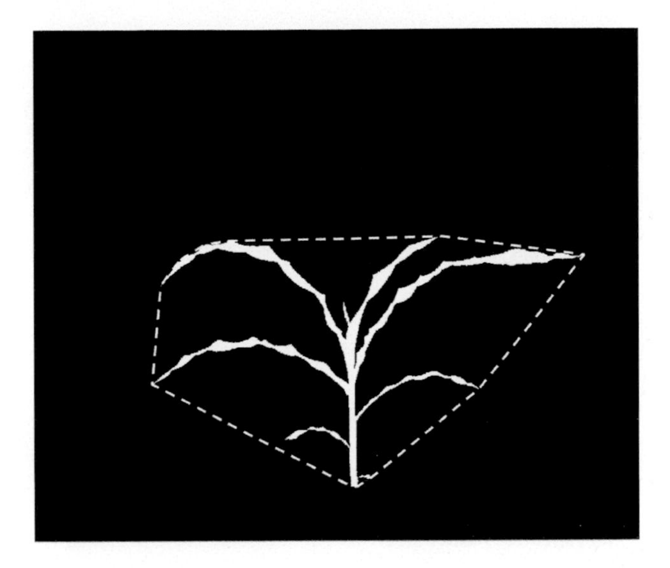

FIGURE 3.22 A segmented plant image with its convex hull.

3.7.1.6 Circularity

The circularity measure for a shape is measured using its area and the length of its boundary. This defines how close the shape is to a circle. For example, leaves may be classified into two types, circular and rectangular, based on their circularity values, computed using the equation:

$$C = 4\Pi \frac{A}{P^2} \tag{3.3}$$

where C is the circularity, and A and P are the area and the perimeter of the segment, respectively.

3.7.1.7 Straightness

The straightness, S, of a shape is computed using its actual (L_e) and apparent length (L_a), ignoring the width factor (assuming that the length is greater than the width). The apparent length of a shape is the total length of its skeleton. The actual length is given by the Euclidean distance between the two end points of the skeleton. The straightness is given by:

$$S = \frac{L_e}{L_a} \tag{3.4}$$

3.7.2 Color Properties

These properties are computed from the original image, but by using the mask for the object of interest, derived using some form of segmentation.

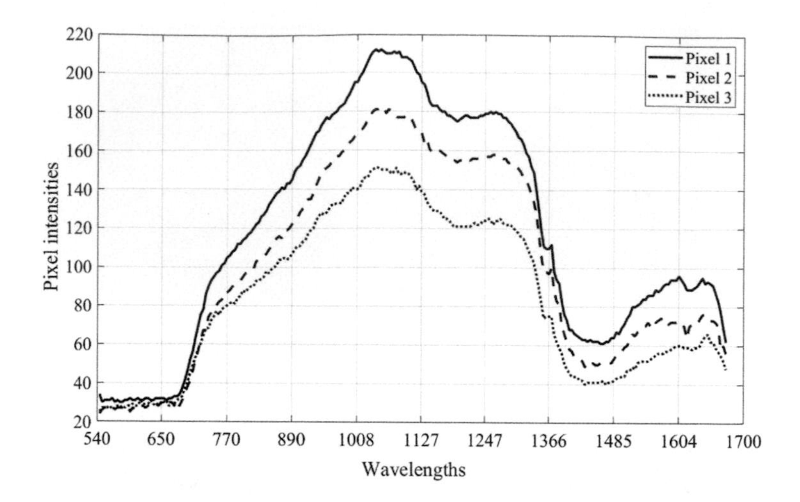

FIGURE 3.23 The spectral reflectance curve for selected plant pixels in Figure 3.5.

3.7.2.1 Average Greenness

The average greenness of a plant can be measured using the excess green index in an RGB image. The value for each pixel is calculated using the expression:

$$Excess\ Green = 2 \times G - R - B \tag{3.5}$$

where R, G, and B are the intensities in the respective color channels. This technique is applicable only in the RGB color space as all the three channels map to color intensities.

3.7.2.2 Normalized Difference Vegetation Index (NDVI)

The *NDVI* is a measure of plant health, based on the amount of sunlight reflected by the plant's pigments at certain wavelengths. The most common photosynthetic pigment, chlorophyll, reflects light in the green spectrum and efficiently absorbs photons in other wavelengths of the EM spectrum. Hence, these pigments appear green in color. Similarly, pigments like carotenoids absorb frequencies in the blue and the green regions and appear orange in color. Figure 3.23 shows the spectral reflectance curve for three randomly selected plant pixels in Figure 3.5 over a wide range of wavelengths (546 nm to 1700 nm), captured with the help of a hyperspectral camera. *NDVI* is computed by:

$$NDVI = \frac{R_n - R_r}{R_n + R_r} \tag{3.6}$$

where R_n is the reflectance in the near-infrared band, and R_r is the reflectance in the red band.

3.8 CONCLUSION

Image analysis plays an important role in the study of phenotypes and their relationship to the plant's genome. It helps to understand a plant's behavior, growth, life cycle events, and responses to external factors like biotic and abiotic stresses by finding and quantifying appropriate phenotypes. The quantified physical features derived from the image of a plant can be combined to establish simple or complex phenotypes (Das Choudhury et al. 2018). Depending on the plant architecture, image type, and the desired phenotype, suitable image analysis algorithms may be employed. For example, to compute the length of a leaf (a phenotypic trait) in a maize plant (exhibiting distichous phyllotactic architecture) from an image (frontal view, RGB image modality) in Figure 3.7, a suitable image-processing pipeline would be: image transformation operations, segmentation, skeletonization, graphical representation, and length computation.

Phenotyping may eventually help to identify the exact set of genes responsible for a specific phenotype. Gene identification and isolation are significant steps in processes like hybridization and genetic engineering, in which the genome is altered to produce a desired change in the plant's genotype and phenotype. However, this relationship between the phenotype and the genotype is not well understood, because it may be a many-to-many relationship. Therefore, the study of a number of phenotypes at the same time is essential, and image-based approaches provide just that. The automatic feature extraction and analysis from images of plants may help in solving the computationally difficult problem of mapping the phenotypes and genotypes. This is an emerging area of research that can be accelerated using image analysis techniques.

REFERENCES

Ali, H., M. I. Lali, M. Z. Nawaz, M. Sharif, and B. A. Saleem. 2017. Symptom bases automatic detection of citrus diseases using color histogram and textual descriptors. *Computers and Electronics in Agriculture* 138:92–104. doi: 10.1016/j.compag.2017.04.008.

Bhargava, A., and A. Bansal. 2018. Fruits and vegetables quality evaluation using computer vision: A review. *Journal of King Saud University – Computer and Information Sciences*. doi: 10.1016/j.jksuci.2018.06.002.

Brown, L. G. 1992. A survey of image registration techniques. *ACM Computing Surveys* 24(4):325–376. doi: 10.1145/146370.146374.

Chapman, S. C. 2007. Use of crop models to understand genotype by environment interactions for drought in real-world and simulated plant breeding trials. *Euphytica* 161(1–2):195–208. doi: 10.1007/s10681-007-9623-z.

Das Choudhury, S., S. Bashyam, Y. Qiu, A. Samal, and T. Awada. 2018. Holistic and component plant phenotyping using temporal image sequence. *Plant Methods* 14:35. doi: 10.1186/s13007-018-0303-x.

Goff, S. A., M. Vaughn, S. McKay et al. 2011. The iPlant collaborative: Cyberinfrastructure for plant biology. *Frontiers in Plant Science* 2:34. doi: 10.3389/fpls.2011.00034.

Gonzalez, R. C., and R. E. Woods. 2018. *Digital Image Processing*, 4th ed. Pearson, New York.

Grillo, O., S. Blangiforti, and G. Venora. 2017. Wheat landraces identification through glumes image analysis. *Computers and Electronics in Agriculture* 141:223–231. doi: 10.1016/j.compag.2017.07.024.

Guerrero, J. M., J. J. Ruz, and G. Pajares. 2017. Crop rows and weeds detection in maize field applying a computer vision system based on geometry. *Computers and Electronics in Agriculture* 142:461–472. doi: 10.1016/j.compag.2017.09.028.

Houle, D., D. R. Govindaraju, and S. Omholt. 2010. Phenomics: The next challenge. *Nature Review. Genetics* 11(12):855–866. doi: 10.1038/nrg2897.

Jiang, B., P. Wang, S. Zhuang, M. Li, Z. Li, and Z. Gong. 2018. Detection of maize drought based on texture and morphological features. *Computers and Electronics in Agriculture* 151:50–60. doi: 10.1016/j.compag.2018.03.017.

Kiani, S., S. Minaei, and M. Ghasemi-Varnamkhasti. 2017. Integration of computer vision and electronic nose as non-destructive systems for saffron adulteration detection. *Computers and Electronics in Agriculture* 46:46–53. doi: 10.1016/j.compag.2017.06.018.

Liu, H., and J. S. Chahl. 2018. A multispectral machine vision system for invertebrate detection on green leaves. *Computers and Electronics in Agriculture* 150:279–288. doi: 10.1016/j.compag.2018.05.002.

Lobet, G., X. Draye, and C. Périlleux. 2013. An online database for plant image analysis software tools. *Plant Methods* 9(1):38. doi: 10.1186/1746-4811-9-38.

Mahner, M., and M. Kary. 1997. What exactly are genomes, genotypes and phenotypes? And what about phenomes? *Journal of Theoretical Biology* 186(1):55–63. doi: 10.1006/jtbi.1996.0335.

Mathworks, Inc. 2019. *MATLAB and Image Processing Toolbox Release 2019a*. Mathworks, Inc., Natick, MA.

Meinke, D. W., J. M. Cherry, C. Dean, S. D. Rounsley, and M. Koornneef. 1998. Arabidopsis thaliana: A model plant for genome analysis. *Science* 282(5389):662–682.

Otsu, N. 1979. A threshold selection method from gray-level histograms. *IEEE Transactions on Systems, Man, and Cybernetics* 9:62–66.

Riedl, M. J. 2001. *Optical Design Fundamentals for Infrared Systems*. Vol. 48. SPIE Press, Bellingham, WA.

Service, P. M., and R. E. Lenski. 1982. Aphid genotypes, plant phenotypes, and genetic diversity: A demographic analysis of experimental data. *Evolution* 36(6):1276–1282. doi: 10.2307/2408159.

Sharif, M., M. A. Khan, Z. Iqbal, M. F. Azam, M. I. U. Lali, and M. U. Javed. 2018. Detection and classification of citrus diseases in agriculture based on optimized weighted segmentation and feature selection. *Computers and Electronics in Agriculture* 150:220–234. doi: 10.1016/j.compag.2018.04.023.

Szeliski, R. 2010. *Computer Vision: Algorithms and Applications*. Springer-Verlag, Berlin, Heidelberg.

Tester, M., and P. Langridge. 2010. Breeding technologies to increase crop production in a changing world. *Science* 327(5967):818–822. doi: 10.1126/science.1183700.

van der Linden, F. M. J. 1996. Creating phyllotaxis: The stack-and-drag model. *Mathematical Biosciences* 133(1):21–50. doi: 10.1016/0025-5564(95)00077-1.

Part II

Techniques

4 Segmentation Techniques and Challenges in Plant Phenotyping

Sruti Das Choudhury

CONTENTS

4.1 INTRODUCTION

An image is a visual representation of a scene that carries large amounts of information. It encodes albedo, pose, and lighting information, among others. Furthermore, the image captures all objects in the field of view. For most applications, however, only one or a few objects are of interest. Separating these objects from the rest is a key computer vision challenge. Objects exhibit many common characteristics that can be used to isolate them. Segmentation, a fundamental problem in computer vision, is the process of partitioning a digital image into multiple homogeneous regions by grouping pixels based on similarity in terms of intensity, texture, or color. In image-based plant phenotyping applications, particularly in high-throughput systems, the objects of interest are often the plant itself or individual parts of a plant, i.e., stem, leaves, flowers, fruits, and roots. The accuracy of computed phenotypes depends on the correctness of the segmentation of the plant and/or its components. In this chapter, we present a comprehensive and critical survey of segmentation techniques developed for above-ground plant phenotyping applications. The image-segmentation algorithms for plant phenotyping are typically developed in specific contexts and evaluated in a structured manner, using suitable datasets. The segmentation framework for plant phenotyping is summarized in Figure 4.1, and will be described next.

4.1.1 Phenotyping Context

From a practitioner's point of view, three main parameters, that play pivotal roles in the choice of segmentation algorithms, are: (a) phenotyping platforms, (b) imaging modalities, and (c) phenotyping categories. They are briefly described below:

- *Phenotyping platforms*: There are two widely used platforms for plant phenotyping research and practice: controlled environment and field-based platforms. The controlled environment plant phenotyping platforms are

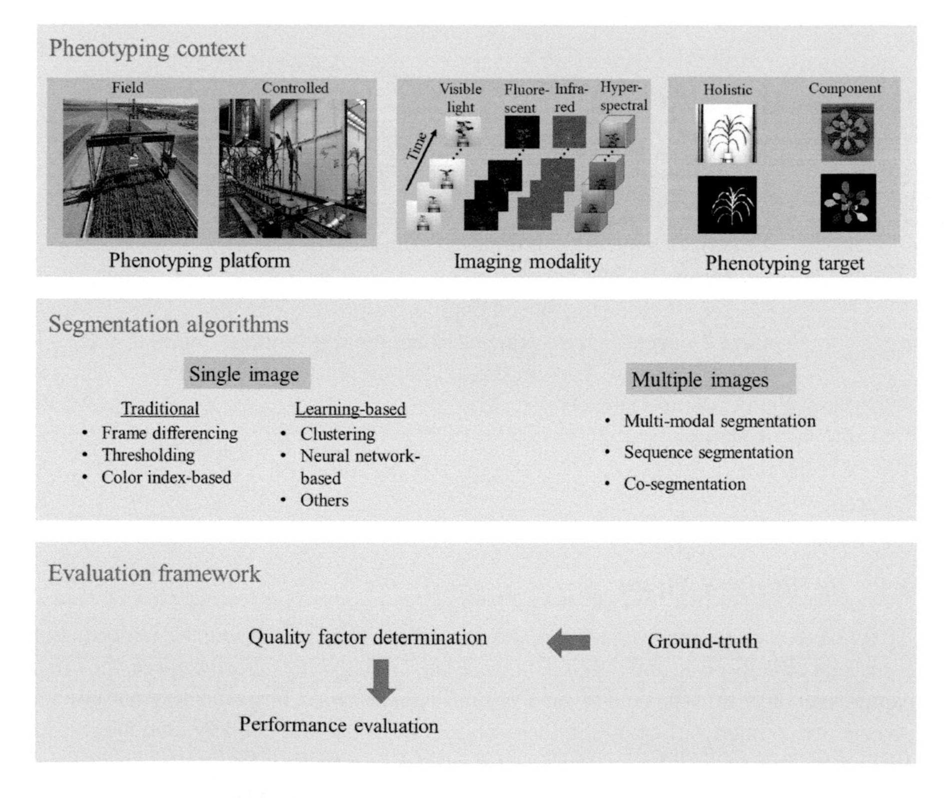

FIGURE 4.1 A general scheme of segmentation for image-based plant phenotyping analysis.

characterized by a homogeneous background that is distinct from the plant, where plants are imaged one at a time under uniform lighting conditions. The field-based plant phenotyping environment is characterized by variations in illumination, cluttered background, and the presence of weeds, occlusions, and shadows, which pose significant challenges.

- *Imaging modalities*: Multimodal imaging is becoming increasingly popular for the extraction of morphological and biophysical phenotyping traits, since different modalities complement one another. For example, visible light encodes morphological characteristics of the plants or their organs (Das Choudhury et al. 2018; Dyrmann 2015; Kumar and Domnic 2019; Lee et al. 2016). In contrast, infrared (IR) images can serve as a proxy for plant's temperature, which, in turn, can be used to detect differences in stomatal conductance, a measure of the plant's transpiration rate and response to water status, reflecting abiotic stress adaptation (Li et al. 2014). Images from different modalities capture information at a disparate range of wavelengths, and hence, require different segmentation strategies.
- *Phenotyping categories*: The choice of segmentation techniques also depends on whether the target phenotypes are holistic or component in nature. Holistic phenotypes are the traits of the plant that are computed by considering the plant as a whole, for instance, the height of the bounding box enclosing the plant to account for the plant's height, or the volume of the convex-hull of the plant as a measure of the plant's biomass. Segmentation for holistic phenotyping analysis simply requires classifying the image into the plant and non-plant pixels. Segmentation for component phenotyping analysis is more complex in comparison, since the individual parts of a plant, e.g., leaves, stems, fruits, and flowers, must first be isolated for their attributes to be computed.

4.1.2 Segmentation Algorithms

Segmentation can be performed using a single image or multiple images of a plant. The multiple images at any timestamp are obtained either by rotating the plant in front of a single camera to capture images from different viewpoints, or by using multiple cameras of the same or different modality. Furthermore, if a plant is imaged over a period of time, the acquired image sequence can be used to segment all the plant images simultaneously. To the best of our knowledge, all segmentation methods used to date for plant phenotyping rely on single image-based segmentation. Hence, this chapter focuses on segmentation algorithms using single images; segmentation based on multiple images is discussed in Section 4.4 as an open problem.

State-of-the-art segmentation techniques used in plant phenotyping can be broadly divided into two categories, namely traditional approaches and learning-based approaches. Traditional segmentation approaches are based on the image-processing pipeline that includes frame differencing, color-based segmentation, edge detection, region growing-based segmentation, and graph-based segmentation. Traditional approaches are easily understandable and more straightforward to implement. They are usually effective in constrained environments, e.g., homogeneous background and uniform lighting conditions, but their performance degrades in complex outdoor scenarios with illumination variations and cluttered background. Since they do not rely on a training phase, they have low computational complexity, and can often be implemented for real-time applications. Learning-based approaches are based on clustering or deep learning techniques. They require the availability of a substantial number of training samples to achieve reliable segmentation. Since they rely on a training phase, they demonstrate better performance compared to traditional approaches in the presence of varying lighting conditions, cluttered background, shadows, and occlusions. It is achieved, however, at the expense of greater computational complexity.

4.1.3 Evaluation Framework

To make systematic and uniform comparisons among different segmentation techniques, public availability of benchmark datasets is indispensable. The performance of the segmentation technique is evaluated by comparing the output of the segmentation algorithm with a human-annotated ground-truth, using a defined quality metric. The examples of quality metrics used in recent literature include *foreground-background dice* (FBD) to determine plant segmentation accuracy (Scharr et al. 2015), *plant-level accuracy* to determine the leaf detection and counting accuracy (Das Choudhury et al. 2018), *symmetric best dice* (SBD), and *difference in count* to evaluate leaf segmentation and leaf counting accuracy (Scharr et al. 2015), *global consistency error, object-level consistency error,* and *precision-recall framework* to determine the leaf segmentation accuracy (Chen et al. 2017; Martin et al. 2004), *Chamfer matching* and *tracking failure* to evaluate the performance of leaf alignment and tracking (Yin et al. 2017), and *mean absolute error* (MAE), and *mean squared error* (MSE) to assess tassel counting performance in the field (Lu et al. 2017).

The rest of the chapter is organized as follows. Section 4.2 presents the state-of-the-art traditional and learning-based segmentation techniques used for plant phenotyping. Section 4.3 summarizes publicly available benchmark datasets, and Section 4.4 discusses the potential open problems. Finally, Section 4.5 concludes the chapter.

4.2 CLASSIFICATION OF SEGMENTATION ALGORITHMS

In this section, we summarize the traditional and learning-based segmentation techniques used for holistic and component plant phenotyping analysis in both controlled environment and field-based phenotyping platforms.

4.2.1 Traditional Approaches

As mentioned before, the traditional techniques use a sequence of image-processing steps to obtain the regions of interest in the image. Some of the most widely used algorithms are briefly described below.

4.2.1.1 Frame Differencing

This approach is simple but effective in constrained environments, where the plant is imaged on a fixed imaging platform with a preset camera location, e.g., high-throughput plant phenotyping systems. Thus, the part of the scene (background without the plant) remains static, while the plants are sequentially transferred from the greenhouse to the imaging cabinets, typically by a movable conveyer belt, for their images to be captured. Given the fixed background image (B) and an image of a plant with the same background (I), the differenced plant image (P) can be computed by simple image subtraction, i.e.,

$$P = I - B.$$

In P, all the non-plant pixels will have a value of 0, and, hence, the plant is effectively segmented. The frame differencing technique is the simplest type of segmentation method, and is particularly suitable for holistic segmentation in a controlled environment, where the part of the image characterizing the plant is significantly different from the homogeneous background in terms of pixel intensity. However, the successful execution of this technique requires the background and foreground images (B and I, respectively) to be aligned with respect to scaling and rotation. Thus, image registration may be needed prior to frame differencing. The registration can be done automatically, using a feature matching approach (Dai and Khorram 1999). The method described by Das Choudhury et al. (2018) used the frame differencing technique to segment the plant from the background. A color-based thresholding step, to remove residual noise, is used to clean up the plant region.

4.2.1.2 Color-Based Segmentation

Color image segmentation is the process of partitioning an image into different regions, based on the color feature of the image pixels. It assumes that an object (or its parts) is homogeneous in its color properties, and hence it can be segmented on the basis of the color features. Since no single color space works best in all types of applications, color-based segmentation algorithms usually examine different color spaces, e.g., *RGB* (red, green and blue), *Lab* (where L indicates lightness, and a and b are color directions: $+a$ is the red axis, $-a$ is the green axis, $+b$ is the yellow axis and $-b$ is the blue axis) and *HSV* (hue, saturation, and value), to determine the most suitable color space for a given application. Color is one of the most common attributes that differentiates a plant from its background (or soil), and different organs of a plant often display different colors, hence, color-based segmentation is widely used in plant phenotyping analysis.

The color-based segmentation first computes an index to measure the greenness of a pixel and then uses a threshold to determine the plant pixels. For example, the normalized difference index (*NDI*) (Woebbecke et al. 1992) is computed from the green (G) and red (R) channels of an image as:

$$NDI = 128 * \left(\frac{G - R}{G + R} + 1 \right).$$

Hamuda et al. (2016) presented a comprehensive survey of the color index-based segmentation approaches, with comparative performance analysis used for plant phenotyping, including excess green index (Woebbecke et al. 1995), excess red index

(Meyer et al. 1998), color index of vegetation extraction (Kataoka et al. 2003), excess green minus excess red index (Meyer et al. 2004), and normalized green–red difference index (Hunt et al. 2005). The method by Zheng et al. (2009) first performed color feature extraction, and then applied the mean-shift algorithm to segment green vegetation for field phenotyping in the presence of variations in illuminations and soil types. The color index-based segmentation formulae are easy to compute and can be adapted to perform better in new environments. Although color-based segmentation algorithms are proven to be effective in indoor laboratory settings with relatively simple backgrounds, they do not show good adaptability in cluttered outdoor environments with varying lighting conditions.

4.2.1.3 Edge Detection and Region Growing-Based Segmentation

Edges are the sharp changes in the intensity values in an image and typically occur at the boundary between two objects. The edge detection-based segmentation methods attempt to identify edges based on analyzing discontinuities in texture, color, and intensities. The most commonly used discontinuity-based edge detection operators include Sobel, Prewitt, Kirsh, Robinson, Marr-Hildreth, Laplacian of Gaussian (LoG), and Canny (Gonzalez and Woods 2017). The edge detection process generates the boundaries of the objects, and the points inside are deemed to be the regions. The main basis for region growing methods is the selection of a set of seed points specific to the given application, followed by the development of the region by iteratively adding the pixels, based on similarity in color, intensity, and texture.

The method by Wang et al. (2018) used the Chan–Vese (C-V) model and an improved Sobel operator to segment overlapping cucumber leaves. The background, such as soil, stems, and membranes, is removed by using a thresholding algorithm based on the relative level of green. Then, contours of complete uncovered leaves are generated using the C-V model. An 8-directional Sobel edge operator is used to detect the edges. Finally, the overlapping leaves are extracted by combining results generated by the C-V model and Sobel edge detector. Experimental analysis on 30 images of cucumber leaves demonstrated better performance than the distance-regularized level set segmentation (Li et al. 2010). Li et al. (2018) proposed a leaf segmentation method for dense plant point clouds, using facet over-segmentation and facet region growing. The method has three steps: (1) point cloud pre-processing, (2) facet over-segmentation, and (3) facet region growing for individual leaf segmentation. The proposed method shows competitive performance based on experimental analysis on 3D point clouds of three different plant species, namely *Epipremnum aureum*, *Monstera deliciosa*, and *Calathea makoyana*, captured by stereo vision and Kinect v2 sensors.

4.2.1.4 Spectral Band Difference-Based Segmentation

Spectral band difference-based segmentation, a variant of the frame differencing technique, is applicable for phenotyping analysis based on hyperspectral or multispectral imagery, where a background image of a plant is not required. Gampa (2019) introduced a spectral band difference-based segmentation method, where two bands of specific wavelengths (770 nm and 680 nm), with a significant contrast in intensity levels, are first enhanced by multiplying each pixel by 2 and then one is subtracted

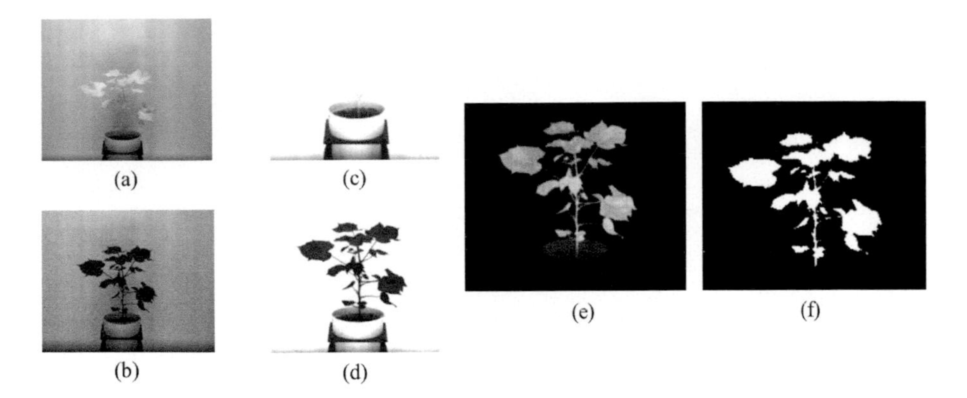

FIGURE 4.2 Illustration of spectral band difference-based segmentation: (a) and (b) hyperspectral images of a cotton plant at wavelengths 770 nm (a) and 680 nm (b); (c) and (d) corresponding enhanced images; (e) image obtained after subtracting (c) from (d); and (f) binary image.

from the other to determine the plant pixels, i.e., the foreground. If I_{770} and I_{680} are the images at the two wavelengths, the foreground image (I_f) is extracted as:

$$I_f = 2 * I_{770} - 2 * I_{680}$$

The resulting foreground image is then binarized, based on Otsu's automatic thresholding technique (Otsu 1979), for use as a mask to segment the plants from all bands of a hyperspectral cube and to generate the reflectance curves for subsequent analysis. The segmentation process is illustrated in Figure 4.2.

4.2.1.5 Shape Modeling-Based Segmentation

A shape-based method for segmenting leaves from low-resolution images, captured by unmanned aerial vehicles (UAVs) flying at high altitudes, is presented by Chen et al. (2019). The method uses two continuous functions, i.e., a midrib function and a width function, for modeling leaf shapes to account for missing edge points resulting from Canny edge detection. The method shows better performance than the Mask R-CNN developed by He et al. (2017b). Figure 4.3 shows the results of leaf segmentation using this method. Chen et al. (2017) transformed each plant image captured by UAVs into a polar coordinate system by using plant centers as the origins. The method carried out leaf shape modeling to segment each leaf, by overcoming occlusions, and finally, transformed the leaf shape model back to Cartesian coordinates to estimate phenotypic traits.

Monitoring flower development can provide useful information for genetic breeding, production management, and yield estimation. Thorp et al. (2016) conducted a study to investigate the differences in flowering dynamics and to estimate seed yield due to different levels of nitrogen and water availability for *Lesquerella*, based on analyzing images captured in the field. The input RGB images were first transformed into hue, saturation, and intensity (HSI) color space, and then the Monte Carlo approach was used to segment *Lesquerella* flowers by addressing uncertainty in HSI parameters.

FIGURE 4.3 Image of sorghum plants captured by UAV at an altitude of 20 m (top); human-annotated ground-truth, where leaves are colored randomly (middle); and leaf segmentation result, obtained using the method of Chen et al. (2019) (bottom). (Image: courtesy Chen et al. (2019)).

4.2.1.6 Graph-Based Segmentation

Graph-based segmentation has recently gained research attention for segmenting plant components, e.g., stem (Das Choudhury et al. 2017) and leaves (Kumar and Domnic 2019), for component phenotyping analysis. A stem segmentation algorithm based on skeleton-graph transformation was introduced by Das Choudhury et al. (2017). In this method, the input plant image is subjected to a two-phase segmentation method, based on the frame differencing technique followed by color thresholding in HSV color space. The binary image thus obtained is skeletonized (i.e., reduced to one-pixel wide lines that preserve the shape topology), using the fast-marching distance transform. The skeleton of the plant is represented by a connected graph where the base of the plant, leaf tips, and junctions are nodes, and the stem segments and leaves are the edges. The stem axis is formed by a linear regression-based curve fitting of all junctions. Kumar and Domnic (2019) proposed a novel graph-based leaf region segmentation algorithm for counting the total number of leaves of rosette plants. Shadow and illumination effects are common obstacles that negatively impact the accuracy of leaf segmentation. Hence, a Weibull distribution (Chang 1982) is used to remove unwanted shadow and light effects, based on a predefined metric of skewness, which indicates uniformity in illumination. The leaf region segmentation algorithm consists of three stages: (a) graph construction, (b) leaf region identification, and (c) removal of the non-leaf region. A circular Hough transform is used over the extracted leaf region to detect circular regions that correspond to individual leaves.

4.2.2 Learning-Based Approaches

Although traditional segmentation techniques perform well in many complex domains, their performance is usually unsatisfactory in the field phenotyping applications in the presence of extensive occlusions and large-scale lighting variations. In addition, traditional segmentation techniques find it challenging to segment individual organs of a plant for component phenotyping analysis. Hence, sophisticated learning-based techniques, including supervised and unsupervised machine learning, and deep learning approaches, in particular, have been investigated.

4.2.2.1 Clustering-Based Segmentation

Clustering is an unsupervised classification technique, whereby a dataset is reorganized into disjoint groups, based on a similarity measure. Clustering-based segmentation logically maps the pixels of an image into a clustering framework, that uses a similarity function to group the homogeneous pixels into different clusters in order to isolate the object of interest in the image. Dadwal and Banga (2012) presented a comprehensive review of clustering-based color image segmentation techniques for detecting ripeness of fruits and vegetables, including k-means, fuzzy c-means, Gustafson-Kessel-Babuska and Gustafson-Kessel possibilistic fuzzy c-means. Ojeda-Magana et al. (2010) performed a comparative analysis of the different clustering-based segmentation techniques to determine the ripeness of bananas and tomatoes.

3D reconstruction of plants is important in plant phenotyping, as some phenotypic traits can be accurately derived only from a 3D representation. The 3D models of the plants can be constructed in many different ways, e.g., multiple views, shape from shading, and direct sensing. Liu et al. (2018) presented an iterative approach to segmenting plant organs, i.e., stem and leaves, from 3D dense point clouds reconstructed from multiview images. The approach is based on spectral clustering, using the Euclidean distance as the basis for similarity. Gélard et al. (2017) used a structure from motion approach to compute 3D point clouds of sunflower plants from multiview RGB images acquired under controlled illumination conditions. They employed clustering to segment the point cloud regions corresponding to the plant organs, namely stem, petioles, and leaves. Jin et al. (2019) proposed a clustering-based algorithm to segment stem and leaves of a maize plant, using light detection and ranging (LiDAR), which has the capability to obtain accurate 3D information, based on laser scanning technology. A median normalized-vectors growth approach is introduced which uses (a) L1-median algorithm to generate the seed for growth tracking in a bottom-up manner, (b) density-based spatial clustering of applications with noise (DBSCAN) theory and (c) Newton's first law to segment the stem first, which naturally separates the leaves. The method described by He et al. (2017a) used structure from motion to reconstruct 3D point clouds of strawberry fruits from multiple views. The point clouds were then converted from the RGB space to the HSV space. The point cloud was segmented, based on thresholding the hue channel into agronomically important external strawberry traits, i.e., achene number, calyx size, height, color, and volume. They employed a clustering algorithm, based on Euclidean distance, to group points corresponding to the same achene for segmenting and counting the total number of achenes. Whereas the majority of the state-of-the-art 3D plant phenotyping methods have considered early growth stages of plants to avoid the complexities due to leaf crossovers and self-occlusions, Das Choudhury et al. (2020) used an iterative space carving approach for voxel-grid reconstruction of maize plants, using multiview images in the advanced vegetative stage. The method used voxel overlapping consistency check and point cloud clustering techniques to segment the 3D voxel-grid of the plant into (a) top leaf cluster (the top part of the plant where multiple newly emerged leaves are incompletely unfolded and surround a single junction occluding each other), (b) individual leaves as separate components and (c) the stalk. Figure 4.4 shows the reconstructed 3D voxel-grid of a maize plant and its segmented components.

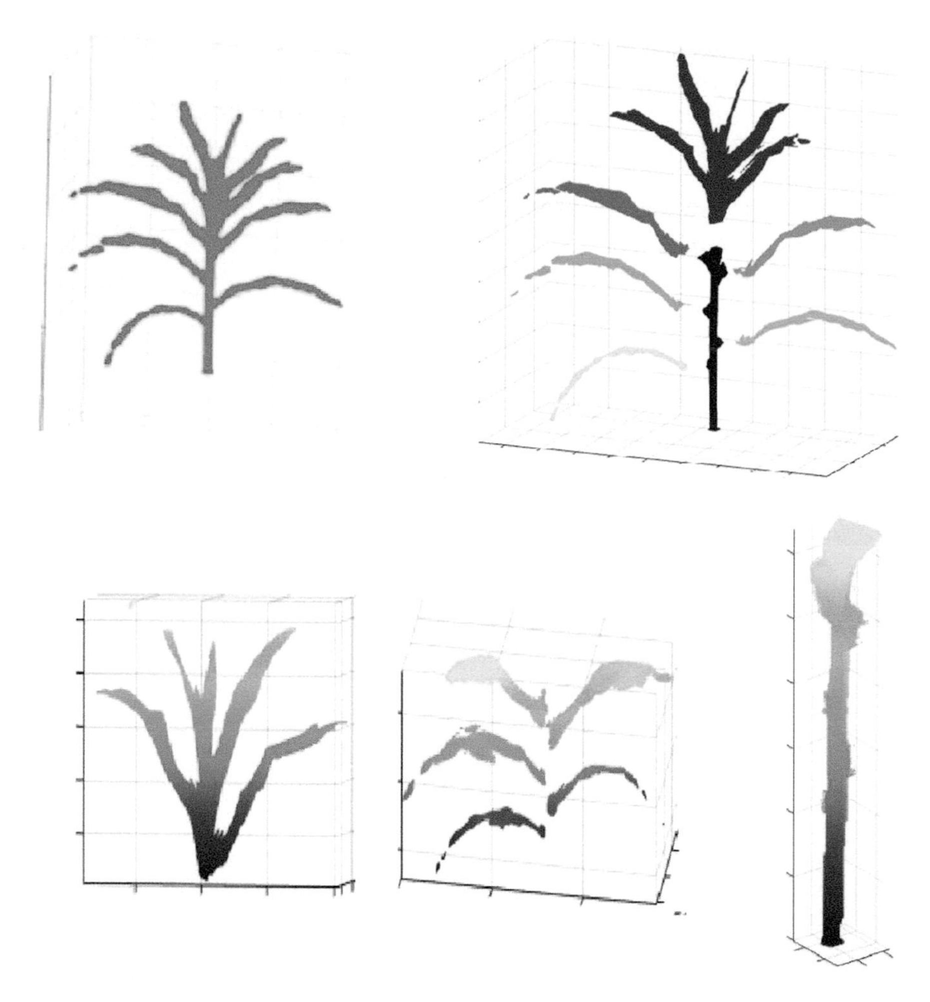

FIGURE 4.4 Illustration of 3D model reconstruction of a plant and segmentation of its components (top row); the reconstructed 3D model of a maize plant and its segmented plant components (bottom row); top leaf cluster, individual leaves, and stalk (left to right).

The color-based segmentation methods often break up a plant into multiple fragments due to color inconsistency between stems and leaves. To address the issue in the context of weed control, Dyrmann (2015) presented a method based on fuzzy c-means and distance transform to segment the plant as a single connected object, while preserving its shape. Fuzzy c-means is a clustering algorithm in which each pixel is categorized into multiple clusters, namely plant and soil clusters, in this case. The author used a distance-dependent threshold to prevent stems, the color of which is different from that of leaves, from being incorrectly grouped into the soil cluster. Experimental results demonstrated that the proposed method was capable of preserving plants' shape better than many other color-based segmentation methods presented in the literature. The Nottingham method, as described in Scharr et al. (2015), used the simple linear iterative clustering super-pixel algorithm (Achanta et al. 2012) to generate super-pixels in *Lab* color space. A region growing algorithm in the super-pixel space was used to segment the plant from the background. The super-pixel with the lowest mean color defined in Lab space was used as the initial seed. The method used the watershed algorithm to segment individual leaves of rosette plants.

4.2.2.2 Neural Network-Based Segmentation

Neural network-based approaches, especially deep learning, have been widely and successfully employed in a wide variety of computer vision tasks. Studies have shown that the response of the human brain to external visual stimulus is layered. The perceptual ability of neural layers is organized to respond hierarchically from low-level information, such as the edges, followed by shape information, up to the final recognition of high-level concepts such as the objects. Convolutional neural networks (CNNs), the most widely used architecture, have demonstrated shift, scale, and distortion invariance (Lecun et al. 1998) in the recognition process. Since CNNs are well suited to extract discriminative and representative features of images, they have been widely used in plant phenotyping tasks, including segmentation.

Aich and Stavness (2017) proposed a leaf counting algorithm that first performs segmentation. In the segmentation stage, convolutional and deconvolutional layers of the network produce the probability that each pixel belongs to the plant. Each of the convolutional and deconvolutional layers is followed by batch normalization and rectified linear unit. Nellithimaru and Kantor (2019) introduced a simultaneous localization and mapping (SLAM) algorithm for counting grapes in a vineyard by combining deep learning with disparity-based 3D reconstruction. The method adapted mask R-CNN (He et al. 2017b) is used to classify stereo images into either grapes, leaves, or branches using instance segregation. A singular value decomposition module is used to characterize the local neighborhood around each 3D point in terms of its 'pointness', 'curveness' and 'surfaceness' and thus to reduce the false positive grape detections and to classify non-grape points into leaves and branches. Segmentation is the key step for the rice panicle phenotyping applications. Since the color, texture, size, and shape of the panicle vary significantly due to water reflections, illumination variations, cluttered background, variations among the rice accessions, and differences in reproductive stages, accurate segmentation of rice panicle is extremely challenging, particularly in the field. Xiong et al. (2017) introduced a novel rice panicle segmentation algorithm, called Panicle-SEG, which uses simple linear iterative clustering to generate candidate regions. Then, a CNN is applied for candidate regions to determine the panicles. Finally, the entropy rate super-pixel algorithm is used to optimize the segmentation.

Most of the existing research in leaf segmentation is focused on single plants in controlled environments, in which the leaves show very little or no occlusion. However, the segmentation of leaves in dense foliage can be an extremely challenging task when applying robotic outdoor inspection on plants in the wild, due to the high degree of occlusion. To address this challenge, Morris (2018) proposed a pyramid CNN that performs boundary detection, followed by the watershed-based segmentation. The boundary detection is formulated as a binary classification problem that generates a confidence map of the probability of a pixel being an edge. The method first performs over-segmentation of leaf boundaries based on the watershed algorithm on the predicted leaf edge, followed by segment merging to generate leaf segments. The Wageningen method, as described by Scharr et al. (2015), also first segments the plant, using a feedforward neural network. Individual leaves are segmented using the watershed algorithm applied to the Euclidean distance map of the plant mask image.

Scharr et al. (2015) presented a comparative study of four leaf segmentation methods: IPK (Pape and Christian 2015), Nottingham, MSU (Yin et al. 2014), and Wageningen for plant and leaf segmentation accuracy (FBD *vs.* SBD). They concluded that high SBD could only be achieved when FBD is also high but that obtaining a high FBD is not a guarantee of good leaf segmentation (i.e., a high SBD). Lu et al. (2016) proposed a two-phase segmentation algorithm, consisting of region proposal generation and color model prediction for jointly segmenting maize plants and tassels in outdoor environments. It first uses a graph-based segmentation algorithm, combined with linear iterative clustering, to generate region proposals. The region proposals are then passed to a neural network-based ensemble model, called the neural network intensity color model, to attach semantic meanings of the crop, tassel, or background to these regions.

4.2.2.3 Other Learning-Based Segmentation Approaches

Simek and Barnard (2015) proposed a probabilistic model for multipart shapes, using piecewise Gaussian processes, and applied the model to segment rosette leaves of Arabidopsis plants. Prior knowledge is added to the model, which contributes robustness to partial occlusions. Singh and Misra (2017) proposed a genetic algorithm-based method for image segmentation, which is used for automatic detection and classification of plant leaf diseases. The algorithm was tested on ten different species, including banana, beans, jackfruit, lemon, mango, potato, tomato, and sapota. Guo et al. (2018) used decision-tree-based pixel segmentation models to segment sorghum heads from RGB images captured by UAVs. The method used a bag-of-features and a majority voting policy to improve segmentation accuracy.

Minervini et al. (2014) proposed a method for automated segmentation of rosette plants from time-lapse top-view images, where each image contained multiple instances of the same plant species. In their method, each instance in the original image was initially located, based on k-means clustering which groups each pixel of the original image as either plant or background, and then segmented, using active contours incorporating color and local texture features. Prior knowledge of plants is also used as the input to the segmentation phase in the form of a plant appearance model built with Gaussian Mixture Models (GMMs). The appearance model is updated incrementally using the previously segmented instances, which significantly reduces the need for prior labeling. The method proposed by Zhou et al. (2015) combined active contours with a visual saliency map for leaf segmentation.

Lee et al. (2016) deployed a bag-of-features model for automatic image classification, to isolate tassels from the cluttered background in a large-scale field experiment. The detected tassel was subjected to consensus-based thresholding for binarization. Pape and Christian (2015) proposed a segmentation (and leaf counting) method, using traditional machine learning. They first extracted a set of geometric features and predicted the number of leaves by using these features as the input to a regression model. They built a machine-learning-based pipeline for edge extraction to accurately detect leaf borders in the presence of overlaps. The extracted leaf border information was further used to reconstruct the detailed leaf shapes in combination with predicted leaf counts. Guo et al. (2013) introduced a decision tree-based segmentation approach for segmenting wheat plants in the presence of varying lighting conditions and a cluttered background in the field.

The IPK method by Pape and Klukas (2014) performed supervised foreground segmentation, based on color 3D histogram cubes, which encoded the probability, for any observed pixel in the given training set, of belonging to either the foreground or the background. The individual leaves of rosette plants were then detected based on supervised feature extraction, using a distance map and skeleton-graph transformation. The MSU method, developed by Yin et al. (2017), presented a framework for simultaneous joint multileaf segmentation, alignment, and tracking. Inspired by crowd detection, where both the number and locations of pedestrians are estimated simultaneously, they used a set of templates to match leaf candidates with the edge map of the last frame in a video. First, the Sobel edge detector was used to generate an edge map and a distance transform image. Then, a set of pre-defined templates were used to find an overcomplete set of transformed leaf templates, using Chamfer matching. The alignment process was formulated as an optimization problem of a designated loss function, using gradient descent. Finally, multileaf tracking was achieved using template transformation, which is also formulated as an optimization problem, using an objective function parameterized by a set of transformation parameters.

4.3 PLANT PHENOTYPING DATASETS

Development and public dissemination of datasets are vital for the advancement of image-based plant phenotyping research, as they provide the computer vision community access to datasets and a common basis for comparative performance analysis of phenotyping algorithms. The performance of an algorithm is typically carried out by comparing the output against human-annotated ground-truth released with the original images, based on a quality metric suitable for the given application. In this section, we briefly summarize the datasets that are publicly disseminated for phenotyping research and are relevant in the context of segmentation.

Aberystwyth leaf evaluation dataset: The dataset consists of top-view time-lapse image sequences of 80 *Arabidopsis thaliana* plants captured every 15 minutes for a period of 35 days, using the Photon Systems Instruments PlantScreen phenotyping platform located at the National Plant Phenomics Centre in Aberystwyth, UK. In addition to the original images, the dataset contains human-annotated ground-truth of a selection of images for evaluation of leaf segmentation, counting, and tracking. It also contains plant-based ground-truth (leaf area, fresh weight, and dry weight) and scanned images of rosette leaves, which are obtained from plants harvested periodically, and from all remaining plants at the end of the image acquisition process. A leaf area scanner was used to measure the area of the rosette leaves (Bell and Dee 2016). The dataset was released with a segmentation evaluation software and distortion correction software. It can be freely downloaded from https://zenodo.org/record/168158#.XaS-KOhKg2y.

CVPPP leaf segmentation and counting challenge (LSC and LCC) dataset: This is a collection of benchmark datasets designed to facilitate the development of algorithms for plant detection and localization, plant segmentation, leaf segmentation, detection, localization and counting, tracking of leaves, and classification and regression of mutants and treatments (Minervini et al. 2015). The Ara2012 and Ara2013 datasets consist of top-view time-lapse images of *A. thaliana* rosettes. The total number of images in Ara2012 and Ara2013 are 150 and 6137, respectively. The tobacco dataset consists of top-view stereo image sequences of tobacco plants (a total number of 165120 images from 80 plants) captured hourly for a period of 18 to 30 days in four different settings, with 20 plants participating in each setting. Figure 4.5

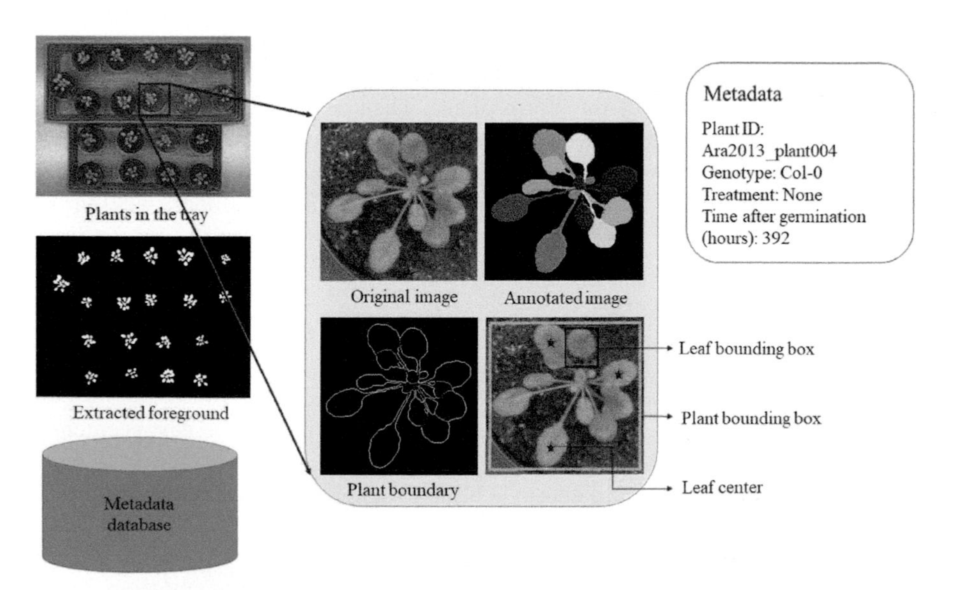

FIGURE 4.5 A symbolic representation of LSC and LCC datasets that depicts the original images with available annotations, ground-truth, and metadata (Minervini et al. 2015).

symbolically depicts the original images, annotated images, the available ground-truth information, and the metadata. The datasets can be freely obtained from https://www.plant-phenotyping.org/datasets-home.

The Ax dataset: The limited availability of training samples adversely impacts the performance of learning-based algorithms for leaf segmentation (Ren and Zemel 2016; Scharr et al. 2015), especially for those using a large number of parameters (e.g., deep network approaches). One common approach to alleviate this problem is dataset augmentation (Giuffrida et al. 2017; Ward et al. 2018). Giuffrida et al. (2017) proposed a generative adversarial network (GAN), trained on CVPPP LSC and LCC datasets, to generate synthetic rosette-shaped plant images. This dataset of artificially generated plant images is named the Ax dataset. The dataset contains 57 RGB images and is released with the total number of leaves in each image as the ground-truth. Figure 4.6 shows some sample images of this dataset. The dataset can be freely downloaded from http://www.valeriogiuffrida.academy/ax.

Joint crop and maize tassel segmentation dataset: The male flower of maize (*Zea mays L.*), i.e., the tassel, is a branched structure grown at the end of the stem that produces pollen to fertilize the female flowers borne on the ear, for reproduction. Accurate segmentation and counting of the tassels are crucial for understanding the diversity of the tassel architecture that is directly linked to grain yield. This dataset contains 229 crop images, 81 tassel images, and 13 background images divided into training and test sets (Lu et al. 2016). Each image in the test set has a manually annotated ground-truth image created with a high level of accuracy, using Photoshop. The crop segmentation test set contains representative images from seven outdoor scenarios, i.e., overcast day, rainy day, cloudy day, sunny day,

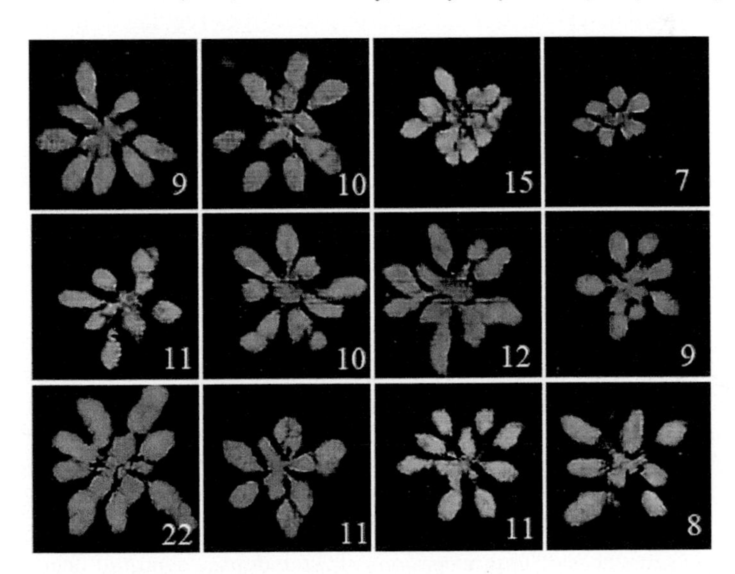

FIGURE 4.6 Sample images from Ax dataset. The numbers in the bottom-right denote the total number of leaves.

FIGURE 4.7 Sample images and corresponding ground-truth from joint crop and maize tassel segmentation dataset.

shadow region, highlight region, and cluttered background. The tassel segmentation dataset is similar to the crop segmentation dataset, but the images with highlighted regions are replaced by images from different cultivars. Figure 4.7 shows some sample images and corresponding ground-truth images from the dataset. The dataset provides a unifying platform for the development and uniform comparisons of maize plant and tassel segmentation methods in realistic field-based plant phenotyping scenarios. The dataset can be freely downloaded from https://github.com/poppinace/crop_tassel_segmentation.

UNL-CPPD: UNL-CPPD is designed to stimulate research in the development and comparison of algorithms for leaf detection and tracking, segmentation, and alignment in maize plants. The dataset is helpful for exploring the temporal variation of component phenotypes regulated by genotypes. It contains a total number of 816 color images of 13 maize plants from two side views captured daily by a visible light camera for 32 days. Each original image is accompanied by a human-annotated ground-truth image, where each leaf is numbered in order of emergence, and an Excel sheet containing information about the co-ordinates of the leaf-tips and leaf-junctions (Das Choudhury et al. 2018). Figure 4.8 shows an example of the ground-truth of a sample image of UNL-CPPD.

Komatsuna dataset: This dataset contains images of komatsuna plants (*Brassica rapa* var. *perviridis*, Japanese mustard spinach) in early growth stages, with a leaf annotation tool to facilitate 3D plant phenotyping analysis, such as leaf segmentation, tracking, and reconstruction (Uchiyama et al. 2017). A set of five komatsuna plants were imaged every four hours for ten days, using an RGB camera (multiview dataset) and an RGB camera fitted with a structured light depth camera (RGBD dataset). The dataset is freely available from http://limu.ait. kyushu-u.ac.jp/~agri/komatsuna/.

Deep phenotyping dataset: This dataset consists of 22 successive top-view image sequences of four accessions of Arabidopsis, captured once daily to study the dynamic growth characteristics of plants for accession classification, using deep learning approaches (Namin et al. 2018). The dataset is

FIGURE 4.8 Ground-truth images of a maize plant from UNL-CPPD: Day 15 (top-left); Day 17 (top-right); Day 19 (bottom-left); and Day 22 (bottom-right).

augmented by rotating each image by 90°, 180°, and 270° to avoid overfitting, while training. It can be freely downloaded from https://figshare.com/s/e18a9782676750595 78f.

MSU dense-leaves dataset: The dataset was introduced for research in automatic detection and segmentation of overlapping leaves in dense foliage (Morris 2018). The dataset consists of 108 original dense-foliage images of trees, vines, and bushes captured under an overcast sky or at night, and their human-annotated ground-truth. Figure 4.9 shows a sample image of this dataset and its human-annotated segmented image (ground-truth). The images are divided into training, validation, and test sets, containing 73, 13, and 22 images, respectively. The dataset is freely available from https://www.egr.msu.edu/denseleaves/.

4.4 OPEN PROBLEMS

The state-of-the-art segmentation techniques used in plant phenotyping rely on single image-based segmentation; hence, many opportunities remain to significantly advance this research field by exploring multiple image-based segmentation to address complex problems. In this section, we briefly describe three significant open

FIGURE 4.9 (a)–(e) Sample images of the MSU Dense-Leaves dataset; and (f) ground-truth of the image shown in (e).

problems in multiple image-based segmentation, i.e., sequence segmentation, co-segmentation, and multimodal segmentation.

Sequence segmentation: For image-based plant phenotyping analysis, plants are imaged at regular intervals to analyze their dynamic growth characteristics. As the plant grows, newer leaves often occlude older leaves. For example, in the case of maize and sorghum plants, environmental factors and normal leaf growth may cause leaves to overlap, resulting in crossovers in an image. These factors make individual leaf segmentation challenging. However, this problem could be addressed by examining the images of the plant during early stages of growth when the leaves did not overlap, and tracking the growth of the leaves over time. Thus, we propose a new problem, called sequence segmentation, which will exploit the growth history of the plants by analyzing image sequences for segmentation instead of analyzing each image separately.

Co-segmentation: The objective of co-segmentation is to simultaneously segment the same or similar objects of the same class from a set of images, by utilizing the consistency information of the shared foreground objects, irrespective of the background. Co-segmentation, introduced by Rother et al. (2006), has been successful in a variety of applications, including image retrieval, image classification, object recognition, tracking, and video content analysis. Co-segmentation techniques attempt to segment either a single foreground or multiple foregrounds from either a pair or a group of images, where the foreground might be present in all the participating images or may intermittently disappear (Kim and Xing 2012). The segmentation problems for plant phenotyping will benefit from using a co-segmentation approach, especially in tracking leaves and flowers, and also for detecting the timing of important events in the plant's life cycle, e.g., germination and the emergence of a leaf, by analyzing image sequences.

Multimodal segmentation: In high-throughput phenotyping systems, plants are typically imaged by multiple cameras in different modalities, e.g., RGB, IR, NIR, hyperspectral, and fluorescence, to capture phenotypic information that are manifested at different wavelengths of the electromagnetic spectrum. Note that the contrast between the plant and the background varies in different modalities. For example, the contrast is lower in IR and NIR images, making the segmentation a challenging task. Therefore, it is more efficient to generate a mask from the RGB image and use it to segment the images from other modalities after aligning them with respect to scaling, rotation, and translation. However, multimodal segmentation is still in its infancy in the domain of image-based plant phenotyping, but has great potential for future work.

4.5 CONCLUSIONS

This chapter presented a survey of segmentation techniques for image-based plant phenotyping analysis. The state-of-the-art segmentation algorithms are discussed in two broad categories, i.e., traditional and learning-based. Traditional approaches use a computing pipeline consisting of image-processing steps. In contrast, learning-based approaches leverage machine-learning algorithms to perform the segmentation. The selection of a segmentation algorithm for a particular application depends on three factors, i.e., phenotyping platform (controlled greenhouse or field), imaging modality (e.g., visible light, NIR, and hyperspectral), and category of phenotypes (holistic or component). This chapter also briefly described the publicly available benchmark datasets, which are indispensable for the development and uniform comparisons of segmentation algorithms. The segmentation techniques for plant phenotyping presented in the literature are based only on single images. We presented open problems regarding segmentation in the context of plant phenotyping analysis, to leverage the additional information that may be obtained from images of the plant acquired through multiple sensors or over time, to efficiently address complex segmentation challenges.

REFERENCES

Achanta, R., A. Shaji, K. Smith, A. Lucchi, P. Fua, and S. Susstrunk. 2012. SLIC superpixels compared to state-of-the-art superpixel methods. *IEEE Transactions on Pattern Analysis and Machine Intelligence* 34(11):2274–2282.

Aich, S., and I. Stavness. 2017. Leaf counting with deep convolutional and deconvolutional networks. In *Proceedings of the IEEE International Conference on Computer Vision Workshop on Computer Vision Problems in Plant Phenotyping* 2080–2089. Venice, Italy.

Bell, J., and H. Dee. 2016. *Aberystwyth Leaf Evaluation Dataset*. Zenodo. doi: 10.5281/zenodo.168158.

Chang, K. T. 1982. Analysis of the Weibull distribution function. *Journal of Applied Mechanics* 49(2):450–451.

Chen, Y., J. Ribera, C. Boomsma, and E. J. Delp. 2017. Plant leaf segmentation for estimating phenotypic traits. In *IEEE International Conference on Image Processing* 3884–3888. Beijing, China.

Chen, Y., S. Baireddy, E. Cai, C. Yang, and E. J. Delp. 2019. Leaf segmentation by functional modelling. In *Proceedings of the International Conference on Computer Vision and Pattern Recognition*, Long Beach, CA.

Dadwal, M., and V. K. Banga. 2012. Color image segmentation for fruit ripeness detection: A review. In *2nd International Conference on Electrical Electronics and Civil Engineering* 190–193. Singapore, Singapore.

Dai, X., and S. Khorram. 1999. A feature-based image registration algorithm using improved chain-code representation combined with invariant moments. *IEEE Transactions on Geoscience and Remote Sensing* 37(5):2351–2362.

Das Choudhury, S., S. Goswami, S. Bashyam, A. Samal, and T. Awada. 2017. Automated stem angle determination for temporal plant phenotyping analysis. In *IEEE International Conference on Computer Vision Workshop on Computer Vision Problems in Plant Phenotyping* 41–50. Venice, Italy.

Das Choudhury, S., S. Bashyam, Y. Qiu, A. Samal, and T. Awada. 2018. Holistic and component plant phenotyping using temporal image sequence. *Plant Methods* 14:35.

Das Choudhury, S., S. Maturu, A. Samal, V. Stoerger, and T. Awada. 2020. Leveraging image analysis to compute 3D plant phenotypes based on voxel-grid plant reconstruction. *Frontiers in Plant Science* under review.

Dyrmann, M. 2015. Fuzzy c-means based plant segmentation with distance dependent threshold. In *British Machine Vision Conference Workshop on Computer Vision Problems in Plant Phenotyping*, Swansea, UK.

Gampa, S. 2019. A data-driven approach for detecting stress in plants using hyperspectral imagery. MS Thesis, University of Nebraska-Lincoln, Lincoln, NE.

Gélard, W., M. Devy, A. Herbulot, and P. Burger. 2017. Model-based segmentation of 3D point clouds for phenotyping sunflower plants. In *12th International Joint Conference on Computer Vision, Imaging and Computer Graphics Theory and Applications* 459–467. Porto, Portugal.

Giuffrida, M. V., H. Scharr, and S. A. Tsaftaris. 2017. ARIGAN: Synthetic Arabidopsis plants using generative adversarial network. In *Proceedings of the IEEE International Conference on Computer Vision Workshop on Computer Vision Problems in Plant Phenotyping* 2064–2071. Venice, Italy.

Gonzalez, R. C., and R. E. Woods. 2017 *Digital Image Processing*, 4th ed. Pearson, New York.

Guo, W., U. K. Rage, and S. Ninomiya. 2013. Illumination invariant segmentation of vegetation for time series wheat images based on decision tree model. *Computers and Electronics in Agriculture* 96:58–66.

Guo, W., B. Zheng, A. B. Potgieter et al. 2018. Aerial imagery analysis – Quantifying appearance and number of sorghum heads for applications in breeding and agronomy. *Frontiers in Plant Science* 9:1544.

Hamuda, E., M. Glavin, and E. Jones. 2016. A survey of image processing techniques for plant extraction and segmentation in the field. *Computers and Electronics in Agriculture* 125:184–199.

He, J. Q., R. J. Harrison, and B. Li. 2017a. A novel 3D imaging system for strawberry phenotyping. *Plant Methods* 13:93.

He, K., G. Gkioxari, P. Dollar, and R. Girshick. 2017b. Mask R-CNN. In *Proceedings of the IEEE International Conference on Computer Vision* 2961–2969. Venice, Italy.

Hunt, E. R., M. Cavigelli, C. S. T. Daughtry, J. E. McMurtrey, and C. L. Walthall. 2005. Evaluation of digital photography from model aircraft for remote sensing of crop biomass and nitrogen status. *Precision Agriculture* 6(4):359–378.

Jin, S., Y. Su, F. Wu, S. Pang, S. Gao, T. Hu, J. Liu, and Q. Guo. 2019. Stem-leaf segmentation and phenotypic trait extraction of individual maize using terrestrial LiDAR data. *IEEE Transactions on Geoscience and Remote Sensing* 57(3):1336–1346.

Kataoka, T., T. Kaneko, H. Okamoto, and S. Hata. 2003. Crop growth estimation system using machine vision. In *Proceedings of the IEEE/ASME International Conference on Advanced Intelligent Mechatronics* 2:1079–1083. Kobe, Japan.

Kim, G., and E. P. Xing. 2012. On multiple foreground cosegmentation. In *IEEE International Conference on Computer Vision and Pattern Recognition* 837–844. Providence, RI.

Kumar, J. P., and S. Domnic. 2019. Image based leaf segmentation and counting in rosette plants. *Information Processing in Agriculture* 6(2):233–246.

LeCun, Y., L. Bottou, Y. Bengio, and P. Haffner. 1998. Gradient-based learning applied to document recognition. *Proceedings of the IEEE* 86(11):2278–2324.

Lee, N., Y. S. Chung, S. Srinivasan, P. S. Schnable, and B. Ganapathysubramanian. 2016. Fast, automated identification of tassels: Bag-of-features, graph algorithms and high throughput computing. In *Knowledge, Discovery and Data Mining Workshop on Data Science for Food, Energy and Water*, San Francisco, CA.

Li, C., C. Xu, C. Gui, and M. D. Fox. 2010. Distance regularized level set evolution and its application to image segmentation. *IEEE Transactions on Image Processing* 19(12):3243–3254.

Li, L., Z. Qin, and H. Danfeng. 2014. A review of imaging techniques for plant phenotyping. *Sensors* 14(11):20078–20111.

Li, D., Y. Cao, X. Tang, S. Yan, and X. Cai. 2018. Leaf segmentation on dense plant point clouds with facet region growing. *Sensors* 18(11):3625.

Liu, J., Y. Liu, and J. Doonam. 2018. Point cloud based iterative segmentation technique for 3D plant phenotyping. In *IEEE International Conference on Information and Automation* 1072–1077. Fujian, China.

Lu, H., Z. Cao, Y. Xiao, Y. Li, and Y. Zhu. 2016. Region-based color modelling for joint crop and maize tassel segmentation. *Biosystems Engineering* 147:139–150.

Lu, H., Z. Cao, Y. Xiao, B. Zhuang, and C. Shen. 2017. TasselNet: Counting maize tassels in the wild via local counts regression network. *Plant Methods* 13:79.

Martin, D., C. Fowlkes, and J. Malik. 2004. Learning to detect natural image boundaries using local brightness, color, and texture cues. *IEEE Transactions on Pattern Analysis and Machine Intelligence* 26(5):530–549.

Meyer, G. E., T. W. Hindman, and K. Lakshmi. 1998. Machine vision detection parameters for plant species identification. In *Precision Agriculture and Biological Quality*, ed. G. E. Meyer and J. A. DeShazer, 3543:327–335.

Meyer, G. E., J. Camargo-Neto, D. D. Jones, and T. W. Hindman. 2004. Intensified fuzzy clusters for classifying plant, soil, and residue regions of interest from color images. *Computers and Electronics in Agriculture* 42(3):161–180.

Minervini, M., M. M. Abdelsamea, and S. A. Tsaftaris. 2014. Image-based plant phenotyping with incremental learning and active contours. *Ecological Informatics* 23:35–48.

Minervini, M., A. Fischbach, H. Scharr, and S. A. Tsaftaris. 2015. Finely-grained annotated datasets for image-based plant phenotyping. *Pattern Recognition Letters* 81:80–89.

Morris, D. 2018. A pyramid CNN for dense-leaves segmentation. In *IEEE International Conference on Computer and Robot Vision* 238–245. Toronto, Canada.

Namin, S. T., M. Esmaeilzadeh, M. Najafi, T. B. Brown, and J. O. Borevitz. 2018. Deep phenotyping: Deep learning for temporal phenotype/genotype classification. *Plant Methods* 14:66.

Nellithimaru, A. K., and G. A. Kantor. 2019. ROLS: Robust object-level SLAM for grape counting. In *CVPR Workshop on Computer Vision Problems in Plant Phenotyping*. Long Beach, CA.

Ojeda-Magana, B., R. Ruelas, J. Quintanilla-Domingvez, and D. Andina. 2010. Color image segmentation by partitional clustering algorithms. In *IECON 2010 – 36th Annual Conference on IEEE Industrial Electronics Society* 2828–2833. Glendale, AZ.

Otsu, N. 1979. A threshold selection method from gray-level histograms. *IEEE Transactions on Systems, Man, and Cybernetics* 9(1):62–66.

Pape, J. M., and K. Christian. 2015. Utilizing machine learning approaches to improve the prediction of leaf counts and individual leaf segmentation of rosette plant images. In *British Machine Vision Conference Workshop on Computer Vision Problems in Plant Phenotyping* 1–12. Swansea, UK.

Pape, J. M., and C. Klukas. 2014. 3-D histogram-based segmentation and leaf detection for rosette plants. In *European Conference on Computer Vision Workshop on Computer Vision Problems in Plant Phenotyping* 8928:61–74. Zurich, Switzerland.

Ren, M., and R. S. Zemel. 2016. End-to-end instance segmentation and counting with recurrent attention. In *IEEE International Conference on Computer Vision and Pattern Recognition* 6656–6664. Las Vegas. NV.

Rother, C., V. Kolmogorov, T. Minka, and A. Blake. 2006. Cosegmentation of image pairs by histogram matching – Incorporating a global constraint into MRFs. In *IEEE Computer Society Conference on Computer Vision and Pattern Recognition* 993–1000. New York, NY.

Scharr, H., M. Minervini, A. P. French et al. 2015. Leaf segmentation in plant phenotyping: A collation study. *Machine Vision and Applications* 27(4):585–606.

Simek, K., and K. Barnard. 2015. Gaussian process shape models for Bayesian segmentation of plant leaves. In *British Machine Vision Conference Workshop on Computer Vision Problems in Plant Phenotyping*, Swansea, UK.

Singh, V., and A. K. Misra. 2017. Detection of plant leaf diseases using image segmentation and soft computing techniques. *Information Processing in Agriculture* 4(1):41–49.

Thorp, K., G. Wang, M. Badaruddin, and K. F. Bronson. 2016. Lesquerella seed yield estimation using color image segmentation to track flowering dynamics in response to variable water and nitrogen management. *Industrial Crops and Products* 86:186–195.

Uchiyama, H., S. Sakurai, M. Mishima, D. Arita, T. Okayasu, A. Shimada, and R. Taniguchi. 2017. An easy-to-setup 3D phenotyping platform for KOMATSUNA dataset. In *Proceedings of the IEEE International Conference on Computer Vision Workshops* 2038–2045. Venice, Italy.

Wang, Z., K. Wang, F. Yang, S. Pan, and Y. Han. 2018. Image segmentation of overlapping leaves based on Chan–Vese model and Sobel operator. *Information Processing in Agriculture* 5(1):1–10.

Ward, D., P. Moghadam, and N. Hudson. 2018. Deep leaf segmentation using synthetic data. In *British Machine Vision Conference*, Newcastle, UK.

Woebbecke, D. M., G. E. Meyer, K. Von Bargen, and D. A. Mortensen. 1992. Plant species identification, size, and enumeration using machine vision techniques on near-binary images. *SPIE Optics in Agriculture and Forestry* 1836:208–219.

Woebbecke, D. M., G. E. Meyer, K. Von Bargen, and D. A. Mortensen. 1995. Shape features for identifying young weeds using image analysis. *Transactions of the American Society of Agricultural Engineers* 38:271–281.

Xiong, X., L. Duan, L. Liu, H. Tu, P. Yang, D. Wu, G. Chen, L. Xiong, W. Yang, and Q. Liu. 2017. Panicle-SEG: A robust image segmentation method for rice panicles in the field based on deep learning and superpixel optimization. *Plant Methods* 13:104.

Yin, X., X. Liu, J. Chen, and D. M. Kramer. 2014. Multi-leaf tracking from fluorescence plant videos. In *IEEE International Conference on Image Processing* 408–412. Paris, France.

Yin, X., X. Liu, J. Chen, and D. M. Kramer. 2017. Joint multi-leaf segmentation, alignment and tracking from fluorescence plant videos. *IEEE Transactions on Pattern Analysis and Machine Intelligence* 40:1411–1423.

Zheng, L., J. Zhang, and Q. Wang. 2009. Mean-shift based color segmentation of images containing green vegetation. *Computers and Electronics in Agriculture* 65(1):93–98.

Zhou, Q., W. Zhicheng, Z. Weidong, and Y. Chen. 2015. Contour-based plant leaf image segmentation using visual saliency. In *International Conference on Image and Graphics* 48–59. Tianjin, China.

5 Structural High-Throughput Plant Phenotyping Based on Image Sequence Analysis

Sruti Das Choudhury and Ashok Samal

CONTENTS

5.1 INTRODUCTION

The complex interaction between genotype and the environment determines the observable morphological and biophysical characteristics of plants, i.e., phenotypes, which influence the yield and acquisition of resources. Image-based plant phenotyping refers to the proximal sensing and quantification of plant traits, based on analyzing their images, captured at regular intervals with precision. The process is generally non-destructive, allowing the same traits to be quantified periodically during a plant's life cycle. However, extracting meaningful numerical phenotypes based on image-based automated plant phenotyping remains a critical bottleneck in the effort to link intricate plant phenotypes to genetic expression.

5.1.1 Plant Phenotyping Computational Framework

The state-of-the-art computer vision-based methods aim to analyze the image sequences of a group of plants (with different genotypes) captured by different types of cameras, i.e., visible light, infrared, near-infrared, fluorescent, and hyperspectral, at regular time intervals from multiple viewing angles under varying environmental conditions, to achieve high-throughput plant phenotyping analysis. In Figure 5.1, we present an abstract computational framework for high-throughput plant phenotyping

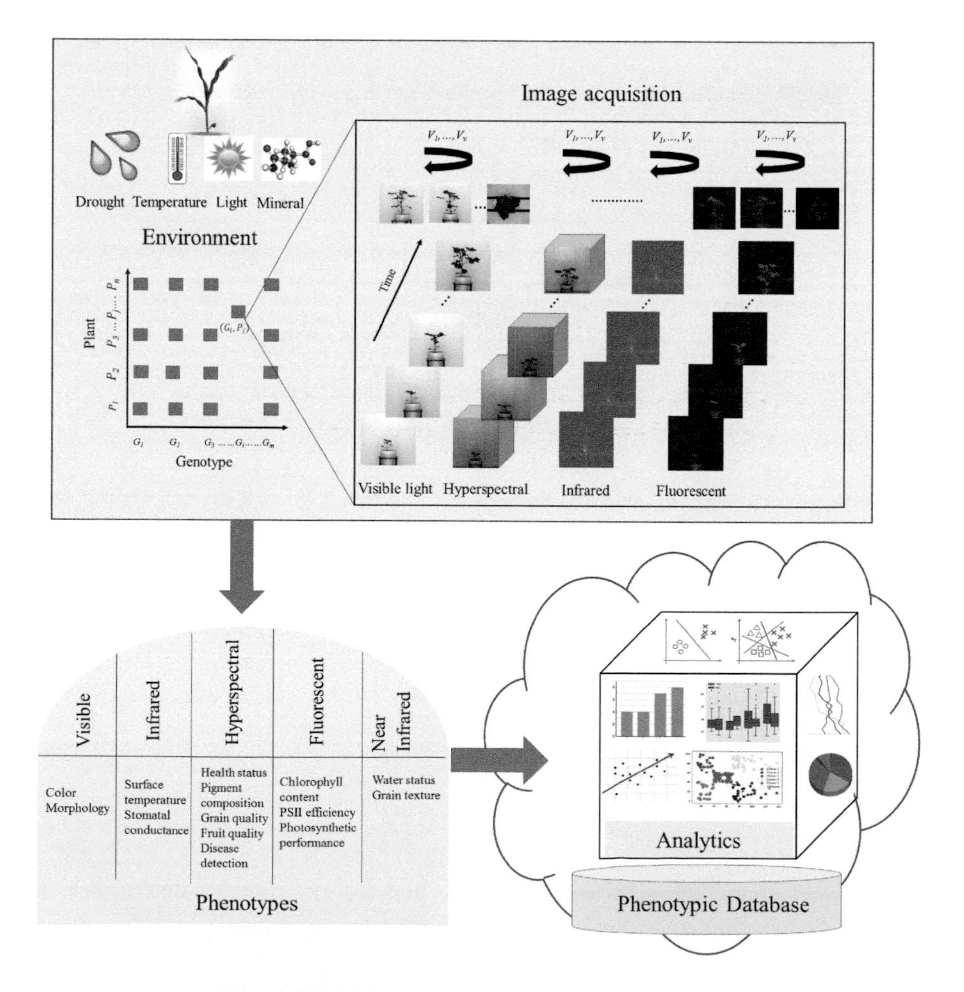

FIGURE 5.1 High-throughput plant phenotyping computational framework.

analysis. The framework facilitates the exploration of computer vision-based plant phenotyping algorithms along multiple dimensions, i.e., multimodal, multiview, and temporal, regulated by genotypes under various abiotic (e.g., drought, salinity, extreme temperature) and biotic stresses (e.g., fungi, arthropods, bacteria, weeds). The visible light images are mainly used to characterize the morphological or structural properties of a plant (Das Choudhury et al. 2018; McCormick et al. 2016; Zhang et al. 2017). Thermal imaging is used as a proxy for the plant's temperature to detect differences in stomatal conductance as a measure of the plant's response to water status and transpiration rate, to reflect abiotic stress adaptation (Li et al. 2014). Fluorescence images are analyzed to measure chlorophyll fluorescence, which helps achieve understanding of the photosynthetic efficiency of plants. A hyperspectral camera simultaneously captures images in hundreds of adjacent narrow wavelength bands. The spectral reflectance curve at each plant pixel, as a function of wavelength, has been used to investigate biophysical and functional properties of plants, e.g., early detection of a plant's susceptibility to stress, identification of parts of a plant affected by a particular disease, stress tolerance mechanisms, and early detection of fruit decay (Lu et al. 2017; Wahabzada et al. 2016).

The phenotypes obtained from an experiment in a high-throughput phenotyping system are stored in a database and then analyzed using statistical methods, data mining algorithms, or visual exploration. A time-series analysis, using cluster purity and angular histograms to investigate genetic influence on temporal variation of phenotypes, is described in Das Choudhury et al. (2017). The method by Das Choudhury et al. (2016) uses analysis of variance (ANOVA) to demonstrate the heritability of holistic phenotypes, i.e., bi-angular convex-hull area ratio and plant aspect ratio. Gampa (2019) developed a classification model using an artificial neural network to classify stressed and non-stressed pixels of a plant, using reflectance curves generated from hyperspectral imagery, and demonstrated temporal propagation of drought stress in plants, based on clustering techniques. Quantitative trait locus (QTL) analysis involves statistical methods to map quantitative phenotypic traits to genomic, molecular markers to explain the genetic basis of the phenotypic variations for improved selective breeding (Martinez et al. 2016). Graphical demonstrations are used to visualize the temporal variation of phenotypes regulated by genotypes in the method developed by Das Choudhury et al. (2018).

5.1.2 Taxonomy of Plant Phenotypes

Figure 5.2 presents a hierarchical classification of aboveground plant phenotypes that can be constructed using the plant phenotyping computational framework shown in Figure 5.1. There are three main categories of phenotypes, namely structural, physiological, and temporal (Das Choudhury et al. 2019). The structural phenotypes characterize a plant's morphology, whereas physiological phenotypes are related to the plant processes that regulate growth and metabolism. Morphological and physiological phenotypes are computed either by considering the plant as a single object (holistic phenotypes) or as individual components, i.e., leaf, stem, flower, and fruit (component phenotypes). Holistic phenotypes are the measurements of attributes of geometrical shapes enclosing the plants, e.g., the height of the bounding box enclosing

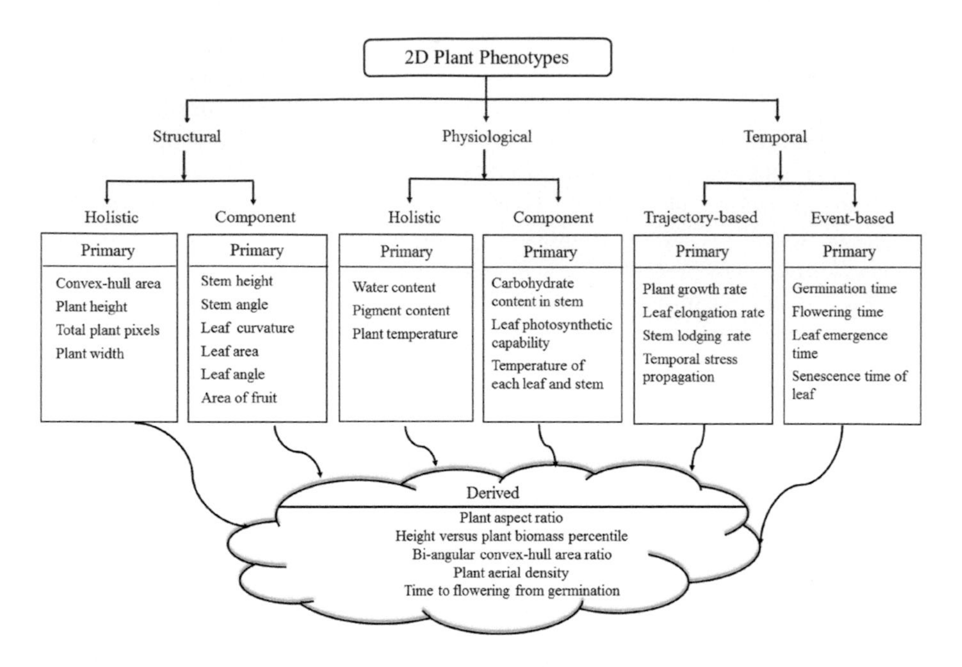

FIGURE 5.2 A taxonomy of 2-dimensional plant phenotypes.

the plant from the side is used to account for plant height, the diameter of the circle enclosing the top-view image of the plant is used to account for plant width, and the area of the convex-hull is a measurement of plant size. Component plant phenotyping analysis is more complex since it requires identifying and tracking individual plant organs that often have similar shape and appearance. The development of effective component phenotypes is important since they have the potential to improve our understanding of plant growth and development at a greater resolution. In literature, many component phenotypes are studied, e.g., rice panicle properties (Al-Tam et al. 2013), leaf length and stem angle (Das Choudhury et al. 2017, 2018), size of maize tassels (Gage et al. 2017), sorghum panicle architecture (Zhou et al. 2019), fruit volume (He et al. 2017), and chlorophyll content of a leaf (Yin et al. 2018).

The genetic regulation of non-uniformity in a plant's growth of its different parts over time has given rise to a new category of phenotype, called the temporal phenotype. Temporal phenotypes provide crucial information about a plant's development over time. Temporal phenotypes, computed by image sequence analysis, are subdivided into two groups, namely trajectory-based and event-based phenotypes. Holistic and component structural and physiological phenotypes are often computed from a sequence of images captured at regular time intervals to demonstrate the temporal variation of phenotypes regulated by genotypes (trajectory-based phenotypes), e.g., plant growth rate, the trajectory of stem angle. Detection of the timing of important events in the life cycle of a plant with reference to an origin (say, planting date), e.g., germination, the emergence of a new leaf, flowering, and fruiting, provide crucial information about the plant's vigor (Agarwal 2017). Whereas the primary phenotypes refer to one specific attribute of a plant, derived phenotypes are computed using two or more primary phenotypes to capture salient traits. Some examples of derived phenotypes include the bi-angular convex-hull area ratio, the plant aspect ratio, the time from germination to flowering, etc. (Das Choudhury et al. 2016).

5.2 PHENOTYPE COMPUTATION PIPELINE

In this section, we present an image-processing pipeline to compute holistic and component structural phenotypes, using 2-D visible light image sequences. Computation of holistic phenotypes is generally easier since it requires the segmentation of the plant and not its parts. However, the determination of the plant's architecture and the isolation of its components are essential to computing the component phenotypes, which increase the complexity of the component phenotyping algorithms. In this chapter, we use a skeleton-graph transformation approach to identify the components of a maize plant. The overall structural phenotype computation pipeline is summarized in Algorithm 1.

ALGORITHM 1: STRUCTURAL PHENOTYPE COMPUTATION

Input: An image sequence of a plant, $P=\{\alpha_{d_1},\alpha_{d_2},\ldots,\alpha_{d_n}\}$, where α_{d_i} denotes the image obtained on day d_i, where $d_i \le d_{i+1}$. n denotes the total number of imaging days. Without loss of generality, we assume $d_i=i$.

Output: $T=\{\{H_1,C_1\},\{H_2,C_2\},\ldots,\{H_n,C_n\}\}$, where H_i and C_i are the sets of holistic and component phenotypes of the plant P for the day d_i, respectively.

1: **for** $j=1{:}n$ **do**
2: $S_j = segment(\alpha_j)$;//Segment the plant
3: $H_j = computeHolisticPhenotypes(\alpha_i)$//Compute holistic phenotypes
4: $W_j = skeleton(S_j)$;//Compute the skeleton W_j of the segmented plant S_j.
5: $Z_j = removeSpur(W_j, threshold)$;//Remove the spurious edges, using a threshold
6: $G_j(V_j,E_j) = determineGraph(Z_j)$;//Represent as a graph; nodes: V_j; edges: E_j
7: $base = determineBase(G_j)$;//The bottom-most node in the graph G_j
8: $J = degreeThreePlusNodes(G_j)$;//All nodes with degree ≥3 are junctions
9: $Stem = getStem(J)$;//Connected path of junctions starting with base
10: $Tips = degreeOneNodes(G_j)$;//Terminal nodes are leaf tips
11: $Leaves = \phi; Interjunctions = \phi$;
12: **for** $e_j \in E_j$ **do**//Label each edge in the graph
13: $[v_{j,1}, v_{j,2}] = vertices(e_j)$;//The two vertices of the edge
14: if $(v_{j,1} \in J) \wedge (v_{j,2} \in J)$ $Interjunctions = Interjunctions \cup \{e_j\}$
15: if $(v_{j,1} \in Tips) \wedge (v_{j,2} \in Tips)$ $Leaves = Leaves \cup \{e_j\}$
16: **end for**
17: //Compute component phenotypes
18: $C_j = computeComponentPhenotypes(Leaves, Stem, Interjunctions)$;
19: **end for**

5.2.1 Segmentation

A three-step segmentation technique, i.e., image subtraction followed by color-based thresholding and connected component analysis, is used to segment the plant (foreground) from the background. Since the imaging chambers of existing high-throughput plant phenotyping systems generally have a fixed homogeneous background, subtracting it from the image is effective in extracting the foreground, i.e., the plant. However, the success of this technique requires the background image and the plant image to be aligned with respect to scale and rotation. Hence, we first use an automated image registration technique, based on local feature detection and matching to address changes in zoom levels (resulting in scale variation) during the image capturing process. The key to feature detection is to find features (e.g., corners, blobs, and edges) that remain locally invariant so that they are detected, even in the presence of rotation and scale changes (Tuytelaars and Mikolajczyk 2008). In this application, the corners of the pots and the frames of the imaging cabinet are used as the local features to align the foreground and the background, based on correspondence. Figure 5.3 (a) shows the background image, Figure 5.3 (b) shows the plant image, and Figure 5.3 (c) shows the extracted foreground after applying image subtraction.

The extracted foreground obtained after image subtraction often retains undesirable noise from the imaging cabinets and the pot due to lighting variations. The noise is eliminated by preserving only the parts that correspond to the plant (green)

FIGURE 5.3 Illustration of the segmentation process. Row 1: (left) the static background image; (middle) the original image; (right) the foreground obtained after applying image subtraction technique. Row 2: (left) the foreground obtained by green pixel superimposition; (middle) the foreground containing green pixels characterizing the plant; and (right) the final binary image mask for the plant.

in the original image. Figure 5.3 (d) shows the result of superimposing the green pixels from the original image shown in Figure 5.3 (b). The green pixels constituting the plant are retained, while noisy pixels, typically other than green, are removed. The resulting foreground, consisting of only green pixels characterizing the plant, is shown in Figure 5.3 (e). Color-based thresholding in HSV (Hue, Saturation, and Value) color space is applied to this image to binarize it. The binary image is subjected to connected-component analysis, followed by morphological operations to fill up any small holes inside the plant image to derive a single connected region for the plant, as shown in Figure 5.3 (f).

5.2.2 View Selection

In high-throughput plant phenotyping, images of plants are usually captured from multiple viewing angles. The junctions and tips of a plant are most clearly visible when the line of sight of the camera is perpendicular to the axis of the leaves. Hence, it is best to use the view that is closest to this optimal view to compute accurate phenotypes. We use the convex-hull as the basis from which to select the best view, i.e., choose the view that has the largest convex-hull area for the plant. Figure 5.4 shows two views of the binary images of a maize plant, enclosed by their convex-hulls. Given that a plant is imaged with m viewing angles each day, the optimal view (OV) is given by:

$$OV = View_j : ConvexHull\left(View_j\right) > ConvexHull\left(View_k\right), 1 \le k \le m, k \neq j$$

where $View_j$ is the jth view of the plant and the function $ConvexHull$ computes the area of the convex-hull of the plant in the image.

5.2.3 Skeletonization

The skeletonization, i.e., the process of reducing a shape to a linear representation, while preserving the main topological and size characteristics, is computed based on

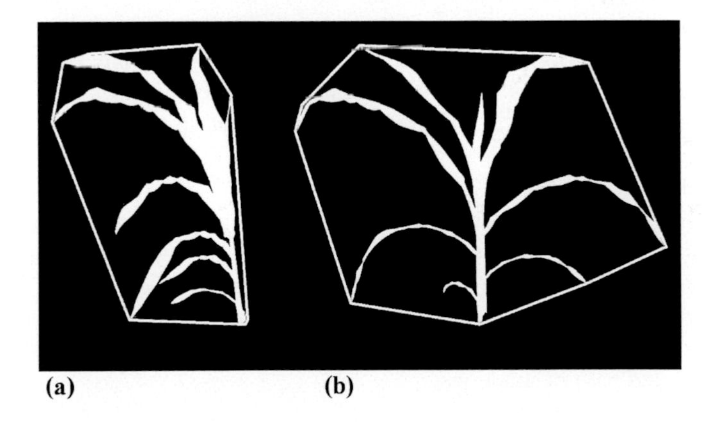

FIGURE 5.4 Illustration of view selection: Binary images of a maize plant in two different views enclosed by convex-hulls.

morphological thinning, geometric methods, or fast marching methods. The morphological thinning- based methods iteratively peel off the boundary, one pixel layer at a time, identifying the points the removal of which does not affect the topology. Although the algorithm is straightforward to implement, it requires extensive use of heuristics to ensure the skeletal connectivity, and hence does not perform well in complex dynamic structures like plants. The geometric methods compute Voronoi diagrams to produce an accurately connected skeleton. However, their performance depends largely on the robustness of the boundary discretization and is computationally expensive. We use fast marching distance transforms to skeletonize the binary image (Hassouna and Farag 2007), due to the robustness of this method to noisy boundaries, low computational complexity, and accuracy in terms of skeleton connectivity. The limiting factor of the skeletonization process is the skeleton's high sensitivity to boundary noise, resulting in redundant spurious branches, or "spurs", which significantly affect the topology of the skeleton graph (Bai et al. 2007). The most common approaches to overcome skeleton instability are based on skeleton pruning, i.e., eliminating redundant skeleton branches. We use thresholding-based skeleton pruning to remove spurious branches. The value of the threshold is experimentally determined using images from the dataset.

5.2.4 Skeleton-Graph Transformation

Graphical representations of skeletons have been investigated in the literature for many computer vision applications, e.g., object recognition (Ruberto 2004), shoot length measurement of submerged aquatic plants (Polder et al. 2010), and detection of leaves and stem (Das Choudhury et al. 2018). Here, we describe a novel skeleton-graph transformation approach, where the skeleton of the plant is represented by a connected graph.

The skeleton of the plant P is represented as the graph, i.e., $P=\{V,E\}$, where V is the set of nodes, and E is the set of edges. The set of nodes, V, is defined by $V=\{B,T,J\}$ where B, T, and J are the base of the plant, the tips of leaves, and the junctions in the stem from which the leaves emerge, respectively. The set of edges E is defined as $E=\{L,I\}$, where L and I represent the leaves and inter-junctions in the plant, respectively. These terms are briefly described below:

- *Base* (B): The point from where the stem of the plant emerges from the ground and is the bottom-most point of the skeleton.
- *Junction* (J): A node at which a leaf is connected to the stem. This is also referred to as a "collar" in plant science. The degree of junctions in the graph is at least 3.
- *Tip* (T): A node with degree 1 is considered to be a tip, the free end of the leaf.
- *Leaf* (L): Leaves connect the leaf tips and the junctions on the stem. An edge with a leaf tip as a node is considered to be a leaf.
- *Inter-junction* (I): An edge connecting two junctions is called an inter-junction. The stem is formed by traversing the graph from the base along a connected path of junctions.

Figure 5.5 illustrates the skeleton-graph transformation for determination of plant architecture. The figure shows the original image (left), its skeleton (middle), and

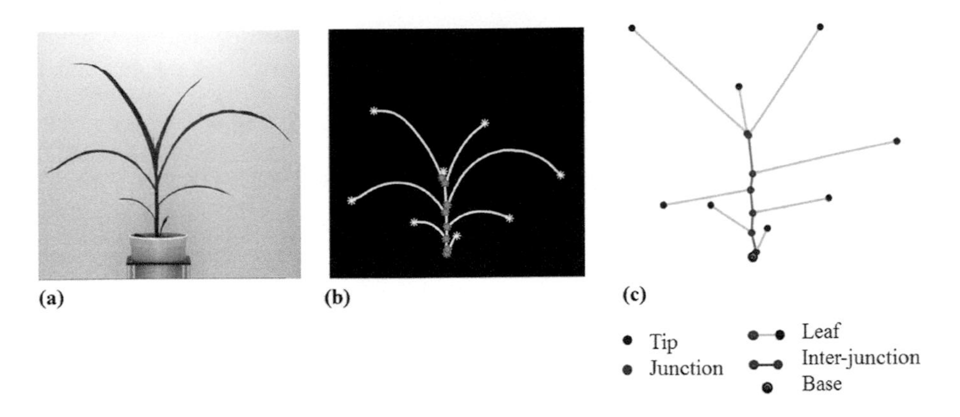

FIGURE 5.5 Illustration of skeleton-graph transformation: (left) the original plant image; (middle) the skeleton image; and (right) the graphical representation of the skeleton.

the graphical representation of the skeleton with all major components of the graph labeled (right).

5.2.5 Detecting Intersecting Leaves using Curve Tracing

In the late vegetative stage, leaves in the upper part of many plants are not fully unfolded, resulting in self-occlusions. Fully-grown leaves often overlap each other due to the plant architecture, their growth stages, or stress conditions. The overlaps result in loops in the skeleton graph and pose a challenge in isolating leaves. In this section, we describe an approach to address the second problem, i.e., leaf crossover, using curve tracing that leverages local continuity. Curve tracing involves the identification, extraction, classification, and labeling of curves in the image. It is used widely in many applications, e.g., road-following in computer games, feature extraction in medical image processing, and object tracking.

Figure 5.6 shows an example of two fully-grown leaves crossing each other at point P, creating a loop. While tracking leaf E_c at P, we have three choices of leaf

FIGURE 5.6 Illustration of leaf detection in the presence of intersecting leaves.

segments to connect it to: E_1, E_2, or E_3. The algorithm chooses the leaf-segment that best maintains continuity at point P with the leaf being tracked, i.e., E_c. This is described by Algorithm 2 below.

ALGORITHM 2: LEAF TRACING AT A CROSSOVER

Input: A leaf segment E_c being tracked at a crossover point P
$E = \{E_1, E_2, \ldots, E_n\}$, a set of other leaf-segments at P
Output: A leaf segment E_d that will be the continuation of E_c

1: $m_c = computeSlope(E_c, P)$;//slope of E_c at crossover point P
2: $M = computeSlopes(E, P)$;//$M = \{m_1, m_2, \ldots, m_n\}$ are slopes for the edges
3: Select E_j: $|m_c - m_j| < |m_c - m_k|$, $1 \le j \le n$.
4: $E_d = E_j$;
5: Return E_d;

5.3 STRUCTURAL PHENOTYPES

In this section, we briefly describe a set of holistic and component phenotypes that are based on the structure of the plant.

5.3.1 Holistic Phenotypes

Three derived holistic phenotypes, namely, bi-angular convex-hull area ratio ($BA_{CH}R$), plant aspect ratio (PAR), and plant areal density (PAD), are defined below.* Their significance in the context of plant phenotyping is also briefly summarized.

5.3.1.1 Bi-Angular Convex-Hull Area Ratio

Bi-angular convex-hull area ratio ($BA_{CH}R$) is defined (Das Choudhury et al. 2018) as:

$$BA_{CH}R = \frac{Area_{CH} \text{ at side view } 0^\circ}{Area_{CH} \text{ at side view } 90^\circ}$$

Where $Area_{CH}$ is the area of the convex-hull of the plant. Figure 5.7 shows the computation of bi-angular convex-hull area ratios for a sequence of plant images. Rows 1 and 3 show the images of two side views (0° and 90°) of a maize plant periodically from Day 7 to Day 24 after emergence. Rows 2 and 4 show the corresponding binary images enclosed by the convex-hulls. It should be noted that the shapes of the convex-hulls vary between the two views of the plant.

Phenotypic significance: Leaves are one of the primary organs of plants, which transform solar energy into chemical energy in the form of carbohydrate through photosynthesis. The total leaf area is associated with the photosynthetic rate (Duncan

* The 2-D holistic phenotype computing software is implemented using OpenCV 2.4.9 and C++ on Visual Studio 2010 Express Edition. It is publicly available at http://plantvision.unl.edu.

FIGURE 5.7 Illustration of bi-angular convex-hull area ratio. Rows 1 and 3 show the original images of a maize plant at two side views. Rows 2 and 4 show the convex-hulls of the plants in Rows 1 and 3, respectively.

1971). Some plants, e.g., maize, alter leaf positioning (i.e., phyllotaxy) in response to light signals perceived through the photoreceptor pigment phytochrome to optimize light interception (Maddonni et al. 2002). $BA_{CH}R$ provides information on plant rotation, due to such shade avoidance, and thus enables the study of phyllotaxy at different phases in the life cycle of plants.

5.3.1.2 Plant Aspect Ratio

Plant aspect ratio (PAR) is defined as

$$PAR = \frac{Height_{BR} \text{ in side view}}{Diameter_{MEC} \text{ in top view}}$$

where $Height_{BR}$ is the height of the bounding rectangle of the plant in the side view, and $Diameter_{MEC}$ is the diameter of the minimum enclosing circle of the plant in the top view. Figure 5.8 shows an illustration of the two components of the plant aspect ratio. Rows 1 and 3 show the images of a maize plant periodically from Day 7 to Day 24 after emergence, from the side view and top view, respectively. Rows 2 and 4 show the binary images of plants enclosed by the bounding box and the minimum bounding enclosing circle of the corresponding plant images shown in rows 1 and 3, respectively.

Phenotypic significance: The plant aspect ratio is a metric for distinguishing between genotypes with narrow *versus* wide leaf extent, after controlling the plant height. The height of the bounding box enclosing a plant from the side view is not affected by the viewing angle. However, the viewing angle does influence the

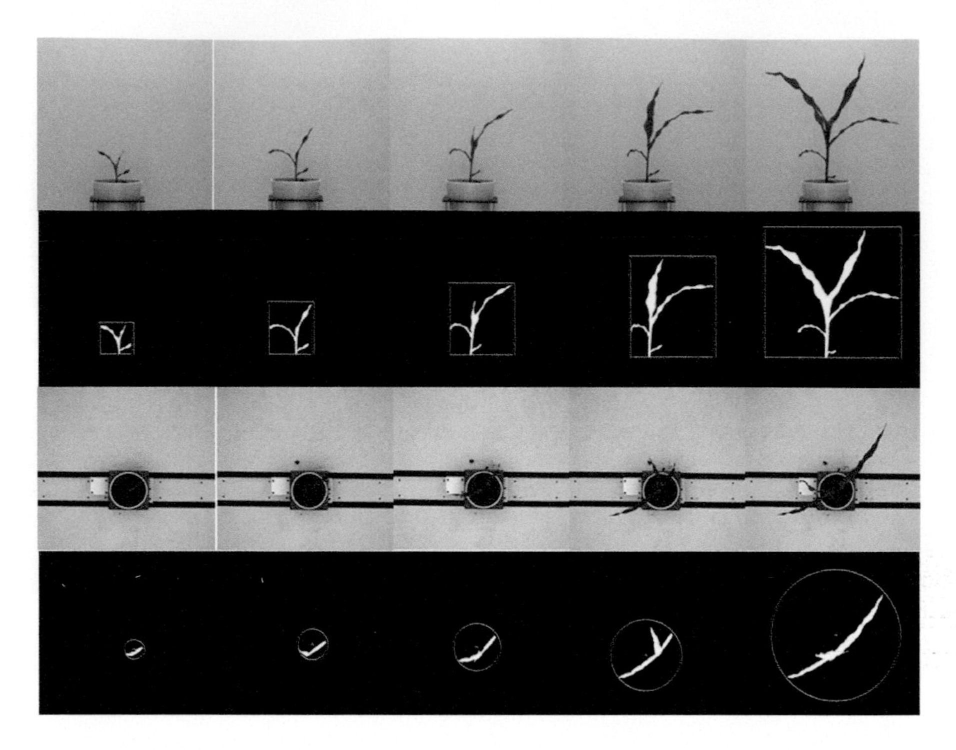

FIGURE 5.8 Computation of plant aspect ratio. Rows 1 and 3 show images of a maize plant from a side view and the top view. Rows 2 and 4 show the bounding rectangle and the minimum enclosing bounding circle of the corresponding images.

computed width of the plant from the image. *PAR*, therefore, uses two invariant measurements of the plant to compute the plant aspect ratio, i.e., the height of the plant from the side view and the diameter of the minimum enclosing circle of the plant from the top view.

5.3.1.3 Plant Areal Density

Plant areal density, *PAD*, is defined as the total number of plant pixels of a side view image to the area of the convex-hull enclosing the plant at the same view, i.e.,

$$PAD = \frac{Plant_{Tpx} \text{ at a side view}}{Area_{CH} \text{ at the same view}}$$

where $Plant_{Tpx}$ denotes the total number of plant pixels, and $Area_{CH}$ is the area of the convex- hull of the plant.

Phenotypic significance: *PAD* provides crucial information about plant growth and development, which vary with environmental conditions. It is also a determining factor for planting density. For maize and sorghum, which are widely used as model genomes for research on the use of grasses as bioenergy crops, *PAD* can be a potential measure of the biomass yield.

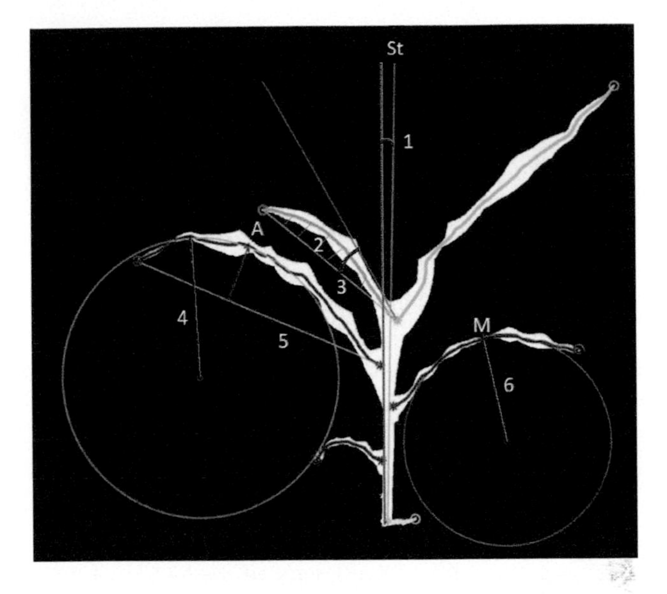

FIGURE 5.9 Component phenotypes. 1: stem angle. 2: integral leaf-skeleton area. 3: leaf-junction angle. 4: radius of apex curvature. 6: radius of mid-leaf curvature. 5: junction–tip distance. Keys: 'A': leaf-apex; 'M': mid-leaf; 'St': stem axis.

5.3.2 Component Phenotypes

In this section, we define several phenotypes applicable for maize, and plants with similar architecture, e.g., sorghum.* Most of these can be extended to plants with different architectures. These component phenotypes are illustrated in Figure 5.9.

1. *Leaf length (LL)*: It measures the length of a leaf, using a polynomial approximation, and is given by:

$$LL = \int_{x_j}^{x_t} \sqrt{1 + \left(\frac{dy}{dx}\right)^2}$$

   where x_j and x_t are the x co-ordinates of the leaf junction and leaf tip, respectively, and $y=p(x)$ is the nth order polynomial that best fits the leaf skeleton, optimizing the least square error.
2. *Inter-junction distance (IJD)*: It is defined as the distance between two adjacent junctions on a stem and is given by:

$$IJD = \sqrt{(x_a - x_b)^2 + (y_a - y_b)^2}$$

   where (x_a, y_a) and (x_b, y_b) are the co-ordinates of two adjacent junctions.

* Leaf detection algorithm and the computation of component phenotypes are implemented in MATLAB R2016a and are publicly available at http://plantvision.unl.edu.

3. *Junction–tip distance*: It is defined as the Euclidean distance between the junction and the tip of a leaf.
4. *Leaf junction angle (LJA)*: Leaf-junction angle, θ, is defined as the angle between the tangent of the leaf at its point of contact with the junction and the junction–tip line. It is measured using

$$LJA(\theta) = \tan^{-1}\frac{m_2 - m_1}{1 + m_1 m_2}$$

where m_1 and m_2 denote the slopes of the tangent to the leaf at its point of contact with the junction and the junction–tip line, respectively.
5. *Integral leaf-skeleton area (ILSA)*: It is defined as the area enclosed by the leaf and the straight line joining the junction and the tip of the leaf.

$$ILSA = \int_{x_t}^{x_j}\left[p(x) - f(x)\right]dx$$

where x_j and x_t are the x co-ordinates of the leaf-junction distance and the leaf-tip distance, respectively, $p(x)$ is the best-fit polynomial for the leaf, and $f(x)$ is the equation of the straight line that connects the junction and the tip of the leaf.
6. *Stem angle*: We define the stem axis as the straight line formed by a linear regression fitting of all the junctions of the stem. The stem angle (*SA*) is defined as the angle between the stem axis and the vertical axis, i.e.,

$$SA = \tan^{-1}(s)$$

where s is the slope of the stem axis.
7. *Leaf curvature*: Leaf curvature encodes significant information about the shape and structure of a leaf. The radius of curvature (R) at any point on the leaf can be derived from the nth order polynomial equation of its skeleton, p, as follows.

$$R = \frac{\left[1 + \left(\frac{dp}{dx}\right)^2\right]^{\frac{3}{2}}}{\left|\frac{d^2p}{dx^2}\right|}$$

The curvature, K, is the reciprocal of the radius of curvature, and is given by:

$$K = \frac{1}{R}$$

8. *Apex curvature*: It is the curvature at the apex of the leaf, i.e., the point on the leaf that is most distant from its junction–tip connecting line.

9. *Mid-leaf curvature*: It is the curvature at the midpoint of the leaf ("mid-leaf") derived from its skeleton.
10. *Growth index*: The total leaf area is measured by the total number of constituent pixels of each leaf in a given image. We introduce a derived component phenotype called the growth index (*GI*), which is defined as the ratio of the total leaf area (T_f) to the height of the stem (S_h) at a given time, i.e.,

$$GI = \frac{T_f}{S_h}.$$

Phenotypic significance: The number of leaves, length of individual leaves, stem height, inter-junction distance, and total leaf area at any phase during a plant's life cycle are the phenotypic expressions that contribute to the interpretation of plant growth and development. They also help in the study of determining plants' responses to environmental stresses. Leaf curvature, leaf-junction angle, and junction–tip distance are measures for the toughness of a leaf. Leaf toughness appears to be an important defense mechanism in maize and sorghum across diverse groups of genotypes. Computer vision-based leaf curvature measurements will replace the manual and tedious process of using mechanical devices, e.g., penetrometers, to measure leaf toughness used in resistance breeding programs and studying phytochemical characteristics of leaves (Bergvinson et al. 1994).

Stem angle, a measure of the deviation of the stem axis from the vertical, can be an early indicator of lodging susceptibility, due to the presence of excessive moisture in the soil, nutrient deficiency, or high planting density (Das Choudhury et al. 2017). Lodging leads to low yield for many grain crops, e.g., maize, wheat, sorghum, and millet. The ratio of integral leaf-skeleton area to the junction–tip distance provides information on leaf drooping, which may be an indicator of nutrient or water deficiency. The method of Das Choudhury et al. (2017) presents a foundational study to demonstrate the temporal variation of the component phenotypes regulated by genotypes, using time-series cluster analysis.

5.4 EXPERIMENTAL ANALYSIS

The detailed experimental analyses of holistic and component structural phenotypes are described in this section. They are performed on two publicly available datasets, i.e., Panicoid Phenomap-1 and UNL-CPPD. We describe the image acquisition process and the datasets used in the experimental analyses in the following sections.

5.4.1 Image Acquisition

The two datasets, i.e., Panicoid Phenomap-1 and UNL-CPPD, are created using the Lemnatec Scanalyzer 3-D high-throughput plant phenotyping system at the University of Nebraska-Lincoln (UNL), USA. Each plant is placed in a metallic carrier on a conveyor belt that moves the plant from the greenhouse to the four imaging chambers successively for capturing images in different modalities. Each imaging chamber has a rotating lifter for up to 360 side view images. The conveyor belt has the capacity to host a maximum number of 672 plants with height up to 2.5 m. It has three automated watering stations with a balance that can add water to a target weight or a specific volume, and records the specific quantity of water added on a daily basis.

5.4.2 Datasets

The Panicoid Phenomap-1 dataset consists of images of 40 genotypes of five categories of panicoid grain crops: maize, sorghum, pearl millet, proso millet, and foxtail millet (Das Choudhury et al. 2016). The images were captured daily by the visible light camera for two side view angles, i.e., 0° and 90° for 27 consecutive days. Panicoid Phenomap-1 contains a total of 13728 images from 176 plants. The dataset is designed to facilitate the development of new computer vision algorithms for the extraction of holistic phenotypes specifically from maize and to encourage researchers to test the accuracy of these algorithms for related crop species with similar plant architectures.

A subset of Panicoid Phenomap-1 dataset consisting of images of the 13 maize plants, called UNL-CPPD, has been created to evaluate the efficacy of component phenotyping algorithms (Das Choudhury et al. 2018). Unlike Panicoid Phenomap-1, UNL-CPPD is released with human-annotated ground-truth, along with the original image sequences. The UNL-CPPD has two versions: UNL-CPPD-I (small) and UNL-CPPD-II (large). UNL-CPPD-I comprises images for two side views: 0° and 90° of 13 maize plants for the first 27 days, starting from germination, that merely exclude self-occlusions due to crossovers. UNL-CPPD-II comprises images for two side views: 0° and 90° of the same 13 plants for a longer duration, i.e., 32 days, to evaluate the component plant phenotyping algorithm in the presence of leaf crossovers and self-occlusions. Thus, UNL-CPPD-I contains a total number of 700 original images, and UNL-CPPD-II contains a total number of 816 original images, including the images contained in UNL-CPPD-I.

Corresponding to each original image, the dataset also contains the ground-truth and annotated image, with each leaf numbered in order of emergence. We release the following ground-truth information in the XML format for each original image of the plant: (a) the co-ordinates of base, leaf tips, and leaf junctions; (b) the status of each leaf, i.e., alive or dead; and (c) the total number of leaves present (which are numbered in order of emergence). The dataset is designed to stimulate research in the development and comparison of algorithms for leaf detection and tracking, leaf segmentation, and leaf alignment of maize plants. The dataset also facilitates the exploration of component phenotypes and investigates their temporal variation as regulated by genotypes. Panicoid Phenomap-1 and UNL-CPPD can be freely downloaded from http://plantvision.unl.edu/.

5.4.3 Holistic Phenotyping Analysis

The experimental analysis of holistic phenotypes is conducted along two dimensions: (a) demonstration of the temporal variation of the phenotypes, as regulated by genotypes, and (b) statistical modeling to analyze the effect of the greenhouse environment on the phenotypes (Das Choudhury et al. 2016, 2018). The study is reported on 31 genotypes of maize plant image sequences of Panicoid Phenomap-1.

Temporal phenotypic behavior: The trajectories of phenotypes over time, regulated by genotype, provide important insights into the functional properties of the plants, and hence, enable us to detect abnormalities in a plant's behavior. Figure 5.10 shows the average plant aspect ratios of the group of plants for the 31 genotypes over 20 consecutive days. The overall trend shows a gradual decrease in aspect ratios, which matches the fact that the width of the plant increases proportionally more than its height. However, the upper and lower boundaries of the range of values for plant aspect ratios for a few genotypes (e.g., PHW52 and PHG39) are very similar to each other. Finally, some genotypes have significantly higher plant aspect ratios (e.g., PHG47) compared to others (e.g., B73). These observations confirm that the plant aspect ratio has the potential to be an effective phenotype.

Environmental effect on phenotypes: In high-throughput plant phenotyping systems, the plants are arranged in rows, often about 1 m apart. The greenhouse environment, i.e., the position of the plants in the greenhouse, may affect the phenotypes. We formulated an experimental design to study the impact of the greenhouse environment on holistic phenotypes, using the Panicoid Phenomap-1 dataset. The study

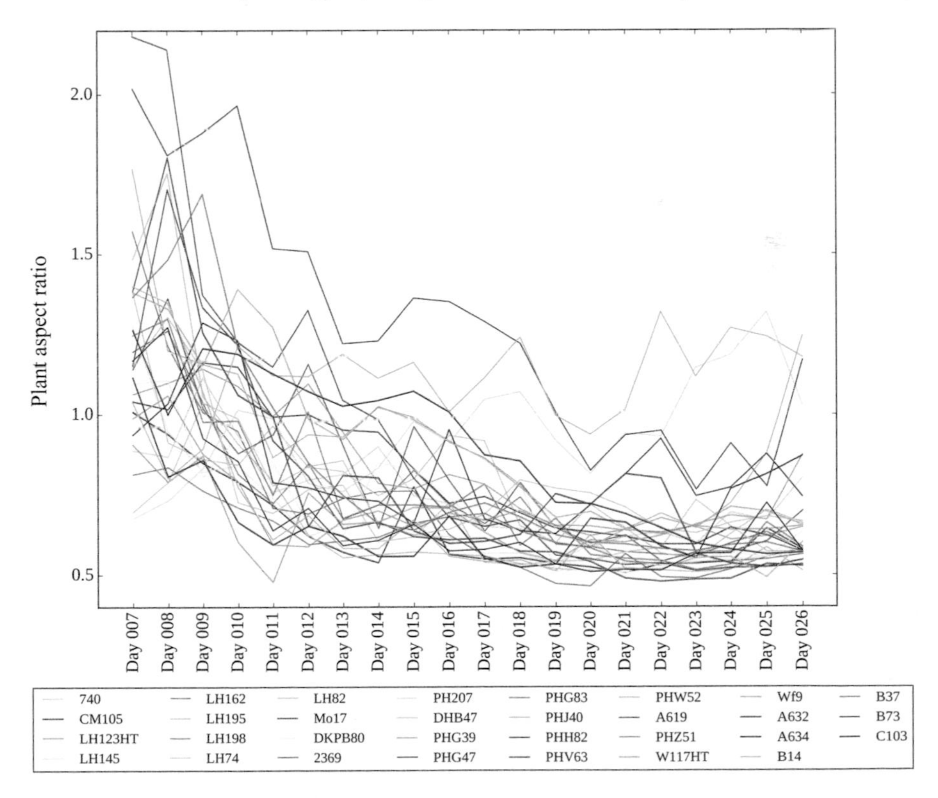

FIGURE 5.10 Illustration of genetic regulation of temporal variation of the plant aspect ratio.

is conducted on 160 plants, where the plants are arranged in 10 rows with 16 plants in each row. The plants in each row are divided into two blocks containing eight plants in each block, totaling 20 blocks. Note that no plant was adjacent to another of the same genotype (Das Choudhury et al. 2018).

To study the effect of genotype and environment on holistic phenotypes, a linear regression model for the phenotypic responses on each day is used. The model for the phenotype values (y) is given below.

$$y_{h,i,j,t} = \mu_{h,t} + \alpha_{h,i,t} + \gamma_{h,\nu(i,j),t} + \epsilon_{h,i,j,t},$$

where $h=1,2,3$ denotes the three types of phenotypes, i.e., plant areal density, bi-angular convex-hull area ratio, and plant aspect ratio, respectively. The subscripts i, j, and t denote the ith block, jth plant in this block, and day t, respectively, and $\nu(i,j)$ stands for the genotype at this pot, which is determined by the experimental design. μ is the mean of the phenotypes on a given day. The parameters α and γ denote the block effect and the genotype effect, respectively, and the error term is denoted by $\epsilon_{h,i,j,t}$.

For plant areal density, the block effect was not significant during the early days of growth, when the plants were small. However, this environment effect became significant as the plants grew larger. It was observed that the plants in the center of the greenhouse have greater areal density than the plants at the edges. This may be attributed to the fact that plants in the center rows experienced light competition from surrounding plants, and as a consequence, responded by increasing in height relative to plants located at the edges of the greenhouse (Das Choudhury et al. 2018). Figure 5.11 shows the estimated greenhouse row effect between the 12th block (in the 6th row) and the first block (in the first row) over time. The circles denote the estimated differences, and the vertical bars denote the corresponding 95% confidence levels. It is evident that the confidence intervals were significantly higher than zero for the last three days, indicating a significant positive effect of the 12th block relative to the first block. The same study was conducted on the other two holistic phenotypes, i.e., bi-angular convex-hull area ratio and plant aspect ratio; however, the block effect was not significant for these two shape-based phenotypes.

5.4.4 Component Phenotyping Analysis

The use of component phenotypes is demonstrated, using three different types of analyses: accuracy of leaf detection, leaf status report, and stem growth performance.

Leaf detection accuracy: Identification of leaves is central to component phenotyping. Table 5.1 summarizes the performance of the leaf detection algorithm for the UNL-CPPD. The plant-level accuracy of leaf detection approach is given by:

$$\text{Accuracy} = \frac{\sum_{i=1}^{n} \frac{N_{d,i} - N_{w,i}}{N_{G,i}}}{n}$$

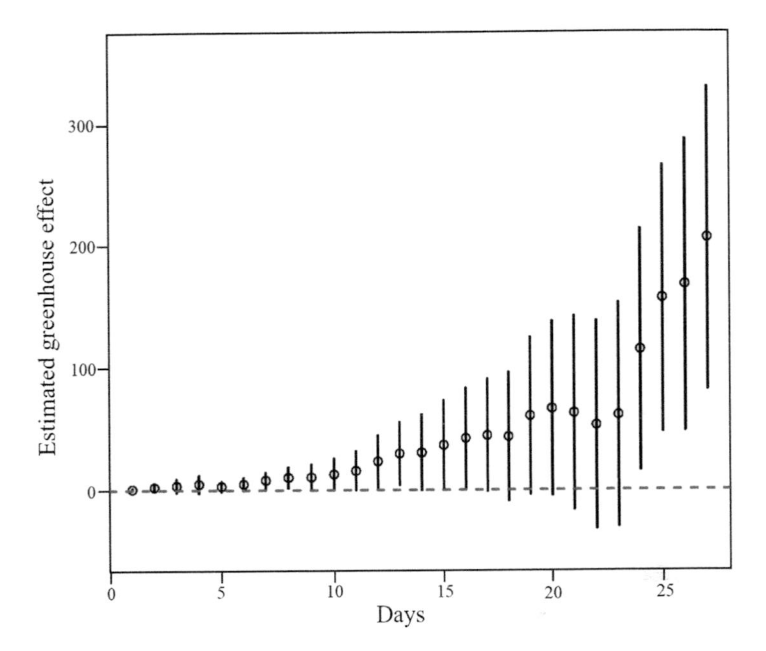

FIGURE 5.11 Estimated greenhouse row effect: the differences (shown as round dots) between the 12th block and the first block over time, with 95% confidence intervals (denoted by the vertical error bars).

where $N_{d,i}$ denotes the number of detected leaves, $N_{w,i}$ denotes the number of leaves that are wrongly detected, and $N_{G,i}$ denotes the ground-truth, i.e., the number of leaves present in the image for the *i*th plant, and *n* denotes the total number of images in the sequence.

The CPPD-II dataset has plants at advanced stages of growth that are more likely to have self-occlusions and leaf crossovers, which degrade the performance of the algorithm. Figure 5.12 shows examples of inaccuracies in leaf detection, where two or more leaves occlude each other, hiding the junctions.

Table 5.1 shows that the average plant-level accuracy for UNL-CPPD-I and UNL-CPPD-II is 92% and 85%, respectively. There are three scenarios in comparing the performance of the leaf detection algorithm for the two datasets.

- *Case 1*: For some plants (e.g., *Plant* 016–20^{+}), the plant-level accuracy for UNL-CPPD-II is higher than that of UNL-CPPD-I. This is attributed to the fact that UNL-CPPD-II contains more images, but the additional images do not suffer from incorrect leaf detection due to self-occlusions.
- *Case 2*: If most of the additional images of UNL-CPPD-II for a sequence have self-occlusions or leaf crossovers, the accuracy is decreased compared to UNL-CPPD-I (e.g., *Plant* 063–32†).
- *Case 3*: The plant-level accuracies for the same genotype for the two datasets are fairly similar (e.g., *Plant* 104–24‡).

TABLE 5.1
Performance Summary of Algorithm 1 on UNL-CPPD

	CPPD-I				CPPD-II			
Plant-ID	No. Leaves	Detected Leaves	False Leaves	Accuracy	No. Leaves	Detected Leaves	False Leaves	Accuracy
001–9	116	93	1	0.79	168	157	5	0.83
006–25	138	136	0	0.98	205	188	5	0.91
008–19	142	140	0	0.98	210	200	9	0.86
016–20⁺	103	86	0	0.83	141	129	0	0.88
023–1	113	101	0	0.89	154	135	8	0.83
045–1	122	120	3	0.96	177	170	6	0.93
047–25	148	142	2	0.94	212	196	5	0.88
063–32†	149	138	0	0.93	214	174	18	0.72
070–11	125	111	0	0.89	177	148	5	0.83
071–8	141	131	0	0.93	199	166	7	0.80
076–24	135	126	2	0.92	191	152	2	0.78
104–24‡	144	140	0	0.97	186	185	0	0.96
191–28	137	111	0	0.96	178	153	5	0.83
Average	132	123	<1	0.92	186	165	≈6	0.85

Key: '+' – an example of plant-level accuracy for UNL-CPPD-II being higher than that of UNL-CPPD-I; '†' – an example of plant-level accuracy for UNL-CPPD-II being lower than that of UNL-CPPD-I; and '‡' – an example of fairly similar plant-level accuracy for UNL-CPPD-I and UNL-CPPD-II.

FIGURE 5.12 Challenges in leaf detection due to self-occlusions.

Leaf status reports: A leaf status report is a graphical representation of the growth of leaves on a plant. It provides the following important information: (a) the day on which a particular leaf emerges, (b) the length of each leaf on a particular day, (c) the total number of leaves present on a particular day, (d) the relative growth rate of each leaf, (e) the day of senescence of a leaf, and (f) the total number of leaves which emerged during the vegetative stage of the life cycle of the plant. Typically, the detected leaves are numbered in order of emergence and tracked over the image sequence.

Figure 5.13 (left) shows the leaf status report of the lengths of the ten leaves of a plant from UNL-CPPD (*Plant* 006-25), starting from emergence until Day 27. From the report, it is evident that Leaf 1 emerged on Day 2, while Leaf 10 emerged on Day 25. Furthermore, the growth rate of the leaves that emerged later is higher than that of the leaves that emerged during the early stage. One possible explanation for this pattern is the reduction in the amount of sunlight received by the older leaves as they grow under

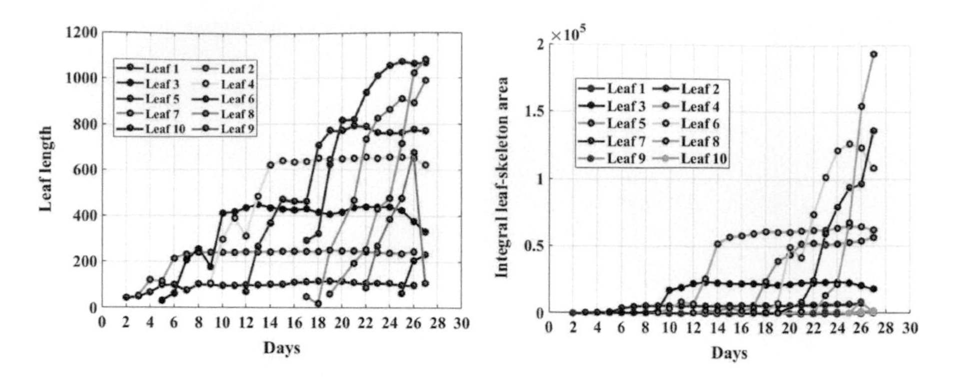

FIGURE 5.13 Illustration of temporal variation of component phenotypes: (left) leaf length; (right) integral leaf-skeleton area.

the shade of upper leaves. It should be noted that, for some days, the length of a leaf decreases, e.g., Leaf 4 was shorter on Day 12 than on the previous day. Some factors that influence this phenomenon include plant rotation, self-occlusion, and the fact that the measurements are made from the 2-D projection of the 3-D leaves. The temporal variation of integral leaf-skeleton area is shown in Figure 5.13 (right). This figure shows that the integral leaf-skeleton area exhibits behavior similar to that of leaf length. Figure 5.14 shows the leaf status report for mid-leaf curvature (left) and apex-leaf curvature (right) for *Plant* 006-25. The mid-leaf curvature and apex-leaf curvature are shown on a log scale for ease of visualization.

Stem growth performance: Figure 5.15 shows a comparative analysis of inter- and intragenotype variation in stem angle as a function of time. Figure 5.15 (right) uses five plants of the same genotype to demonstrate the intra-genotype variation, while Figure 5.15 (left) uses five plants of different genotypes to demonstrate the inter-genotype effect on the stem angle. The difference in the temporal behavior of intra- and intergenotype stem angles confirms that the stem angle is likely to be controlled by a genotypic effect.

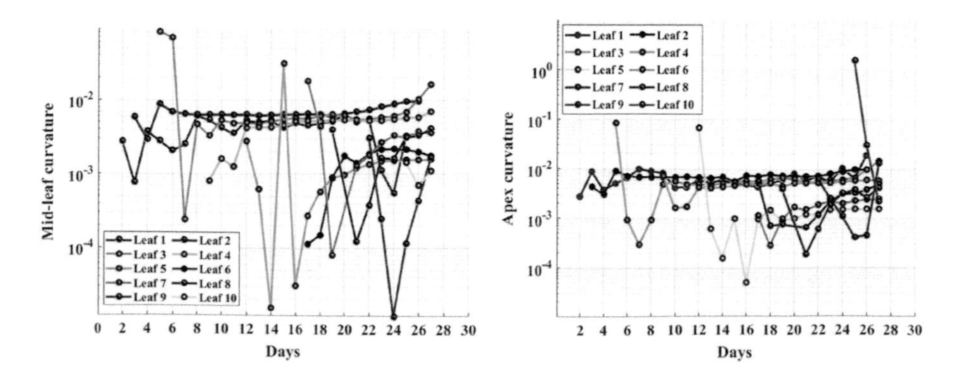

FIGURE 5.14 Illustration of temporal variation of component phenotypes: (left) mid-leaf curvature; (right) apex-leaf curvature.

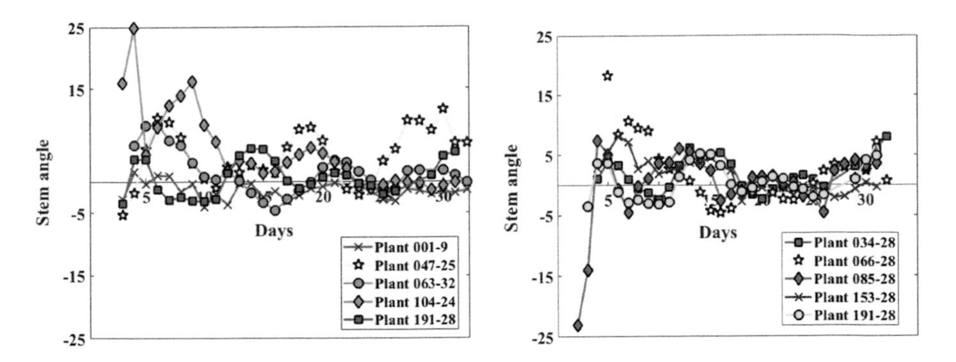

FIGURE 5.15 Comparative analysis of inter- (left) and intragenotype (right) variation of stem angles.

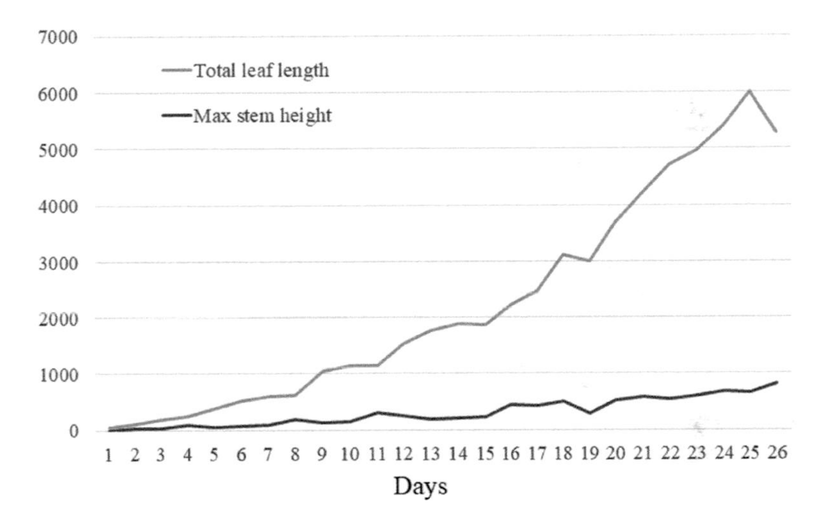

FIGURE 5.16 Illustration of the relative growth rate of total leaf length and stem height of a plant.

Figure 5.16 shows the relative growth rate of the total leaf area and the stem height of a plant over different days.

5.5 CONCLUSION

A plant phenotyping computational framework and a phenotypic taxonomy to advance research in high-throughput image-based plant phenotyping analysis are presented in this chapter. In addition, it provides a discussion of an image-processing pipeline to compute holistic and component structural phenotypes from 2-D visible light image sequences. Three holistic phenotypes, along with their significance to plant science, are discussed. The emergence timing of each leaf, size of each leaf, the total number of leaves present at any development stage, the growth rate of each leaf, leaf curvature, leaf area, and stem angle are the component phenotypic expressions that best represent plant vigor. The chapter also introduces a novel algorithm to detect leaves and stem of the maize plants using a skeleton-graph transformation approach to compute these component phenotypes. The algorithm uses curve tracing to address the challenge of intersecting leaves. Detailed experimental analyses are performed on Panicoid Phenomap-1 and UNL-CPPD to demonstrate the temporal variation of the holistic and component phenotypes in maize, regulated by the environmental and genetic variation, with a discussion on their significance in the context of plant science.

The prerequisite of the skeleton-graph transformation approach for component phenotyping analysis is the selection of the optimal view at which the line of sight of the camera is perpendicular to the orientation of the leaves. Since the proposed method deals with 2-D image sequences, its performance is limited by self-occlusion caused by leaf crossovers. In addition, the leaves are tracked manually from emergence until senescence for graphical demonstration of leaf life-history. Therefore, future work will consider the 3-D reconstruction of the plant from multi-view image sequences to overcome the limitation due to self-occlusion. An automatic leaf- tracking approach, in the presence of self-occlusion and view variations, will also be explored in future work.

REFERENCES

Agarwal, B. 2017. Detection of plant emergence based on spatio temporal image sequence analysis. MS Thesis, University of Nebraska, Lincoln, NE.

Al-Tam, F., H. Adam, A. D. Anjos, M. Lorieux, P. Larmande, A. Ghesquière, S. Jouannic, and H. R. Shahbazkia. 2013. P-TRAP: A panicle trait phenotyping tool. *BMC Plant Biology* 13:122.

Bai, X., L. J. Latecki, and W.-Y. Liu. 2007. Skeleton pruning by contour partitioning with discrete curve evolution. *IEEE Transactions on Pattern Analysis and Machine Intelligence* 29(3):449–462.

Bergvinson, D. J., J. T. Arnason, R. I. Hamilton, J. A. Mihm, and D. C. Ewell. 1994. Determining leaf toughness and its role in maize resistance to the European com borer (Lepidoptera: Pyralidae). *Journal of Economic Entomology* 87(6):1743–1748.

Das Choudhury, S., V. Stoerger, A. Samal, J. C. Schnable, Z. Liang, and J.-G. Yu. 2016. Automated vegetative stage phenotyping analysis of maize plants using visible light images. In *KDD workshop on Data Science for Food, Energy and Water*, San Francisco, CA.

Das Choudhury, S., S. Goswami, S. Bashyam, A. Samal, and T. Awada. 2017. Automated stem angle determination for temporal plant phenotyping analysis. In *ICCV workshop on Computer Vision Problems in Plant Phenotyping* 41–50. Venice, Italy.

Das Choudhury, S., S. Bashyam, Y. Qiu, A. Samal, and T. Awada. 2018. Holistic and component plant phenotyping using temporal image sequence. *Plant Methods* 14:35.

Das Choudhury, S., A. Samal, and T. Awada. 2019. Leveraging image analysis for high-throughput plant phenotyping. *Frontiers in Plant Science* 10:508.

Duncan, W. G. 1971. Leaf angles, leaf area, and canopy photosynthesis. *Crop Science* 11(4):482–485.

Gage, J. L., N. D. Miller, E. P. Spalding, S. M. Kaeppler, and N. de Leon. 2017. Mar. TIPS: A system for automated image-based phenotyping of maize tassels. *Plant Methods* 13:21.

Gampa, S. 2019. A data-driven approach for detecting stress in plants using hyperspectral imagery. MS Thesis, University of Nebraska, Lincoln, NE.

Hassouna, M. S., and A. A. Farag. 2007. Multistencils fast marching methods: A highly accurate solution to the eikonal equation on Cartesian domains. *IEEE Transactions on Pattern Analysis and Machine Intelligence* 29(9):1563–1574.

He, J. Q., R. J. Harrison, and B. Li. 2017. A novel 3D imaging system for strawberry phenotyping. *Plant Methods* 13:93.

Li, L., Z. Qin, and H. Danfeng. 2014. A review of imaging techniques for plant phenotyping. *Sensors* 14(11):20078–20111.

Lu, Y., Y. Huang, and R. Lu. 2017. Innovative hyperspectral imaging-based techniques for quality evaluation of fruits and vegetables: A review. *Applied Sciences* 7(2):189.

Maddonni, G. A., M. E. Otegui, B. Andrieu, M. Chelle, and J. J. Casal. 2002. Maize leaves turn away from neighbors. *Plant Physiology* 130(3):1181–1189.

Martnez, P., D. Robledo, S. T. Rodrguez-Ramilo, M. Hermida et al. 2016. *Turbot (Scophthalmus maximus) Genomic Resources: Application for Boosting Aquaculture Production*. Academic Press, San Diego, CA.

McCormick, R. F., S. K. Truong, and J. E. Mullet. 2016. 3D sorghum reconstructions from depth images identify QTL regulating shoot architecture. *Plant Physiology* 172(2):823–834.

Polder, G., H. L. E. Hovens, and A. Zweers. 2010. Measuring shoot length of submerged aquatic plants using graph analysis. In *Proceedings of the ImageJ User and Developer Conference* 172–177. Mondorf-les-Bains, Luxembourg.

Ruberto, C. D. 2004. Recognition of shapes by attributed skeletal graphs. *Pattern Recognition* 37(1):21–31.

Tuytelaars, T., and K. Mikolajczyk. 2008. Local invariant feature detectors: A survey. *Foundations and Trends in Computer Graphics and Vision* 3(3):177–280.

Wahabzada, M., A.-K. Mahlein, C. Bauckhage, U. Steiner, E.-C. Oerke, and K. Kersting. 2016. Plant phenotyping using probabilistic topic models: Uncovering the hyperspectral language of plants. *Scientific Reports* 6:22482.

Yin, X., X. Liu, J. Chen, and D. M. Kramer. 2018, June. Joint multi-leaf segmentation, alignment, and tracking for fluorescence plant videos. *IEEE Transactions on Pattern Analysis and Machine Intelligence* 40(6):1411–1423.

Zhang, X., C. Huang, D. Wu, et al. 2017. High-throughput phenotyping and QTL mapping reveals the genetic architecture of maize plant growth. *Plant Physiology* 173(3):1554–1564.

Zhou, Y., S. Srinivasan, S. V. Mirnezami, A. Kusmec, Q. Fu, L. Attigala, M. G. Salas Fernandez, B. Ganapathysubramanian, and P. S. Schnable. 2019. Semiautomated feature extraction from RGB images for sorghum panicle architecture GWAS. *Plant Physiology* 179(1):24–37.

6 Geometry Reconstruction of Plants

Ayan Chaudhury and Christophe Godin

CONTENTS

6.1 INTRODUCTION

Modeling the interesting and complex geometry of plants has been a center of attention of research for biologists and mathematicians for decades (Prusinkiewicz and Lindenmayer 1990; Godin and Sinoquet 2005). Numerous approaches have been proposed in order to mathematically model the geometrical structure of plants in a robust manner. Whereas one motivation for studying plant geometry is to better understand the structure of plants from a mathematical perspective, realistic modeling of plants has become an active area of research in computer graphics, in particular, due to the explosion of plant phenotyping platforms (Fahlgren et al. 2015). We are fast approaching the routine availability of intelligent imaging-based systems for automated modeling of realistic phenotypic models of plants. Due to the non-invasive nature and automated analysis techniques used, imaging-based systems are becoming extremely popular (Das Choudhury et al. 2018). The reconstructed 3D model of a plant from multiple view images mimics the original plant and can be used to extract desired biological features of the plant non-invasively (Chaudhury et al. 2018). Numerous systems have been proposed to study the effect of different environmental conditions on various types of plant species as well as on their genetically modified versions. These types of systems are extremely useful to quantitatively analyze different experimental outcomes in the field of crop science research. For example, one might want to study the effect of different types of lighting conditions on different genetically modified cultivars of a particular species, or study the growth pattern of a plant species over a certain period, or track the development of individual leaves in order to study how leaf shape changes over time. With the advent of imaging-based technologies, such analyses are becoming possible, and achievable in a fully automated manner.

Although the notion of *realistic* plant models in computer graphics and animation does not necessarily imply the actual biological relevance of the models to *real* plants, the ultimate goal of synthetic plant modeling is to mimic the appearance of the original plant, in terms of both the geometrical structure and of the texture realism effect. Synthetic modeling of plants can be beneficial to two broad types of research studies. First, this could be the quintessential tool to perform simulation experiments. Accurate modeling of the plant's geometry can help us to build models that can be used to simulate the behavior of the actual plant (Yi et al. 2017). For example, assume that we intend to study the mechanical effect of wind on a particular plant. If we have constructed a 3D model of a plant in such a way that the mathematical representation and interactions between different organs are well defined in terms of plant dynamics, we can simulate the effect of wind on different organs and can simultaneously model the behavior of the plant as a whole in a biologically plausible manner. Similarly, we might be able to simulate the growth of a plant over time from its synthetic model. The ultimate goal is to study specific aspects of the plant behavior in a fully virtual manner without the need for experimentation with a real plant. This is comparable to a flight simulator, where the flying is virtually simulated without the expensive process of actual flying. Similar types of modeling of chemical bonding structures are also studied in drug development research. Basically, the goal is to minimize the overheads and luxury (and limitations) of working on a real object and replace it with a synthetic model.

The second type of application of synthetic models includes quantitative analysis of different types of biological traits of plants. The models can help us to compute phenotypes, such as internode distances, insertion and divergence angles of individual branches, length and diameter of each branch, extraction of branching points, etc., in a fully automated manner. Also, studying properties, like plant growth (e.g., volume and surface area), changes in leaf shape, and inflorescence of flowers, is possible with the help of synthetic models. However, one of the biggest challenges associated with these types of analyses is to accurately model the geometry of the plant under consideration. Due to the complex structure of plants, the problem is extremely challenging. For example, self-occlusion results in incomplete or missing data and directly affects the quality of geometry reconstruction in occluded regions; the reconstruction of very thin branches is a common problem in this context. Various types of computer vision and machine-learning algorithms are reported in the literature to solve these types of problems. Still, most of the techniques are task-specific and lack the robustness and generality to be applied to different plant species. In general, geometric modeling of plants can be broadly categorized into three types of approaches (Stava et al. 2014). The first type of approach is the rule-based procedural modeling approach. This type of approach is based on a set of predefined rules to define complex objects by successively replacing them with parts of a simple initial object. A classic example of this type of approach is the L-system (Prusinkiewicz and Lindenmayer 1990). Although rule-based procedural approaches have been widely used in plant modeling, it is hard to use the framework to work on real data and reconstruct the original plant geometry. Also, the implementation of a procedural model is very cumbersome. However, due to the recent advances in the implementation framework of L-system-based modeling, using high-level programming languages, the approach has become more flexible to carry out (Boudon et al. 2012; Karwowski and Prusinkiewicz 2003). Figure 6.1 shows an example of a synthetic plant generated by procedural modeling technique in the L-Py framework (Boudon et al. 2012). The second type of modeling approach is based on real data (e.g., 2D images or 3D point clouds), called *data-driven* modeling. The goal of this

FIGURE 6.1 An example of a synthetic plant model generated in L-Py, using a procedural model.

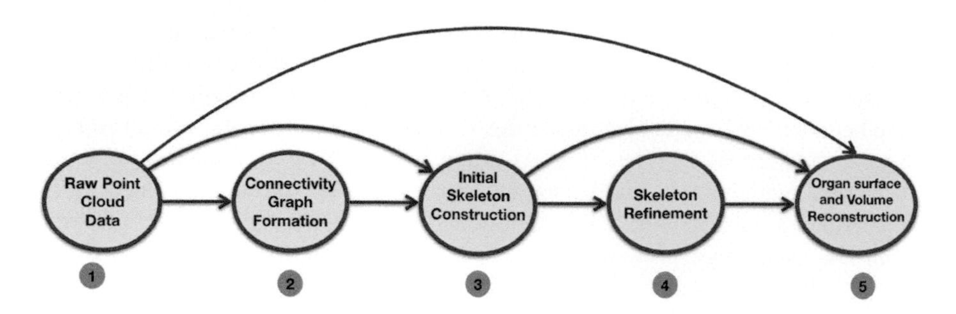

FIGURE 6.2 The general pipeline of data-driven modelling. Each circle represents a block of the pipeline and the arrows represent possible transitions of the workflow.

type of approach is mostly focused on the biological relevance of the model to reconstruct the geometry of the original plant from the input data. Finally, the third type of approach is based on input from the user and may be used to refine the results of the previous two approaches (Prusinkiewicz et al. 2001). This approach is mostly used in computer graphics and animation applications, where the user can interactively edit the geometry, according to the needs of the application.

In this chapter, we focus on the state-of-the-art techniques used to reconstruct the plant geometry from real data (*data-driven modeling* techniques). Many works in this area are published in the literature, focusing on different aspects of data-driven modeling techniques. These techniques can be clustered into different categories. Each of these categories constitutes a building block of a pipeline, which can be thought of as a general sequence of steps for data-driven modeling. In a broader sense, the pipeline consists of the following main building blocks: (1) Raw point cloud extraction, (2) connectivity graph formation, (3) initial skeleton construction, (4) skeleton refinement, and (5) organ surface and volume reconstruction. The pipeline is visually represented in Figure 6.2, where each circle represents a building block (or a *state*). In general, many of the techniques follow all the steps in the pipeline in a sequential manner, although some algorithms skip certain steps. Basically, the underlying idea is that the pipeline can be followed only in the forward direction, and skipping states is allowed. The possible transitions are indicated by the arrows in the figure. The states are sequentially numbered at the bottom, and we use the following convention to denote the steps followed by a particular algorithm. For example, if an algorithm takes raw point cloud data as input, a graph is constructed from the data, followed by skeletonization of the input data using the graph, and finally a particular organ of the plant is reconstructed, we denote the transition sequence as: $1 \rightarrow 2 \rightarrow 3 \rightarrow 5$. In the next section, we discuss the building blocks of the pipeline with respect to the current state-of-the-art techniques.

6.2 DATA-DRIVEN MODELING OF PLANTS

6.2.1 Step 1: Raw Point Cloud Data Extraction

A point cloud is a set of data points, which are the 3D point coordinates at the surface of an object. Point clouds can be generated using different types of techniques. One of the initial (now primitive) techniques for point cloud generation was the manual *digitization strategy by contact* (Godin et al. 1999), where the user-clicks on the surface of the object (plant) are recorded and stored as a point cloud. In recent years, two main techniques have been widely used in generating point cloud data, namely the back-projection technique to generate 3D data from a sequence of overlapping 2D images around the plant (Reche et al. 2004; Shlyakhter et al. 2001), and laser scanner technology to record point coordinates at the surface of an object (Xie et al. 2016; Yin et al. 2016; Zhang et al. 2014). Whereas 2D image-based techniques generate 3D data using *structure from motion* techniques, laser scanning is an efficient way to obtain 3D data directly. Laser scanning is a state-of-the-art technology to perform non-invasive 3D analysis for plant phenotyping. Depending on the need of the application, the resolution of the point spacing in the generated point cloud is possible to control. The 3D model of the plant can be generated by registering multiple overlapping scans.

6.2.2 Step 2: Connectivity Graph Formation

The raw point cloud data does not contain any topological connectivity information on the points. One of the very first steps in geometric reconstruction is the construction of a graph from the raw point cloud data. A typical way to construct the graph is to connect the points by using the nearest-neighbor strategy (e.g., a *kd*-tree search). Such a graph is also known as the Riemannian graph. The points represent vertices of the graph, and the edges are denoted by the connection of the vertices. The edge weight is determined by the distance between the vertices connected by the edge. Different types of distance measures can be used (e.g., Euclidean distance, geodesic distance, etc.). Graphs having edge weights are called weighted Riemannian graphs. The Riemannian graph structure is used as a basis of geometric reconstruction by many techniques (Livny et al. 2010; Xu et al. 2007; Yan et al. 2009).

6.2.3 Step 3: Initial Skeleton Construction

A skeleton is a thin structure obtained from an object shape, which encodes the topology and basic geometry of the original object. Skeletons are a compact representation of the shape and are extremely useful in various types of shape analysis tasks (Cornea et al. 2007). Say $\mathcal{P} \in \mathbb{R}^3$ is a point set. A ball of radius r centered at $p \in \mathcal{P}$ is defined as $\mathcal{S}(p) = \{q \in \mathbb{R}^3, d(p,q) < r\}$, where $d(p,q)$ is the distance between p and q in $\mathbb{R}^3$. A maximal ball is defined as a ball $\mathcal{S}(p) \subset \mathcal{P}$, which is not completely included in any other ball included in $\mathcal{P}$. The skeleton is the set of centers of all maximal balls included in $\mathcal{P}$. Extracting an accurate skeleton from a shape is an extremely challenging task. By *accurate*, it means that, ideally, the skeleton should pass through the exact centerline of the shape. In other words, within the local neighborhood of the shape, the distances from the skeleton point to the enclosing shape boundaries should be the same. Although 2D skeleton extraction is a well-studied problem in the literature, 3D skeletonization of point cloud data is still an open problem. The task is even more challenging in the case of plants, because of their complex geometry. Moreover, the problems of non-uniform point density and missing data make the task extremely complicated.

Different types of techniques have been proposed for skeleton extraction from point cloud data. Some of these approaches are based on Riemannian graph construction, and some of them compute the skeleton from the point cloud data, along with the sequence of 2D images. Whereas these skeletons are constructed in the form of a graph, there might be loops in the graph due to factors like close branches, missing data, etc. We will briefly discuss recent advances in skeletonization techniques, along with their strengths and limitations.

A semi-automatic method of skeletonization, described by Xu et al. (2007), is one of the most notable works on skeleton extraction from point cloud data. The workflow transition sequence is $1 \rightarrow 2 \rightarrow 3$, as shown in Figure 6.2. The goal is to reconstruct the original plant model, along with *completing* the missing information caused by self-occlusion. Initially, assuming that the root of the tree is located at the bottom, all the points in the scanned data are connected as a weighted Riemannian graph, where the weight is defined as the edge length. Then, the shortest path from each vertex in the graph to the root vertex is computed, using Dijkstra's Shortest Path algorithm. The whole point cloud is then *clustered*, based on the quantized shortest path lengths and graph adjacency information. The center of each cluster is assigned as a skeleton point (node), which is then connected to each of the other nodes, according to their spatial locations. This idea produces a basic skeleton of the plant in the form of a graph. However, due to occlusion and missing data, there might exist components (or sub-graphs) that are not connected to the main skeleton. In order to solve this problem, a breadth-first traversal is initiated from the root node. During the traversal, a cone of a certain angle is projected along the direction vector, from the parent node to the current node. This process is performed for every node of the graph. If any point, other than the nodes of the skeleton graph, falls within the current volume of the cone, it is attached to the current node of the skeleton graph and becomes a node of the graph. Then the above-mentioned process is repeated by considering the newly added point as the root node and searching for more points closer to the current point. This ensures that disconnected parts are joined to the main skeleton in a meaningful way. This approach is reported to work well on deciduous trees (when the trees have shed all their leaves) when evaluated in a qualitative/visual manner. In order to extract the skeleton in the presence of leaves, some heuristics are used to synthesize the skeleton near the tree crown, where most of the leaves are located. Finally, a parametric curve smoothing (Hermite curve) is performed to incorporate a ``smooth'' transition at the branching points, to make the skeleton more realistic.

One problem associated with the technique is that the skeleton points do not maintain the *centeredness* criterion, resulting in a *zigzag* structure near the branching points (we will discuss this issue in Section 6.2.4). This type of result does not support the biological relevance of the skeleton (although, visually, it might look good at a distance) and leads to incorrect geometric reconstruction by the data-driven modeling pipeline. Also, the skeleton might contain *loops*, resulting from close branches of the tree; this problem was later handled by Yan et al. (2009) by transforming the skeleton graph into a tree structure, using some heuristics.

Motivated by these types of problems, Bucksch et al. (2009) proposed a skeletonization technique to compute tree parameters, with the emphasis on botanical accuracy of the measurements of the parameters (height, length, and diameter of branches, etc.). Instead of constructing the Riemannian graph from discrete points, the method exploits the voxel structure and applies some local heuristics to infer the connectivity of the skeleton points (the state transition sequence is $1 \rightarrow 3$ in Figure 6.2). Initially, the input point cloud is organized in an octree structure by subdividing the point cloud into octree cells, where each cell contains a certain number of points. Then, a graph is formed from the octree representation by forming vertices from the cells and connecting the vertices by edges, using a nearest-neighbor strategy. However, simply considering the center of gravity of each octree cell as a graph vertex and joining the vertices by edges might lead to incorrect tree topology, because of the inevitable noise and varying point density of typical real-world laser scans. To solve this problem, a heuristic method is used. For any two adjacent octree cell centroids and the midpoint of the line connecting them, three planes are passed through these points, perpendicular to the line joining the points. The median values of the squared distances of the points in each cell to the corresponding planes (d_1 and d_2, respectively) are computed. Next, the median values of the squared distances of all the points belonging to both of the cells to the plane passing through their midpoint are computed (d_3). Then, based on a threshold on these values, the inference is made as to whether the cells are connected to each other. The idea is that, if d_3 is sufficiently smaller than d_1 and d_2, then it is likely that the points in the cells are more "scattered" and there is a connection between the two cells (unlike the case where the points are "clustered" around a small neighborhood in each cell, and thus it is likely that the cells are independent of each other). Finally, the centeredness of the skeleton is achieved by embedding the octree graph into the point cloud by a *graph-embedding* technique (Pascucci et al. 2007). Experimental results demonstrate the effectiveness of the technique under bad sampling conditions and for accurate skeleton geometric reconstruction for various types of cases. However, the method is based on many heuristic assumptions and is not robust enough to be useful for a range of applications.

For a different type of approach, *particle flow*-based techniques build the skeleton from the sequence of input photographs (state transition sequence $1 \rightarrow 3$ in Figure 6.2). The motivation of a particle simulation-based approach is based on the process of transport and exchange of energy, water, and sustenance between roots, branches, and leaves of a plant (Neubert et al. 2007; Rodkaew et al. 2003).

Initially, a standard segmentation algorithm is used to extract the tree from the background of the input photograph. Then, some random (or user-defined) seed pixels are chosen, which are denoted as *particles*. By choosing the lowermost pixel at the root of the tree as the target point, a path is traced, starting from each seed pixel to the target. The path is created by the standard *Livewire* segmentation technique (Mortensen and Barrett 1995), which is based on Dijkstra's Shortest Path algorithm, using the edge information of the image. This idea is motivated by the fact that each particle is assumed to contain some energy, and the transportation rule directs each particle towards a target. The 2D skeleton, or branching structure, is computed by tracing the paths of the particles at the end of the simulation. Neubert et al. (2007)

called the combination of all the paths the *attractor graph* in an image. The final attractor graph is obtained by combining the direction vectors of the attractor graphs for each image plane. Next, a 3D voxel model of the tree is computed by assuming a parallel projection model and back-projecting the sequence of images (similar to the well-known *space carving* approach). Each voxel is assigned to a density estimate, and the 2D attractor graphs are used to compute the direction field of the particles (see Neubert et al. (2007) for details). These voxels and the density information are used to initialize the positions of another set of particles (which are not the same as the seed pixels used in the 2D image). Then, the process of path tracing from each particle to the target point is repeated (as done before in the 2D image). However, a different strategy is used in the 3D case. Unlike the *Livewire* approach, the fundamental laws of Newtonian mechanics are used to compute the path. The position of a particle is updated, based on certain rules of force and velocity of a particle (Neubert et al. 2007), and nearby particles, within some distance threshold, are joined together to form a new particle. The trace of the particles forms the skeleton of the plant.

Although the technique is motivated by the energy transport phenomena of real plants, the skeleton geometry is not guaranteed to follow the actual geometry of the input data. The choice of seed points plays an important role in forming the skeleton structure. Random placement of seed points may result in skeleton construction at the wrong places. The attractor graph computation has inherent limitations. For the 2D attractor graph computation, the *Livewire* technique is used to compute the branches of the tree. However, the generic *Livewire* technique is mainly designed to follow the edges of the image, but not specifically to follow the branching structures. For computing the 3D attractor graph, the flow of the particles by the rules of Newtonian mechanics is not supported by the branching geometry of the original plant.

Space colonization algorithm (Palubicki et al. 2009; Runions et al. 2007) has been successful in modeling the complex branching structure of plants. The main idea of the algorithm is to generate a random point cloud at first, to indicate the free space where the plant can grow. Then, starting from the lowermost point (which will be the root of the reconstructed plant), the aim is to "*eat-up*" the points (or particles) in the cloud in an iterative manner. It is assumed that each bud (undeveloped or embryonic shoot located at the tip of a stem) is surrounded by a spherical *occupancy zone* of a certain radius (also called the *kill distance*) within a conical perception volume of a certain angle and radius. Initially, each particle is labeled as *unprocessed*. At every iteration, a bud searches in its perception volume to find the closest particles. If there is more than one particle within the volume, the branch direction is set to the average direction of all the particles from the current particle, and these particles are labeled as *processed*. The same operation is continued in every iteration until there is no particle left in the cloud that is labeled as *unprocessed*. Although this idea can produce visually pleasing skeletal structures, it is not fully supported by the biological constraints of plants. The "*eat-up*" process might result in the creation of a branching structure in the wrong direction, and there is no mechanism to constrain the process to the original branching structure. The space colonization algorithm was proposed to generate synthetic plants, and no real dataset was used to reconstruct the original plant geometry. Preuksakarn et al. (2010) later exploited the space colonization approach and applied it to point cloud data instead of a random (or user-defined) set

FIGURE 6.3 An example of a plant model (right) generated by a space colonization algorithm using real data (left).

of points, thus improving the result regarding the biological relevance of the reconstructed plant. Figure 6.3 shows an example of a plant model generated by the space colonization algorithm on real data.

The overall technique is a local optimization-based strategy. In each iteration, a local neighborhood is considered, and no backtracking is possible to *correct* a wrong move or to refine the particle search from a global point of view. Although the space colonization approach can produce incorrect branching geometry, it can perform well if prior knowledge of the plant geometry is embedded along with the algorithm; we discuss this type of improved method in the next stage of the pipeline.

6.2.4 Step 4: Skeleton Refinement via Local and Global Optimization

With the motivation to *improve* or *refine* the skeleton structure obtained in the previous step, different types of techniques are proposed in the literature. Refinement refers to deforming the skeleton according to the botanical rules in order to follow the geometry of the original input point cloud data. While some of the skeleton refinement techniques are optimization-based, others are *ad hoc* and locally heuristic in nature. The general goal of all these types of approaches is to make the skeleton biologically relevant.

Livny et al. (2010) proposed a series of optimization techniques to improve the imperfect skeletal structure (contaminated by noise, outliers, and missing data). The method is based on the initial Riemannian graph structure and the particle flow method (state transition sequence $1 \rightarrow 3 \rightarrow 4$ in Figure 6.2). Initially, a minimum weight-spanning tree is computed from the graph by Dijkstra's Shortest Path algorithm. This is called the initial Branch Structure Graph (BSG). In the context of BSG, the following attributes are defined. A *simple branch* is defined as the type of branch which does not contain any junction point (where two or more branches meet). A *branch chain* is defined as the set of edges corresponding to a *simple branch*. Each node of the BSG is assigned a weight, which is equal to the size of its subtree. At this stage, the BSG hardly looks like a real tree structure due to the lack of *smoothness* in the branching structure, and to noise and occlusions. A number of biological constraints are proposed on the BSG structure to *refine* the branching geometry, and a series of global optimization strategies are used to apply the constraints on the BSG.

For skeleton refinement, the following criterion is imposed: the *branch chains* should be smooth, which ensures that the bending angles should be small. This is achieved by using the notion of an *orientation field* (Neubert et al. 2007) on the BSG vertices. Consider a BSG denoted as $\mathcal{G}$ in Figure 6.4. An orientation field is a map associating each vertex $v \in \mathcal{G}$ with a vector $\overrightarrow{\mathcal{O}_v} \in \mathbb{R}^3$. This vector represents an estimated direction for moving each vertex, v, so as to smooth-out the skeleton. Given a vertex $v \in \mathcal{G}$ having parent v_p and the vector $\overrightarrow{v_p v}$ along the edge $e(v_p,v)$, each $\mathcal{O}_v$ is computed to minimize the following two factors: i) the change in direction with the vector, $\overrightarrow{v_p v}$, and ii) the change in direction with the orientation, $\mathcal{O}_{v_p}$, of the parent node. The minimization is performed simultaneously over all the vertices of the graph $\mathcal{G}$ in a least squares fashion (Livny et al. 2010). This leads to an optimal orientation $\mathcal{O}_v^*$ at each vertex. Then, the algorithm updates the vertex positions on the basis of the computed orientation $\overrightarrow{\mathcal{O}_v}$. Further refinement is obtained by performing a similar type of optimization, considering the center of the edges in the skeleton graph. Finally, allometric rules are used to reconstruct the branch thickness.

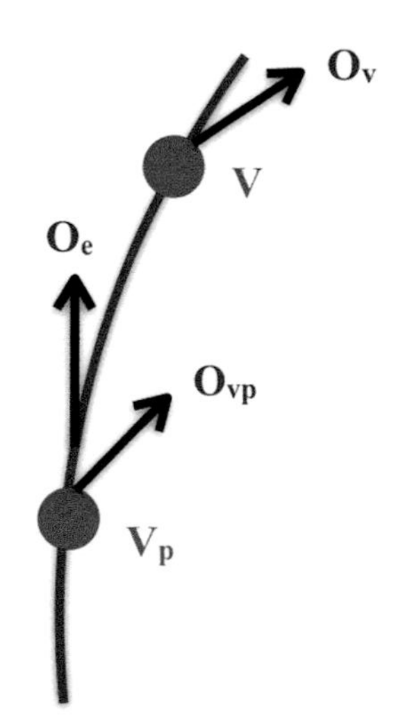

FIGURE 6.4 Illustration of concepts in a branch structure graph (BSG).

The optimization technique does not incorporate prior botanical knowledge of plants into the framework and thus suffers from botanical inconsistency in the model. For example, the skeleton points might get shifted away from the original branch point cloud after the optimization. Also, there is no strategy for the skeleton points to follow the centerline of the branches.

In a similar line of work (state transition sequence $1 \rightarrow 3 \rightarrow 4$ in Figure 6.2), Wang et al. (2016) proposed a combined local and global optimization technique to refine an existing skeleton, especially to handle the case of occlusion/missing data. Given an input point cloud $\mathcal{P} = \{p_i\}$ in the 3D point cloud format, obtained from terrestrial laser scanning (having missing/incomplete points), firstly, a rough skeleton is extracted from the data, using a standard skeletonization technique. The missing parts of the data result in disconnected components in the skeleton, which are connected together by a spanning tree-based approach. However, this approach does not necessarily support the actual topology of the structure. In the next step, skeletonization of the input data is performed again by *contracting* the original point cloud to the previously obtained skeleton points, using the Laplacian contraction method of Cao et al. (2010). The idea of the Laplacian contraction method is to first *contract* or *collapse* the input point cloud to a 1D structure of points, using a Laplacian smoothing technique Au et al. (2008). An energy function is defined, which consists of two terms: the first term removes geometric details along the surface normal direction, and the second term preserves the shape of the original point cloud Au et al. (2008). The advantage of performing a two-stage skeletonization process is to exploit the strengths of both types of techniques. Whereas the skeleton obtained in the first step roughly *completes* the missing parts of the data, the contraction technique in Cao et al. (2010) cannot handle the missing data problem. On the contrary, although rough skeletonization does not guarantee the preservation of the topology of the object shape, the contraction technique described in Cao et al. (2010) performs better in this aspect. Further processing is performed on the skeleton to fine-tune the result by minimizing an objective function, which takes into account the branch dominant direction and point density factors. In the optimization process, the skeleton point cloud $\mathcal{S} = \{s_j\}$ is pushed to move by means of a repulsive force $\mathbf{F}_r$ away from the original point cloud, whereas the original point cloud is forced to contract toward the skeleton point cloud by means of a constraint force $\mathbf{F}_s$. The optimal skeleton points are obtained by solving the following energy minimization problem:

$$\min_{\mathcal{S}} \sum_{p_i \in \mathcal{P}} \mathbf{F}_r\left(p_i, \mathcal{S}\right) + \lambda \mathbf{F}_s\left(p_{i,\mathcal{S}}\right)^2,$$

where λ controls the trade-off between the two terms (Wang et al. 2016).

In order to prevent large displacement of the points during the optimization process, the points are constrained to move very short distances, and an iterative procedure is used to refine the displacements. Figure 6.5 shows some results from the technique.

In recent work, Wu et al. (2019) developed a skeleton refinement technique specifically for maize plants, focusing on local issues like centeredness and maintaining

FIGURE 6.5 Handling missing data in skeletonization, by Wang et al. (2016). Two rows show two examples. In each row, the left-hand image is the raw point cloud, the central image is the reconstructed 3D model, and the right-hand image is the photograph of the original tree. (Figure re-used with permission of IEEE).

the geometry of the original point cloud. As with the previous approach, the initial skeleton is obtained by the Laplacian-based contraction method of Cao et al. (2010). In order to adaptively sample the contracted skeleton points, principal component analysis is performed to determine the intersection points of the tree branches (junctions), which are used as keypoints. The sampling is performed at a higher density near the keypoint regions. After connecting the skeleton points by edges, using the nearest-neighbor strategy, a final refinement step is performed in order to alleviate the problem of a *zigzag* structure in the skeleton (Figure 6.6). Assuming that the points near the junctions are mainly responsible for the *zigzag* structure, a least squares straight line is fitted locally to the stem skeleton points, except for the points near the junction. This removes the *zigzag* locally. For the leaf vein skeleton points, first, a tangent line is constructed through the adjacent points. Then, a cutting plane perpendicular to the tangent line is projected onto the line, and the central point of the intersection is considered to be the refined skeleton point. The model is based on the geometry of the maize plant and is not reported to work on other types of species.

In general, the above-mentioned skeletonization algorithms have their strengths and limitations, based on the type of application. There is no algorithm that can handle any type of data in a robust manner. Basic skeletonization techniques discussed in the first stage of the pipeline can be used for general purpose applications, whereas the refined skeletons may suit better for specific cases. Local optimization techniques tend to focus on the issues like centeredness criteria, whereas the global optimization techniques focus on maintaining the overall shape of the skeleton. Ideally, the trade-off between the two types of optimization techniques is the crucial factor in the performance of the skeletonization process.

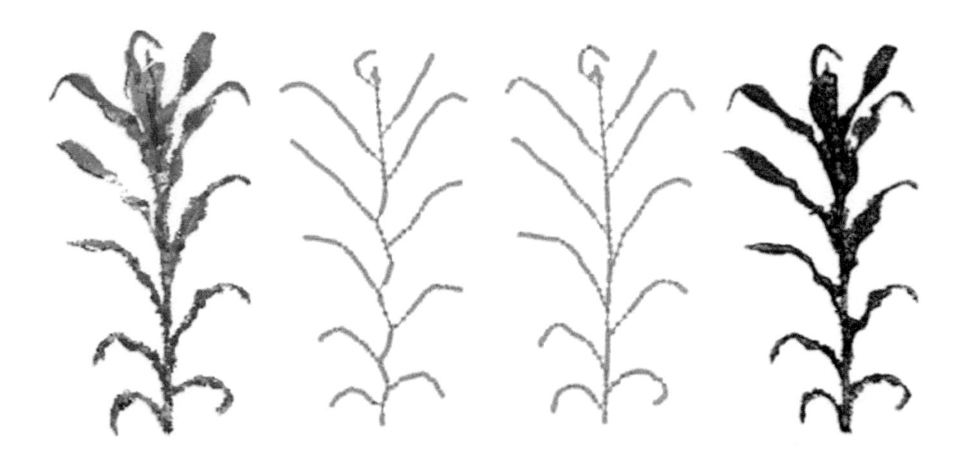

FIGURE 6.6 Skeletonization approach of Wu et al. (2019). The left-hand image is the original plant, and the central image is the skeleton extracted by the Laplacian contraction (Cao et al. 2010), which suffers from the zigzag structure. The right-hand image is the result of the proposed skeletonization algorithm that solves this problem. In the right-most image, the skeleton points are embedded in the original plant for visualization purposes. (Distributed under the terms of Creative Commons Attribution License).

Apart from the above-mentioned skeletonization algorithms, which are mostly focused on plants, some other types of general skeletonization techniques have also been successful for plants to some extent. Some notable works in this aspect include those of Tagliasacchi et al. (2009) and (Huang et al. (2013). The technique proposed in Tagliasacchi et al. (2009) is based on the notion of a generalized rotational symmetry axis. The local neighborhood of each point on the object is assumed to be cylindrical, and surface normal information is exploited to compute the skeleton. For handling the (non-cylindrical) regions, where different (cylindrical) object parts join, a spatial coherence strategy is used. The L1-median skeleton technique (Huang et al. 2013) extended the idea of classical L1-median statistics by introducing a regularization term in the energy function. For a recent survey of 3D skeletonization algorithms, interested readers are encouraged to read Tagliasacchi et al. (2016).

6.2.5 Step 5: Organ Surface and Volume Reconstruction

Whereas skeletonization of plant point cloud data provides the basic geometry of the branching structure of plants, reconstruction of different organs in terms of volumetric (e.g., for branch thickness) and surface (e.g., for leaf, flower) reconstruction is the ultimate goal of plant modeling. There are different types of techniques available to achieve this. In the subsequent sections, we discuss recent approaches of volume and surface reconstruction techniques by categorizing the algorithms; for most of these algorithms, the transition sequence is $1 \rightarrow 5$ in Figure 6.2.

6.2.5.1 Reconstruction Using Real Plant Parts

The main idea of this type of technique is to dissect a plant into different parts, followed by scanning of individual parts in an offline manner, and then to reconstruct the original tree geometry using these scanned parts. The technique performs skeletonization as an intermediate step, and thus the state transition sequence is $1 \rightarrow 3 \rightarrow 5$ in Figure 6.2).

Yin et al. (2016) proposed a method to reconstruct plants consisting of mainly leaves. The motivation for the work is to handle the case of occlusion explicitly. Due to problems of heavy self-occlusion, inaccessible parts, diverse topologies, slim petioles, and complex foliage geometry, it is extremely hard to obtain a complete model of the plant *via* conventional acquisition techniques (such as using a laser scanner to scan the whole plant from multiple views and then reconstructing the overlapping views to obtain a full 3D model). A two-step approach is proposed to handle this problem. In the first step, the plant is scanned from several overlapping views, which are then registered to obtain a *coarse* 3D point cloud model of the plant. In the next step, the plant is disassembled into disjoint parts. Each of these parts is scanned at a finer level of detail, and a mesh is reconstructed for each part. Then, with some user assistance, each part is placed near the corresponding part in the coarse 3D model, and point set registration is performed to fit the part in the coarse model. This process is repeated for all the parts, which results in *improving* the coarse 3D model to obtain a complete 3D model of the plant.

Mesh reconstruction of the individual plant parts is performed as follows. After dissecting the kth part from the plant, it is scanned to obtain a point cloud $\mathcal{S}_k$. The point cloud is converted into a skeleton *via* the L1-median skeleton method Huang et al. 2013), and the points are resampled to the desired number of points in a uniform manner. Then, for each skeleton point p_i, a slicing plane perpendicular to the skeleton curve is computed, and the original points from $\mathcal{S}_k$ are projected to the closest slicing plane. This forms a set of cross-sectional slices (s_i) along the skeleton. Now, the points in this model are classified as leaf or stem (or petiole) points by a segmentation technique proposed by Li et al. (2013), which is an extension of the classical graph-cut segmentation technique (Boykov et al. 2001) for the 3D case. Then each slice s_i in the stem is approximated to by a circular Non-Uniform Rational Basis Spline (NURBS) curve, and the slices in the leaf are approximated to by closed NURBS curves (except for the tip of the leaf). The thickness of the closed curves is modeled as gradually decaying values, from the center to the boundary. Finally, the profile curves are joined together to form a manifold mesh $\mathcal{M}_k$ by a *sweeping reconstruction* technique (Yin et al. 2014).

Once all the individual parts are reconstructed, each of these part meshes $\mathcal{M}_k$ is registered to the coarse plant model $\mathcal{P}$. A joint global and local geometric reconstruction method is used to perform the registration task. The global step of the registration starts with an interactive procedure. The user places a part close to its actual location in the coarse model. Then a point-by-point correspondence set is obtained by the nearest-neighbor strategy. The part mesh is transformed (rotation and translation) toward the full model, using this coarse correspondence set. The optimal set of transformation parameters is obtained as

$$\underset{r_i, t_i}{\operatorname{argmin}} \left(D\left(\mathcal{M}_k\left(r_i, t_i\right), \mathcal{P} \right) + \lambda L\left(\mathcal{S}_k\left(t_i\right) \right) \right)$$

where the first term $D(\cdot,\cdot)$ is the data fitting term, which transforms $\mathcal{M}_k$ by the transformation parameters (r_i,t_i), the second term $L(\cdot)$ is the regularization term, which translates the skeleton curve by t_i, and λ is the weighting factor of the regularization (or smoothness) term. The data term computes the weighted sum of the inner correspondence distances, and the smoothness term ensures the uniform point distribution along the curve skeleton with preservation of the skeleton length.

In the local registration step, a similar type of objective function is created. The objective functions are minimized by a standard framework (the Broyden-Fletcher-Goldfarb-Shanno solver).

In a similar line of work, Xie et al. (2016) proposed a semi-automatic modeling approach, using scanned tree parts. The method is designed for computer graphics applications, where the user can choose certain parts of the tree from a number of parts stored in the database, and then the algorithm automatically constructs the whole plant model by *joining* these parts in a biologically meaningful manner (state transition sequence $1 \rightarrow 5$ in Figure 6.2).

Initially, 15 different tree species are cut into about 200 tree parts. Each of these parts is scanned separately to obtain a point cloud of each tree part, which is stored in a database. Next, the user chooses to reconstruct a particular tree from a template photograph and places the required tree parts from the database in the 3D space according to the photograph. This represents the basic structure of the tree, which remains in a disconnected or incomplete stage.

In the next phase, the *tree cuts* are connected to form a realistic tree in a biologically relevant manner. Given a tree cut, first, the Euclidean distances from the endpoint of the current tree cut to the endpoints of all the cuts are computed. The tree cut with the minimum distance is considered to be the nearest object from the current tree cut. Now, the transformation parameters (rotation and translation) between the contours (a contour is assumed to be closed, and thus forms a "*loop*") of the two tree cuts are estimated to obtain point-to-point correspondence between the two parts. In order to estimate the branch surface between the two parts, an interpolation strategy is used. A Hermite curve is computed between the center points of the two contours, and a number of *loops* are generated along the curve by keeping the loop centers perpendicular to the curve direction. Incorporation of additional bifurcations (apart from the bifurcation data in the database) is introduced into the model by means of some user assistance. The user gives input for the position of a bifurcation, and the algorithm connects different tree parts associated with the bifurcation in an automatic manner, using an interpolation strategy. The branching diameters and angles are estimated using standard *allometry rules* to add realism to the reconstruction process.

One drawback of the above approach is that the method demands a number of tree parts to build the database. Also, it needs some user assistance in order to work properly. Although modeling the missing branch parts is approximated to by parametric curves, it is not fully supported by the biological relevance of the structure. The technique is not fully data-driven in the sense that the input point cloud data of the whole plant is not used, and thus the algorithm does not focus on reconstructing the original plant geometry.

6.2.5.2 Hybrid Approach

The hybrid approach is based on a combined framework of procedural (i.e., the type of modeling that is based on a set of predefined rules to produce the output) and data-driven modeling techniques (state transition sequence $1 \rightarrow 5$ in Figure 6.2). The *Inverse Procedural Model* approach (Stava et al. 2014) is a hybrid approach where the main idea is to estimate the parameter of a procedural model that best represents the input scan data, instead of defining rules that generate a tree model, as is typically done in classical procedural modeling techniques. Each set of parameters defines the general architecture of a particular species, and, by varying the parameters, different species under different environmental conditions can be modeled. The model is based on a set of 24 parameters that describe the geometry of a tree. These parameters include branching angle, internode distance, number of lateral buds, etc. which are based on geometric and environmental factors acting on the plant. Several biological assumptions/constraints are employed in the modeling process to represent different entities of the tree mathematically. For example, the apical bud is assumed to be located at the tip of the plant shoot. The orientation of the apical bud with respect to the shoot is modeled as the following polar and azimuthal angle in a spherical coordinate system:

$$\theta \sim \mathcal{N}(0, \sigma_v),$$

$$\phi \sim \mathcal{U}(0, 2\pi),$$

where σ_v is the apical angle variance parameter, and $\mathcal{N}$ and $\mathcal{U}$ are normal and uniform distributions, respectively. So, by varying σ_v, it is possible to generate different branching structures. In this way, the parametric model of the tree is defined by using the selected set of parameters. In order to find the parameters that maximize the similarity measure between the input data and the model, a number of trees are generated by perturbing the discrete values of the parameters empirically by the user. Then, each of these trees is compared with the input tree by a similarity measure, based on three types of distances: *shape distance*, *geometric distance*, and *structural distance* (refer to Section 6.3 for details). However, an infinite number of tree models can be generated by using different combinations of the parameters. The best set of parameters corresponding to the best matching tree is obtained using a Monte Carlo technique.

Recently, Guo et al. (2018) proposed a hybrid approach to reconstruct plant geometry with the aim of alleviating the need for tuning a large number of parameters, as in the previous approach. The technique extends the space colonization technique of Preuksakarn et al. (2010) by introducing an L-system-type rule-based framework in the model.

Initially, a technique is proposed to generate high-quality 3D point cloud data from a sequence of overlapping images around a plant. Next, a parametric model is fitted to the point cloud by using a rule-based method. The model is based on five parameters to represent the skeletal structure of the tree: internode length (l), roll angle (ϕ), branching angle (ψ), growth units (ρ), and diameter coefficient (γ). Starting from the root/bottom of the plant, the model simulates a set of rules by making certain assumptions. Basically, the rules make use of the points in a biologically meaningful manner. From the seed locations, the following set of rules are used:

$$\text{Seed}(\mathbf{p},\mathbf{v}) \rightarrow A(\mathbf{p},\mathbf{v}),$$

$$A(\mathbf{p},\mathbf{v}) \rightarrow \left\{\text{Metamer}(\mathbf{p}',\mathbf{v}')*\right\} A(\mathbf{p}'',\mathbf{v}''),$$

$$\text{Metamer}(\mathbf{p},\mathbf{v}) \rightarrow \text{Internode}(\mathbf{p},\mathbf{v}) L(\mathbf{p}',\mathbf{v}'),$$

$$L(\mathbf{p},\mathbf{v}) \rightarrow A(\mathbf{p}',\mathbf{v}'),$$

where $\mathbf{p}$ is the seed position and $\mathbf{v}$ is the orientation, which indicates the growth direction of each bud. Initially the growth direction is upward and is adjusted, according to the branching structure and roll angles at each stage. $A(\mathbf{p},\mathbf{v})$ is the *kernel* of the growth process, which results in a chain of metamers (or a growth unit) represented by {*} in the rules. At any particular stage, the position of a metamer is computed from the previous metamer, orientation, and length. The orientation is also updated by the phyllotaxy (roll) angle. In the implementation level, the data points are selected as in the technique of the space colonization algorithm (Palubicki et al. 2009; Preuksakarn et al. 2010). Finally, standard allometric rules are applied to compute branch diameter and length.

6.2.5.3 Other Types of Approaches

Zhang et al. (2014) proposed a technique by first constructing the visible parts of the scanned input point cloud data, and then synthesizing the non-visible parts, using a shape prediction model. The approach is a combination of geometric reconstruction at different hierarchical levels (from fine to coarse) and attempts to model the multiscale aspect of tree reconstruction.

Initially, all the points in the input point cloud data are labeled as *unlabeled*. Then, starting from a user-specified point, an initial cylinder is fitted to the local neighborhood. Once a set of points is fitted within a cylinder, these points are labeled as *processed* and are not considered anymore for processing. An iterative process is continued to fit successive cylinders until no cylinder can be fitted any more to the data (*cylinder marching* process). This process is based on the assumption that branch shapes are cylindrical with gradually varying radii. If the number of connected cylinders goes beyond a threshold value, the cylinders constitute a single branch of the tree. At this stage, visible branches of the tree are obtained from the fitted cylinders. To reconstruct the non-visible branches of the tree (due to occlusion), the following heuristic is used. At the end of the cylinder marching process, any remaining unlabeled scattered points are classified as tree crown points, which mostly belong to the leaves at the tip of the branches. Then, the classical particle flow technique is used to model the non-visible branches by considering the crown points as the source and the nearest main branch point as the destination. The directions of the flow of the particles are computed, using the conical search operation of the space colonization algorithm, as discussed earlier. Finally, some standard techniques are applied to produce the texture effect, mesh model, and leaf configurations. One drawback of the technique is that a lot of heuristics are used in the whole process. Also, tuning of parameters is very important to obtain the desired result.

6.2.5.4 Flower Geometry Reconstruction

Although the goal of volume reconstruction is mainly to model the branching system of the plant, reconstruction of organs, like flowers and leaves, requires surface-based techniques. Because of the complicated thin structure of the petals and their tight configurations along with a high level of occlusions, geometric reconstruction of flowers is important, as well as a difficult problem. Unlike the case of branching structure reconstruction, there are only a handful of techniques that take into account the real data to model flower geometry.

Recently, Yan et al. (2014) proposed a model to reconstruct the flower geometry from a single photograph. The model is based on the assumption that the 3D shapes of individual petals are roughly the same. Initially, modeling flowers with a single layer of petals is considered, and then the idea is extended for flowers having multiple layers of petals. Initially, the petals of a single-layer flower are segmented in an interactive way by the well-known *GrabCut* technique (Rother et al. 2004). Individual petals are located from the segmented image, using the following technique. To begin with, the center of the flower is located by user assistance. The contour of the segmented flower is traversed, and the distance from each contour point to the center is computed. *Valleys* in the sequence of the distance values indicate the intersection points between adjacent petals. The contour of each petal is identified by simulating the classical particle flow technique, where a particle is traced from the intersection point toward the flower center by following the edge of the petal (similar to the *Livewire* technique discussed earlier). The tip of each petal is located at the midpoint of the contour.

In the next phase, the underlying surface of the petals is modeled by a 3D cone, assuming the flower structure to be of conical shape. The 3D cone is fitted to the extracted petals, where the apex of the cone is positioned at the center, and the base is aligned with the petal tips. This cone is used as a special case of a *surface of revolution*, which can be defined by rotating a curve around the main axis of the flower. By an iterative procedure, each petal contour is fitted to the surface of the revolution framework, and the template of the individual petal is obtained. Once the petal contour is obtained, a 3D mesh is constructed by sampling points enclosed in the contour and triangulating these points. However, at this stage, the mesh does not reflect the geometry of the bending of each petal. In order to facilitate this feature, the mesh needs to be bent to approximate the petal shape. This deformation *drags* the vertices on the boundary of the mesh, so that the differences between the observed contours and the projections of the corresponding mesh are minimized. This is performed by minimizing the following energy:

$$\mathcal{E} = \lambda_{geo}\mathcal{E}_{geo} + \lambda_{con}\mathcal{E}_{con},$$

where $\mathcal{E}_{geo}$ is the geometric preserving energy that maintains the geometric features of the mesh, $\mathcal{E}_{con}$ is the contour fitting energy that drags boundary vertices to their new locations, and λ_{geo} and λ_{con} are the weighting factors for the two terms, respectively. The deformation is performed in an iterative manner.

In a similar type of work, Zheng et al. (2017) presented an algorithm to dynamically track the shape of a blooming flower over time. The idea is based on the reconstruction of flower geometry and deformation of the model over time *via* a template model.

During the blooming stage, the flower is scanned over the time span $1 \ldots T$ to obtain a sequence of point clouds $\mathbb{Q} = \mathbb{Q}_{1:T}$. During the early stage, interior petals are completely occluded, whereas during the latter stage, exterior petals decay, bend, and twist heavily. So, none of the scans can actually represent the whole flower geometry by itself. Initially, one intermediate frame is selected manually, which is considered to be the *key frame*, and the geometry $\mathbb{M}$ of the full flower is reconstructed from this frame in an interactive manner (assuming that the petals are at least partially visible in the selected key frame). The main idea of the technique is to use $\mathbb{M}$ as the template, and fit this to the point cloud in the adjacent frames by deformation. The *track and fit* operation is formulated as an Expectation Maximization (EM) framework, where the E-step computes the correspondence between the template and the captured point cloud, and the M-step updates the vertex locations, based on the computed correspondence of the vertices. The tracking is performed both forward and backward in time, which yields a series of 3D model sequences $\mathbb{M}_{1:T}$, which represent the blooming process.

Let $\mathbb{Q}_t$ represents the captured data at time t and $\mathbb{M}_t$ represents the deformed template mesh model (which we are seeking) of the original template model $\mathbb{M}$. In the EM framework, $\mathbb{Q}_t$ is assumed to be the observation of a Gaussian Mixture Model (GMM), the centroids of which are the vertices of the unknown mesh $\mathbb{M}_t$. The problem is formulated as a Maximum A Posteriori (MAP) problem to obtain the deformed mesh vertices $\mathbb{M}_t^*$., which is obtained by maximizing the probability of the observation:

$$\mathbb{M}_t^*. = \underset{\mathbb{M}_t}{\operatorname{argmax}}\, p\left(\mathbb{Q}_t \middle| \mathbb{M}_t\right) p\left(\mathbb{M}_t\right),$$

where $p\left(\mathbb{Q}_t \middle| \mathbb{M}_t\right)$ is the likelihood term, and $p\left(\mathbb{M}_t\right)$ is the prior term. During the E-step, vertex correspondences between $\mathbb{M}$ and $\mathbb{Q}$ are computed. Instead of computing the correspondence using the point cloud as a whole, these are classified into different parts (petals), where each part $\mathbb{Q}^k \in \mathbb{Q}$ corresponds to a part $\mathbb{M}^k \in \mathbb{M}$. The classification is performed by the standard GMM clustering strategy. Assuming that a point $q_j \in \mathbb{Q}$ is normally distributed around $m_i \in \mathbb{M}$ as $q_j \sim \mathcal{N}\left(m_i, \sigma_i\right)$ with covariance σ_i, the probability of q_j given m_i is

$$p\left(q_j \middle| m_i\right) = \frac{1}{\sqrt{\left(2\pi\right)^3 \left|\sigma_i\right|}} \exp\left(-\frac{1}{2}\left(q_j - m_i\right)^T \sigma_i^{-1}\left(q_j - m_i\right)\right).$$

The probability that a point q_j belongs to $\mathbb{Q}^k$ is computed as

$$p\left(q_j \in \mathbb{Q}^k\right) = \frac{\sum_{m_i \in \mathbb{M}^k} \phi_i p\left(q_j \middle| m_i\right) v_i}{\sum_{m_i \in \mathbb{M}} \phi_i p\left(q_j \middle| m_i\right) v_i},$$

where $\phi_i = \sum_{q_j \in \mathbb{Q}^k} p(q_j|m_i)$, and $v_i \in \{0,1\}$ is called the *visibility term*, which is set to 1 if the point is visible (within a threshold distance from the mesh); otherwise, it is set to 0.

Now, for each part (petal) k, the correspondence between $\mathbb{Q}^k$ and $\mathbb{M}^k$ is computed as a point correspondence matrix $\mathbb{Z} : \mathbb{M}^k \rightarrow \mathbb{Q}^k$, where each element $\mathbb{Z}_{ij} \in \mathbb{Z}$ is computed as

$$\mathbb{Z}_{ij} = \frac{\phi_i p(q_j|m_i) v_i}{\sum_{m_i \in \mathbb{M}^k} \phi_i p(q_j|m_i) v_i}.$$

Once the correspondences of vertex locations are estimated for each petal, the task is now to optimize the vertex locations in $\mathbb{M}$, so that $\mathbb{M}$ fits better to the data $\mathbb{Q}$ (the M-step). This is formulated as the following stochastic minimization problem

$$\underset{\mathbb{M}}{\operatorname{argmin}} \left(-\log p(\mathbb{M}|\mathbb{Q},\mathbb{Z}) - \log p(\mathbb{M}) \right).$$

The first term in the above equation is the data term, where the goal is to minimize the distances between the mesh vertices and their corresponding points. Along with an additional penalty, to ensure that the mesh follows the contour of the petal, the data term is written as

$$-\log p(\mathbb{M}|\mathbb{Q},\mathbb{Z}) = \sum_k \left(\omega_1 D(\mathbb{Q}^k, \mathbb{M}^k) + \omega_2 D(\mathbb{Q}_b^k, \mathbb{M}_b^k) \right),$$

where $\mathbb{Q}_b^k$ and $\mathbb{M}_b^k$ are the boundary points, which are not taken into consideration at the initial stage of the blooming process since the petals are not sufficiently separated at this stage, and ω_1, ω_2 are the weighting factors for the two terms, respectively. The second term of the stochastic minimization formulation is the prior term, which regularizes the solution. This term is modeled as a combination of three different types of priors, as follows:

$$-\log p(\mathbb{M}) = \mathcal{E}_{\text{shape}} + \mathcal{E}_{\text{collision}} + \mathcal{E}_{\text{root}}.$$

The first term is used to preserve the flower shape. Using some user intervention, a reliable template mesh is created, and the deformation of $\mathbb{M}^k$ is constrained by this template mesh. In the second term, avoidance of the penetration of two petals is ensured. Whenever penetration is detected, the point is moved to another location so that the surface normal directions are not violated. The third term ensures that the root (or the base) point of each petal does not change during deformation. Finally, the energy is minimized by an alternate optimization strategy, using different types of nonlinear least square solvers.

One of the drawbacks of the method lies in its limitation in modeling fine geometric details. Also, large deformation of petals will result in poor registration of the point cloud, which will affect the overall accuracy of the reconstructed model.

6.3 EVALUATION TECHNIQUES

So far, we have discussed different types of modeling techniques that aim at reconstructing the geometry of plants from real data. However, an important aspect of reconstruction is to measure the quality of the solution. From the computer graphics perspective, the results are analyzed mostly on qualitative or visual assessment. However, in biological applications, quantitative analysis of the results is extremely important. Given a reference plant structure (the ground-truth), different types of measures are proposed in the literature to quantify the similarity between the reconstructed model and the reference model. A naive approach is to measure quantities like crown volume, total branch length, etc. and then to compare the two models by the difference between these quantities. However, this type of method cannot truly assess the quality of the reconstruction of tree topology. Ferraro and Godin (2000) proposed an advanced metric to compare 2D tree structures by means of *edit distance*. The distance is computed as the minimum cost of a sequence of edit operations that transforms an initial tree into a target tree. Boudon et al. (2014) adopted the idea of 2D edit distance into the 3D case to compare two trees. An optimization framework is proposed to compute geometric and structural similarities of two trees. Initially, the scales (or the resolutions) of the scan data of two trees are homogenized. Then, similar elements of the trees are determined by adopting the Hausdorff distance between their skeleton curves. Structural similarity between the two trees is obtained by computing the similarity of the organization of the edges.

Stava et al. (2014) proposed a similarity measure by introducing two different types of distance measures along with the edit distance (Ferraro and Godin 2000). *Shape distance* is used to measure the difference between the overall shapes of the trees. Different types of descriptors (e.g., height, radius, etc.) are evaluated in the shape distance function as:

$$\delta = 1 - \exp\left[1 - \frac{a\left(\xi_{\tau_i} - \xi_{\tau_j}\right)^2}{2\sigma^2}\right]$$

where ξ_{τ_i} and ξ_{τ_j} are the descriptors of the two trees, and σ is a normalization factor, which is set empirically. Another type of measure, the *geometric distance*, computes the difference between the global branching structure of the two trees from some sample points over the trees. Several parameters, like branch angle, and branch slope with respect to the horizontal plane, etc. are computed for all the sample points over the tree, and the descriptor is defined as the weighted mean and variance of these samples. The geometric distance is computed using these descriptors. The final similarity measure $\bar{\delta}\left(\tau_i, \tau_j\right)$ of two trees τ_i and τ_j is computed as the sum of the three measures, along with the weighting factors that control the influence of each measure.

On a different type of evaluation strategy, Guo et al. (2018) used two types of evaluation metrics: *model-based* and *scan-based*. In the model-based strategy, synthetic plant models are used as ground-truth, and a number of views are generated from the model. Then, the similarity distance, as proposed by Stava et al. (2014), is used as the evaluation metric. In the scan-based strategy, the comparison is made directly on the real data. For every point in the reconstructed model, the distance to the nearest point in the actual scan data is computed. The mean and standard deviation of the distances are used as performance measurements. If the values are lower, then the performance is treated as better, whereas higher values indicate worse performance.

6.4 CONCLUSION

In this chapter, we have elaborately studied the topic of geometric reconstruction of plants from real-world data. A number of state-of-the-art techniques are discussed, along with their strengths and limitations. We infer, from a thorough literature survey, that only a handful of techniques take into account the actual biological relevance of the results. Also, not much work has focused on exploiting prior botanical knowledge in the modeling framework. Given the huge varieties of plant species, there is a lot of scope for geometric modeling to accurately reconstruct the plant geometry in a robust manner. Ideally, the ultimate goal of geometric reconstruction will be to develop a generalized algorithm that will be able to handle a diverse variety of species for real-time phenotyping analyses.

ACKNOWLEDGMENT

This work is supported by the Robotics for Microfarms (ROMI) European project.

REFERENCES

Au, O. K., C. Tai, H. Chu, D. Cohen-Or, and T. Lee. 2008. Skeleton extraction by mesh contraction. *ACM Transactions on Graphics* 27(3):44.

Boudon, F., C. Pradal, T. Cokelaer, P. Prusinkiewicz, and C. Godin. 2012. L-Py: An L-system simulation framework for modeling plant architecture development based on a dynamic language. *Frontiers in Plant Science* 3:76.

Boudon, F., C. Preuksakarn, P. Ferraro, J. Diener, P. Nacry, E. Nikinmaa, and C. Godin. 2014. Quantitative assessment of automatic reconstructions of branching systems obtained from laser scanning. *Annals of Botany* 114(4):853–862.

Boykov, Y., O. Veksler, and R. Zabih. 2001. Fast approximate energy minimization via graph cuts. *IEEE Transactions on Pattern Analysis and Machine Intelligence* 23(11):1222–1239.

Bucksch, A., R. C. Lindenbergh, and M. Menenti. 2009. Skeltre – Fast skeletonisation for imperfect point cloud data of botanic trees. In *Proceedings of the Eurographics Workshop on 3D Object Retrieval* 13–20. Munich, Germany.

Cao, J., A. Tagliasacchi, M. Olson, H. Zhang, and Z. Su. 2010. Point cloud skeletons via Laplacian-based contraction. In *Proceedings of the Shape Modeling International Conference* 187–197. Aix-en-Provence, France.

Chaudhury, A., C. Ward, A. Talasaz, A. G. Ivanov, M. Brophy, B. Grodzinski, N. P. A. Huner, R. V. Patel, and J. L. Barron. 2018. Machine vision system for 3D plant phenotyping. *IEEE/ACM Transactions on Computational Biology and Bioinformatics* 2009–2022.

Cornea, N. D., D. Silver, and P. Min. 2007. Curve-skeleton properties, applications, and algorithms. *IEEE Transactions on Visualization and Computer Graphics* 13(3):530–548.

Das Choudhury, S., S. Bashyam, Y. Qiu, A. Samal, and T. Awada. 2018. Holistic and component plant phenotyping using temporal image sequence. *Plant Methods* 14:35.

Fahlgren, N., M. A. Gehan, and I. Baxter. 2015. Lights, camera, action: High-throughput plant phenotyping is ready for a close-up. *Current Opinion in Plant Biology* 24:93–99.

Ferraro, P., and C. Godin. 2000. A distance measure between plant architectures. *Annals of Forest Science* 57(5):445–461.

Godin, C., E. Costes, and H. Sinoquet. 1999. A method for describing plant architecture which integrates topology and geometry. *Annals of Botany* 84(3):343–357.

Godin, C., and H. Sinoquet. 2005. Functional-structural plant modeling. *The New Phytologist* 166(3):705–708.

Guo, J., S. Xu, D. Yan, Z. Cheng, M. Jaeger, and X. Zhang. 2018. Realistic procedural plant modeling from multiple view images. *IEEE Transactions on Visualization and Computer Graphics* 26(2):1372–1384.

Huang, H., S. Wu, D. Cohen-Or, M. Gong, H. Zhang, G. Li, and B. Chen. 2013. L1-medial skeleton of point cloud. *ACM Transactions on Graphics* 32:65.

Karwowski, R., and P. Prusinkiewicz. 2003. Design and implementation of the L+C modeling language. *Electronic Notes in Theoretical Computer Science* 86(2):134–152.

Li, Y., X. Fan, N. J. Mitra, D. Chamovitz, D. Cohen-Or, and B. Chen. 2013. Analyzing growing plants from 4D point cloud data. *ACM Transactions on Graphics* 32(6):6.

Livny, Y., Y. Feilong, M. Olson, B. Chen, H. Zhang, and J. El-Sana. 2010. Automatic reconstruction of tree skeletal structures from point clouds. *ACM Transactions on Graphics* 29:151.

Mortensen, E. N., and W. A. Barrett. 1995. Intelligent scissors for image composition. In *Proceedings of the 22nd Annual Conference on Computer Graphics and Interactive Techniques* 191–198. Los Angeles, CA.

Neubert, B., T. Franken, and O. Deussen. 2007. Approximate image-based tree-modeling using particle flows. *ACM Transactions on Graphics* 26(3):88.

Palubicki, W., K. Horel, S. Longay, A. Runions, B. Lane, R. Mech, and P. Prusinkiewicz. 2009. Self-organizing tree models for image synthesis. *ACM Transactions on Graphics* 28(3):3.

Pascucci, V., G. Scorzelli, P. Bremer, and A. Mascarenhas. 2007. Robust on-line computation of Reeb graphs: Simplicity and speed. *ACM Transactions on Graphics* 26(3):58.

Preuksakarn, C., F. Boudon, P. Ferraro, J. B. Durand, E. Nikinmaa, and C. Godin. 2010. Reconstructing plant architecture from 3D laser scanner data. In *Proceedings of the 6th International Workshop on Functional-Structural Plant Models* 12–17. Davis, CA.

Prusinkiewicz, P., and A. Lindenmayer. 1990. *The Algorithmic Beauty of Plants.* Springer-Verlag.

Prusinkiewicz, P., L. Mündermann, R. Karwowski, and B. Lane. 2001. The use of positional information in the modeling of plants. In *Proceedings of the 28th Annual Conference on Computer Graphics and Interactive Techniques* 289–300. Los Angeles, CA.

Reche, A., I. Martin, and G. Drettakis. 2004. Volumetric reconstruction and interactive rendering of trees from photographs. *ACM Transactions on Graphics* 23(3):720–727.

Rodkaew, Y., P. Chongstitvatana, S. Siripant, and C. Lursinsap. 2003. Particle systems for plant modeling. In *Proceedings of the Plant Growth Modeling and Applications* 210–217. Beijing, China.

Rother, C., V. Kolmogorov, and A. Blake. 2004. GrabCut: Interactive foreground extraction using iterated graph cuts. *ACM Transactions on Graphics* 23(3):309–314.

Runions, A., B. Lane, and P. Prusinkiewicz. 2007. Modeling trees with a space colonization algorithm. In *Proceedings of the Eurographics Workshop on Natural Phenomena* 63–70. Prague, Czech Republic.

Shlyakhter, I., M. Rozenoer, J. Dorsey, and S. Teller. 2001. Reconstructing 3D tree models from instrumented photographs. *IEEE Computer Graphics and Applications* 21(1):53–61.

Stava, O., S. Pirk, J. Kratt, B. Chen, R. Mech, O. Deussen, and B. Benes. 2014. Inverse procedural modeling of trees. *Computer Graphics Forum* 33(6):118–131.
Tagliasacchi, A., T. Delame, M. Spagnuolo, N. Amenta, and A. Telea. 2016. 3D skeletons: A state-of-the-art report. *Computer Graphics Forum* 35(2):2.
Tagliasacchi, A., H. Zhang, and D. Cohen-Or. 2009. Curve skeleton extraction from incomplete point cloud. *ACM Transactions on Graphics* 28(3):71.
Wang, Z., L. Zhang, T. Fang, X. Tong, P. T. Mathiopoulos, L. Zhang, and J. Mei. 2016. A local structure and direction-aware optimization approach for three-dimensional tree modeling. *IEEE Transactions on Geoscience and Remote Sensing* 54(8):4749–4757.
Wu, S., W. Wen, B. Xiao, X. Guo, J. Du, C. Wang, and Y. Wang. 2019. An accurate skeleton extraction approach from 3D point clouds of maize plants. *Frontiers in Plant Science* 10:248.
Xie, K., F. Yan, A. Sharf, O. Deussen, H. Huang, and B. Chen. 2016. Tree modeling with real tree parts examples. *IEEE Transactions on Visualization and Computer Graphics* 22(12):2608–2618.
Xu, H., N. Gossett, and B. Chen. 2007. Knowledge and heuristic based modeling of laser scanned trees. *ACM Transactions on Graphics* 26(4):19.
Yan, D., J. Wintz, B. Mourrain, W. Wang, F. Boudon, and C. Godin. 2009. Efficient and robust reconstruction of botanical branching structure from laser scanned points. In *Proceedings of the IEEE International Conference on Computer Aided Design and Computer Graphics* 572–575. New Orleans, LA.
Yan, F., M. Gong, D. Cohen-Or, O. Deussen, and B. Chen. 2014. Flower reconstruction from a single photo. *Computer Graphics Forum* 33(2):439–447.
Yi, L., H. Li, J. Guo, O. Deussen, and X. Zhang. 2017. Tree growth modeling constrained by growth equations. *Computer Graphics Forum* 37:239–253.
Yin, K., H. Huang, H. Zhang, M. Gong, D. Cohen-Or, and B. Chen. 2014. Morfit: Interactive surface reconstruction from incomplete point clouds with curve driven topology and geometry control. *ACM Transactions on Graphics* 33(6):202.
Yin, K., H. Huang, P. Long, A. Gaissinski, M. Gong, and A. Sharf. 2016. Full 3D plant reconstruction via intrusive acquisition. *Computer Graphics Forum* 35(1):272–284.
Zhang, X., H. Li, M. Dai, W. Ma, and L. Quan. 2014. Data-driven synthetic modeling of trees. *IEEE Transactions on Visualization and Computer Graphics* 20(9):1214–1226.
Zheng, Q., X. Fan, M. Gong, A. Sharf, O. Deussen, and H. Huang. 2017. 4D reconstruction of blooming flowers. *Computer Graphics Forum* 36(6):405–417.

7 Image-Based Structural Phenotyping of Stems and Branches

Fumio Okura, Takahiro Isokane, Ayaka Ide, Yasuyuki Matsushita, and Yasushi Yagi

CONTENTS

7.1 INTRODUCTION

The structure of plant shoots (i.e., leaves and stems) is an important target for plant phenotyping and artificial intelligence (AI)-aided cultivation. Whereas it should be possible to reconstruct the shoot structure using image analysis, difficulties arise due to heavy occlusions, or to structural complexity. In particular, determining the hidden skeleton structures of the branches of plants remains one of the most challenging topics in the computer vision community. Skeleton estimation is actively being studied for human pose estimation in the field of computer vision (Cao et al. 2017). A recent trend in image-based human skeleton estimation relies on prior knowledge of the relationships among joints to extract and interpret joints from images. Unlike human bodies, the branch structures of plants are less rigidly organized, and the number of joints and their connections are generally unknown. In addition, heavy occlusions due to leaves make the problem harder. These aspects pose a unique challenge when estimating the structure of plants.

In this chapter, as shown in Figure 7.1, we introduce an approach for the estimation of 3D branch (stem) structures of plants from multiview images using a deep

FIGURE 7.1 Estimation of partly hidden branches. From multiview plant images, our method infers the branch structure. An explicit graph structure (red lines) can be derived from the combination of a deep learning-aided 3D reconstruction technique.

learning-aided 3D reconstruction technique (Isokane et al. 2018). To estimate the branching paths hidden under leaves, we cast the estimation problem as an image-to-image (I2I) translation problem (Isola et al. 2017), which converts an image from one domain to another domain. In this context, we converted an input image of a leafy plant to a map that represents the branch structure, in which each pixel contains a prediction of the probability of branch existence. After undertaking an I2I translation using deep neural networks, we transformed the 2D results into a 3D branching shape and structure, using a volumetric 3D reconstruction approach.

First, we will briefly review the prior literature for 3D modeling of plant shoots and I2I translation, the key component of our 3D reconstruction technique.

7.2 A BRIEF REVIEW OF PLANT SHOOT MODELING

Although the context is slightly different from plant phenotyping, 3D modeling of plant shoots has been actively studied in the computer graphics community because of their importance as a rendering subject, and because the manual modeling of plants is notably time-consuming. Interactive 3D modeling methods for trees using user-provided hints, e.g., lines, were proposed in the early 2000s (Boudon et al. 2003; Okabe et al. 2005). Growth models for trees that include branching rules are occasionally used (Galbraith et al. 2004; Palubicki et al. 2009; Streit et al. 2005), and some of these are augmented with realistic textures (Livny et al. 2011). These approaches generally rely heavily on branch models (or rules), and the resulting structure cannot deviate much from the presumed models.

In contrast, 3D plant reconstruction methods based on observations of real-world trees and plants, that use photographs (Shlyakhter et al. 2001) or 3D scans (Livny et al. 2010; Xu et al. 2007), have shown potential for automatic tree and plant modeling. Several reconstruction approaches that use multiview images have been proposed (Neubert et al. 2007; Reche-Martinez et al. 2004; Tan et al. 2007). These approaches have been further extended to include single-image-based methods (Argudo et al. 2016; Tan et al. 2008), to provide greater applicability. The main focus of most image-based modeling approaches has been to generate 3D tree models that provide a suitable silhouette or volume, and they do not necessarily aim to recover the structure of every branch. When using 3D reconstruction for plant phenotyping, it is essential to estimate biologically correct branching patterns.

FIGURE 7.2 Occlusions in a plant photograph. Although it may be possible for a human to imagine hidden branch structures, inferring them based on image analysis is one of the most challenging issues in computer vision.

Using multiview images of bare trees, previous approaches have achieved geometric reconstructions of branch structures (Lopez et al. 2010; Stava et al. 2014; Zhang et al. 2016). One study calculated a time-space segmentation of dense point clouds for scenes with mild occlusions, which were captured from various angles using active stereo (Li et al. 2013). The leaves of plants or trees, however, make 3D reconstruction considerably more difficult due to occlusion; an example is shown in Figure 7.2. Note that for relatively small plants, Quan et al. used an interactive approach (i.e., involving manual labor) to define unobserved branches (Quan et al. 2006).

7.3 A BRIEF REVIEW OF IMAGE-TO-IMAGE (I2I) TRANSLATION

I2I translation aims at transferring contextual or physical variations between the source and target images. A similar technique had been actively studied before the invention of deep learning. Some traditional I2I translation techniques have been referred to as texture synthesis or texture transfer when the approach focused on the textures. Many approaches have been proposed for modifying the color distribution of an image, such as color transfer methods. More generally, such techniques are referred to as image style transfer. The early works on texture synthesis include image analogies (Hertzmann et al. 2001) and image quilting (Efros and Freeman 2001), which were proposed in the early 2000s. Commonly, these approaches divide the image into small patches and transfer the change based on correspondences between patches (Darabi et al. 2012; Lefebvre and Hoppe 2005). Using patch-wise computation, texture and color transfer can be effectively unified to transfer an entire scene from a photograph (Okura et al. 2015). PatchMatch (Barnes et al. 2009) dramatically speeds up the patch correspondence estimation; this has made texture synthesis popular in consumer applications, *via* its implementation in Adobe Photoshop.

More recent I2I translation methods benefit from deep learning, such as convolutional neural networks (CNNs) with encoder-decoder architectures or generative adversarial networks (GANs). Pix2Pix (Isola et al. 2017), which uses a conditional GAN, shows impressive performance on a wide variety of translation tasks. In this context, CycleGAN (Zhu et al. 2017) has shown the possibility of I2I translation without the use of paired training images, by connecting two GAN architectures. The inventors of deep I2I translation, who developed Pix2Pix and CycleGAN, have also devised image quilting, one of the earliest approaches to texture transfer. Pix2Pix employs rules that imitate the effect of traditional texture synthesis in its loss function. This provides a key lesson for deep learning approaches in that Pix2Pix is merely a good optimizer, and it is, therefore, important to leverage the domain knowledge of *before-deep-learning* to develop better methods for the use of deep learning.

7.4 3D MODELING OF PARTLY HIDDEN BRANCHES

Our method takes as input multiview images of a plant and generates a probabilistic 3D branch structure in a 3D voxel space. Our method begins by estimating a 2D probabilistic branch existence map in each of the multiview images, based on an altered I2I translation method. Once the probabilistic branch existence map is computed for each view, they are merged in a 3D voxel space, using the estimated camera poses, based on a structure-from-motion method (Wu 2013), to yield a probabilistic 3D branch structure. Finally, an explicit 3D branch structure is generated by a particle flow simulation, which is inspired by a traditional tree modeling approach (Neubert et al. 2007), using the probabilistic 3D branch structure. In the following sections, we explain the individual steps of the proposed method.

7.4.1 Estimation of 2D Branch Probability Using Bayesian I2I Translation

From a leafy plant image, we first estimated a pixel-wise 2D branch existence probability. The major challenge is to infer the branch structure that is hidden under leaves and cannot be directly observed by a camera, potentially not from any of the selected viewpoints. At this stage, we adopted a Pix2Pix approach (Isola et al. 2017) to I2I translation as a means to derive a statistically valid prediction of the existence of branches in the multiview images. In this context, we trained a Pix2Pix network, using image pairs of a leafy plant and its corresponding label map, describing the branch region.

Pix2Pix is a state-of-the-art approach for I2I translation that uses conditional GAN architecture, as illustrated in Figure 7.3. Given training pairs of input and output image domains, a generator creates an output candidate. A discriminator, which forms a CNN, estimates whether the pair of input and output images is real or fake (i.e., generated by the generator network). The training process is called adversarial learning because these two networks are *adversarially* optimized; i.e., the generator is optimized to mislead the discriminator, which learns to detect images created by the generator. As with the original Pix2Pix, we used U-Net (Ronneberger et al. 2015) as the generator network. It consists of encoder and decoder CNNs, as well as skip connections, that allow data to bypass deeper enhancement layers to retain

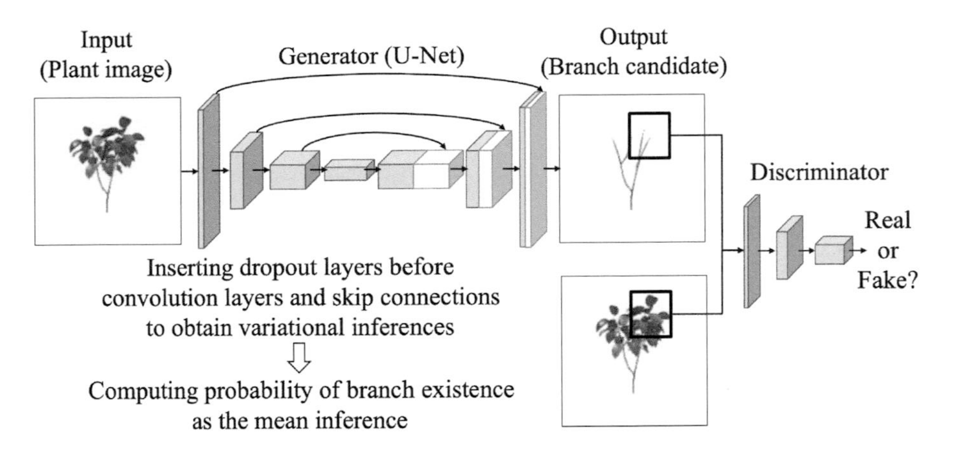

FIGURE 7.3 An illustration of Pix2Pix architecture. We constructed a Bayesian extension for this implementation to generate probabilistic branch existence as output via a Monte Carlo sampling of multiple variational inferences.

the original content. The discriminator network in Pix2Pix is called PatchGAN. Although CNNs usually make decisions for a whole image, PatchGAN outputs real or fake decisions for every image patch (a small region of an image). This idea imitates traditional texture synthesis, which optimizes the output image on the basis that every image patch in the output image should be plausible.

We intend to use the generator output as an input image for a multiview volumetric 3D reconstruction pipeline. In such cases, directly using estimates generated by I2I translation networks may not be a good idea. The generator network creates one output (branch) image that corresponds to an input (leafy) image. When you imagine branches partly hidden under leaves, it is possible to roughly estimate where the branches are located, but it is generally difficult to designate accurate branch paths. Since the real world includes a level of uncertainty, it is often inadequate to employ deterministic approaches – if branch paths created by the generator are slightly different from the actual branch location, it can cause missing or duplicate branches during the 3D reconstruction stage. We, therefore, extended the Pix2Pix framework to obtain a probability of the branch existence for each pixel. When the generator cannot confidently decide the branch location, the probability map will be blurred and will allow completion by other views, using this inference with greater confidence. We used the original I2I translation approach in a Bayesian deep learning framework (Gal and Ghahramani 2016; Neal 2012) with Monte Carlo sampling. This strategy was implemented by inserting dropout layers before the convolution layers in the Pix2Pix encoder-decoder network. To further increase inference variation, dropout was also applied to skip connections. Multiple variational inferences yielded the degree of uncertainty, which cannot be obtained by a single inference.

As such, we obtained the mean of the inferences from the I2I translation network. By treating the variational inferences as stochastic samples, each pixel in the mean inference can be regarded as the probability of branch existence in the range of [0,1]. For the ith image I_i, the branch probability $B_{2D_i}:\mathbb{R}^2 \rightarrow [0,1]$ at pixel x_{2D} is written as:

$$B_{2D_i}(\mathbf{x}_{2D}) = \frac{1}{n_v}\sum_v \hat{B}_{2D_{i,v}}(\mathbf{x}_{2D}),$$

$$\hat{B}_{2D_{i,v}} = \pi_v(I_i),$$

where π_v denotes the Pix2Pix translation from an image I_i to the corresponding branch existence $\hat{B}_{2D_{i,v}}$ with the vth variation of random dropout patterns. The probability map B_{2D_i} for each viewpoint i is then obtained by marginalizing the individual samples $\hat{B}_{2D_{i,v}}$ over random trials v.

7.4.2 Probabilistic 3D Branch Structure Generation

Once the view-wise probability maps $\{B_{2D_i}\}$ are obtained, our method estimates a 3D probability map B_{3D} of the branch structure defined in the 3D voxel coordinates, using a traditional computed tomography (CT) method (Brooks and Di Chiro 1975). We used multiview input images to estimate the camera positions and intrinsic parameters by a structure-from-motion method (Wu 2013). The probability of the branch existence B_{3D} at voxel $\mathbf{x}_{3D}$ can be computed as a joint distribution of $\{B_{2D_i}\}$, by assuming their independence as

$$B_{3D}(\mathbf{x}_{3D}) = \prod_i B_{2D_i}(\theta_i(\mathbf{x}_{3D})),$$

in which θ_i represents a projection from the voxel to the ith image coordinates. While it is possible that none of these views will convey branch structure information in its entirety, due to heavy occlusion, this aggregation effectively recovers the branch structure in a probabilistic framework.

7.4.3 Branch Path Generation Using Particle Flows

The probabilistic 3D branch structure can be converted to an explicit representation of 3D branch paths that can be used for structural analysis applications (e.g., counting branches and measuring their lengths) for plant phenotyping. Inspired by a conventional tree modeling approach (Neubert et al. 2007), we developed a branch structure generation method, using particle flows, as shown in Figure 7.4. The resulting 3D model is represented by a graph that consists of nodes and edges that correspond to joints and branches.

First, we generated particles proportionally to the 3D probability map B_{3D}. The root position of the plant was also set to the bottom-most point that had a high probability of being part of the branches. With these settings, starting from a random distribution of particles, the particle positions were iteratively updated using the following rules:

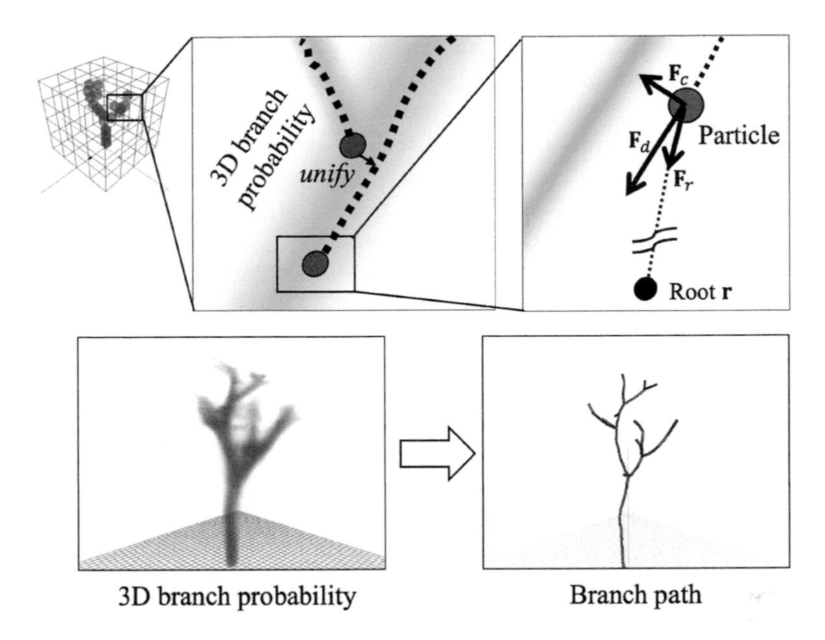

FIGURE 7.4 Branch path generation using particle flows.

$$\mathbf{p}_{t+1} \leftarrow \mathbf{p}_t + \mathbf{F}(\mathbf{p}_t),$$

$$\mathbf{F}(\mathbf{p}_t) = \lambda_c \mathbf{F}_c(\mathbf{p}_t) + \lambda_d \mathbf{F}_d(\mathbf{p}_t) + \lambda_r \mathbf{F}_r(\mathbf{p}_t).$$

Here, $\mathbf{F}_c(\mathbf{p}_t)$ and $\mathbf{F}_d(\mathbf{p}_t)$ represent normalized vectors towards and parallel to the stream of branch probability, respectively. $\mathbf{F}_r(\mathbf{p}_t)$ represents the unit direction to the root point of the plant. When multiple particles get close to each other, they are unified. The traces of the particles are recorded in a tree graph as vertices and edges, and unification of particles is treated as a joint.

While the flow simulation generates a lot of branch candidates, these are then simplified and refined to yield the final structure. The refinement process involves smoothing of the branch paths, unification of the branch points, and deletion of subtrees located in areas with low probabilities.

7.5 EXPERIMENTAL RESULTS

We conducted experiments using simulated and real-world plant images, and assessed the quality of the reconstructions, both quantitatively and qualitatively. For Pix2Pix network training, we used images rendered from ten synthetic plants. The plant models were created by changing the parameters of a self-organizing tree model (Palubicki et al. 2009) that was implemented in L-studio (Prusinkiewicz et al. 1999), an L-system-based plant modeler. For each plant, we rendered images of the plants with and without leaves viewed from 72 viewpoints. Each image was flipped for data augmentation. As a result, we created $10 \times 72 \times 2 = 1440$ pairs of leafy and branch image pairs.

In this chapter, we describe the experimental result using simulated plant models, which contain the ground-truth branches, for quantitative evaluation. The accuracy of plants generated using the proposed method is assessed using geometric error, which evaluates the Euclidean distances between 3D points in the generated 3D branches and the ground-truth branches. We assessed the geometric error by densely sampling 3D points on both generated and the ground-truth branches. Let $\mathbf{g} \in \mathcal{G}$ and $\mathbf{t} \in \mathcal{T}$ be generated and true 3D branch points, respectively. The geometric error is defined as a bidirectional Euclidean distance $d(\mathcal{G}, \mathcal{T})$ between the two-point sets, written as:

$$d(\mathcal{G},\mathcal{T}) = \frac{1}{2}\left(\frac{\sum_{\mathcal{G}} \mathbf{g} - N_{\mathcal{T}}(\mathbf{g})}{|\mathcal{G}|} + \frac{\sum_{\mathcal{T}} \mathbf{t} - N_{\mathcal{G}}(\mathbf{t})}{|\mathcal{T}|} \right),$$

where $N_{\mathcal{T}}(\mathbf{x})$ and $N_{\mathcal{G}}(\mathbf{x})$ are functions used to acquire the nearest point to $\mathbf{x}$ from $\mathcal{G}$ and $\mathcal{T}$, respectively, and $|\mathcal{G}|$ and $|\mathcal{T}|$ denote the numbers of points in each point set.

Using these error metrics, we assessed the method using three different settings: volumetric reconstruction using 1) the ground-truth branch regions in multiview images that are not occluded by leaves, 2) the original I2I translation results (without Bayesian extension), and 3) the proposed method, with Bayesian extension of Pix2Pix, as the 2D branch probability map $\{B_{2D_i}\}$.

Table 7.1 summarizes the accuracy evaluation of the four settings with varying numbers of views for 3D reconstruction. The geometric error is defined only up to scale, because our branch recovery method is also up to scale, like most multiview 3D reconstruction methods. Note that we set the scale of the coordinates so that the height of simulated plants ranged from 100 to 200 units. The proposed approach generates accurate branch structures across all the view settings, but the accuracy decreases with a smaller number of views. Figure 7.5 shows a few results for simulated plants using 72 images. When the method is applied to plant models where their species (i.e., texture and shape of leaves) are different from the simulated plants used for training, the proposed method still generates sufficiently accurate branch structures.

Figure 7.6 shows the results of the proposed method using the images of real-world plants. In this experiment, we used the same trained model for the I2I translation that

TABLE 7.1
The Accuracy of 3D Branch Structure (Averaged over Five Simulated Plants) The Error is Defined on a Relative Scale

		Geometric Error (Euclidean Distance)			
Number of Views		72	36	12	6
2D branch map $\{\boldsymbol{B}_{2D_i}\}$	1) Visible branch region	4.20	3.98	3.59	18.05
	2) I2I translation	1.76	2.26	2.40	14.95
	3) Proposed	**1.69**	**1.74**	**2.14**	**14.53**

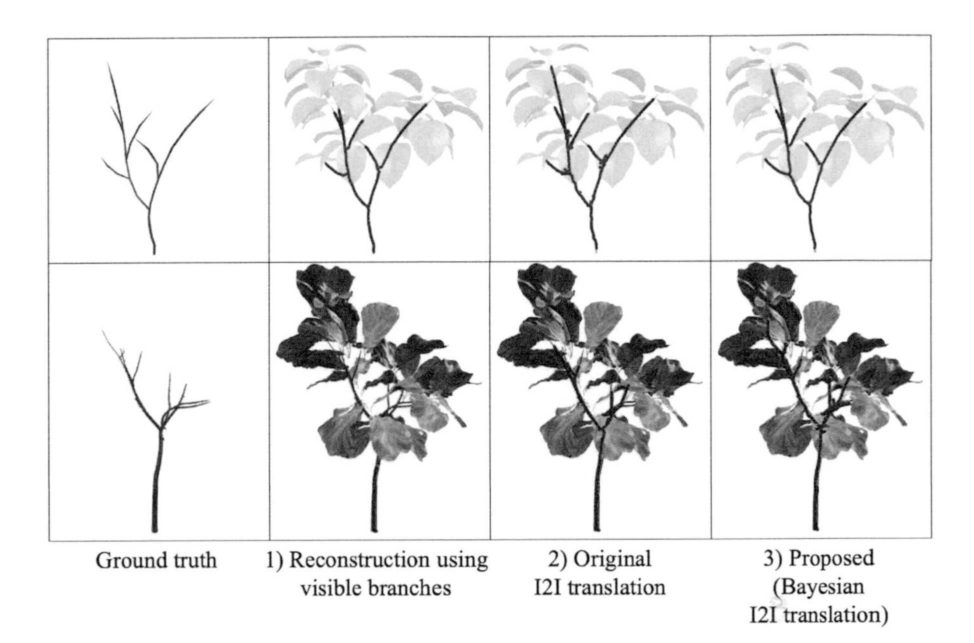

FIGURE 7.5 Results using simulated plants. In comparison to other settings, the proposed approach generates accurate and stable branch structure.

FIGURE 7.6 Results using real plants. Using input images from which the foreground plants are extracted, the proposed approach generates convincing branch structures.

was used in the simulation experiment. To avoid unmodeled factors in the experiment, we first manually extracted the plant regions from the images, which can also be achieved using chroma-keying. The proposed method qualitatively yields convincing branch structures, even though the I2I translation network was trained using simulated plants.

7.6 CONCLUSION

This chapter presents a plant modeling approach that utilizes I2I translation to estimate branch structures of 3D plants from multiview images, even if the branches are occluded under leaves. The combination of Bayesian I2I (leafy to branch image) translation and 3D aggregation generates the branch existence probability in a 3D voxel space, resulting in a probabilistic model of a 3D plant structure. We have shown that explicit branch structures can be generated from the probabilistic representation *via* particle flow simulation. The experimental results generated using simulated plants show that the proposed approach is able to accurately recover the branch structure of a plant. We have also shown that the Bayesian extension of I2I translation is effective in obtaining a stable estimate of branch structure in the form of probability, compared with a non-Bayesian one.

From the viewpoint of computer vision and the image analysis community, plants are interesting yet challenging targets, owing to their complexity (e.g., mutual occlusions among leaves and branches, thin branches, and complex structures). However, AI-powered plant reconstruction techniques can lead the application of leaf-wise and branch-wise analysis of growing plants. It may be a game-changing technology for plant phenotyping and cultivation in the future, when unified with other plant phenotyping techniques and with greater knowledge of plant science.

ACKNOWLEDGMENTS

This work was supported in part by JST PRESTO Grant Number JPMJPR17O3.

REFERENCES

Argudo, O., A. Chica, and C. Andujar. 2016. Single-picture reconstruction and rendering of trees for plausible vegetation synthesis. *Computers & Graphics* 57:55–67.

Barnes, C., E. Shechtman, A. Finkelstein, and D. Goldman. 2009. PatchMatch: A randomized correspondence algorithm for structural image editing. *ACM Transactions on Graphics* 28(3):24.

Boudon, F., P. Prusinkiewicz, P. Federl, C. Godin, and R. Karwowski. 2003. Interactive design of bonsai tree models. *Computer Graphics Forum* 22(3):591–599.

Brooks, R. A., and G. Di Chiro. 1975. Theory of image reconstruction in computed tomography. *Radiology* 117(3 Pt 1):561–572.

Cao, Z., T. Simon, S. Wei, and Y. Sheikh. 2017. Realtime multi-person 2D pose estimation using part affinity fields. In *Proceedings of the IEEE Conference on Computer Vision and Pattern Recognition* 7291–7299. Honolulu, HI.

Darabi, S., E. Shechtman, C. Barnes, D. B. Goldman, and P. Sen. 2012. Image melding: Combining inconsistent images using patch-based synthesis. *ACM Transactions on Graphics* 31(4):82.

Efros, A. A., and W. T. Freeman. 2001. Image quilting for texture synthesis and transfer. In *Proceedings of the 28th Annual Conference on Computer Graphics and Interactive Techniques* 341–346. Los Angeles, CA.

Gal, Y., and Z. Ghahramani. 2016. Dropout as a Bayesian approximation: Representing model uncertainty in deep learning. In *International Conference on Machine Learning* 1050–1059. New York, NY.

Galbraith, C., L. Muendermann, and B. Wyvill. 2004. Implicit visualization and inverse modeling of growing trees. *Computer Graphics Forum* 23(3):351–360.

Hertzmann, A., C. E. Jacobs, N. Oliver, B. Curless, and D. H. Salesin. 2001. Image analogies. In *Proceedings of the 28th Annual Conference on Computer Graphics and Interactive Techniques* 327–340. Los Angeles, CA.

Isokane, T., F. Okura, A. Ide, Y. Matsushita, and Y. Yagi. 2018. Probabilistic plant modeling via multi-view image-to-image translation. In *Proceedings of the IEEE Conference on Computer Vision and Pattern Recognition* 2906–2015. Salt Lake City, UT.

Isola, P., J.-Y. Zhu, T. Zhou, and A. A. Efros. 2017. Image-to-image translation with conditional adversarial networks. In *Proceedings of the IEEE Conference on Computer Vision and Pattern Recognition* 1125–1134. Honolulu, HI.

Lefebvre, S., and H. Hoppe. 2005. Parallel controllable texture synthesis. *ACM Transactions on Graphics* 24(3):777–786.

Li, Y., X. Fan, N. J. Mitra, D. Chamovitz, D. Cohen-Or, and B. Chen. 2013. Analyzing growing plants from 4D point cloud data. *ACM Transactions on Graphics* 32(6):157.

Livny, Y., F. Yan, M. Olson, B. Chen, H. Zhang, and J. El-Sana. 2010. Automatic reconstruction of tree skeletal structures from point clouds. *ACM Transactions on Graphics* 29(6):151.

Livny, Y., S. Pirk, Z. Cheng, F. Yan, O. Deussen, D. Cohen-Or, and B. Chen. 2011. Texture-lobes for tree modelling. *ACM Transactions on Graphics* 30(4):53.

Lopez, L. D., Y. Ding, and J. Yu. 2010. Modeling complex unfoliaged trees from a sparse set of images. *Computer Graphics Forum* 29(7):2075–2082.

Neal, R. M. 2012. *Bayesian Learning for Neural Networks*. Springer Verlag, New York, NY.

Neubert, B., T. Franken, and O. Deussen. 2007. Approximate image-based tree-modeling using particle flows. *ACM Transactions on Graphics* 26(3):88.

Okabe, M., S. Owada, and T. Igarashi. 2005. Interactive design of botanical trees using free-hand sketches and example-based editing. *Computer Graphics Forum* 24(3):487–496.

Okura, F., K. Vanhoey, A. Bousseau, A. A. Efros, and G. Drettakis. 2015. Unifying color and texture transfer for predictive appearance manipulation. *Computer Graphics Forum* 34(4):53–63.

Palubicki, W., K. Horel, S. Longay, A. Runions, B. Lane, R. Měch, and P. Prusinkiewicz. 2009. Self-organizing tree models for image synthesis. *ACM Transactions on Graphics* 28(3):58.

Prusinkiewicz, P., R. Karwowski, R. Měch, and J. Hanan. 1999. L-Studio/Cpfg: A software system for modeling plants. In *International Workshop on Applications of Graph Transformations with Industrial Relevance* 457–464. Kerkrade, Netherlands.

Quan, L., P. Tan, G. Zeng, L. Yuan, J. Wang, and S. B. Kang. 2006. Image-based plant modeling. *ACM Transactions on Graphics* 25(3):599–604.

Reche-Martinez, A., I. Martin, and G. Drettakis. 2004. Volumetric reconstruction and interactive rendering of trees from photographs. *ACM Transactions on Graphics* 23(3):720–727.

Ronneberger, O., P. Fischer, and T. Brox. 2015. U-Net: Convolutional networks for biomedical image segmentation. *International Conference on Medical Image Computing and Computer-Assisted Intervention* 234–241. Munich, Germany.

Shlyakhter, I., S. Teller, M. Rozenoer, and J. Dorsey. 2001. Reconstructing 3D tree models from instrumented photographs. *IEEE Computer Graphics & Applications* 21(1):53–61.

Stava, O., S. Pirk, J. Kratt, B. Chen, R. Měch, O. Deussen, and B. Benes. 2014. Inverse procedural modelling of trees. *Computer Graphics Forum* 33(6):118–131.

Streit, L., P. Federl, and M. C. Sousa. 2005. Modelling plant variation through growth. *Computer Graphics Forum* 24(3):497–506.

Tan, P., G. Zeng, J. Wang, S. B. Kang, and L. Quan. 2007. Image-based tree modeling. *ACM Transactions on Graphics* 26(3):87.

Tan, P., T. Fang, J. Xiao, P. Zhao, and P. Quan. 2008. Single image tree modeling. *ACM Transactions on Graphics* 27:108.

Wu, C. 2013. Towards linear-time incremental structure from motion. In *International Conference on 3D Vision* 127–134. Seattle, WA.

Xu, H., N. Gossett, and B. Chen. 2007. Knowledge and heuristic-based modeling of laser-scanned trees. *ACM Transactions on Graphics* 26(4):19.

Zhang, D., N. Xie, S. Liang, and J. Jia. 2016. 3D tree skeletonization from multiple images based on PyrLK optical flow. *Pattern Recognition Letters* 76:49–58.

Zhu, J. Y., T. Park, P. Isola, and A. A. Efros. 2017. Unpaired image-to-image translation using cycle-consistent adversarial networks. In *Proceedings of the IEEE International Conference on Computer Vision* 2242–2232. Venice, Italy.

8 Time Series- and Eigenvalue-Based Analysis of Plant Phenotypes

Sruti Das Choudhury, Saptarsi Goswami, and Amlan Chakrabarti

CONTENTS

8.1 INTRODUCTION

Time series modeling has been an active research field for many decades, and is of fundamental importance in a wide variety of application domains. A time series is an ordered sequence of values of a variable measured at regular time intervals, e.g., sales data, stock prices, exchange rates in finance, weather forecasts, biomedical measurements (e.g., blood pressure and temperature), biometrics (e.g., face, fingerprint, and gait recognition), and locations in particle tracking in physics. Time series analysis is carried out to understand the underlying structure that produces the observed data for forecasting, monitoring, and even feedback and feedforward control (Montgomery et al. 2015). Since time series modeling enables the study of past observations to develop an appropriate model that describes the inherent temporal structure of the event, we extend its application to plant phenotyping for the study of ontogenetical traits that record the development history of a plant.

High-throughput plant phenotyping based on image analysis has drawn the attention of researchers in recent times to help achieve higher yields of higher quality crops with minimum resource utilization. The process of image-based plant phenotyping is non-destructive, and hence, phenotypes can be extracted at multiple time points during a plant's life cycle. Plants are continually growing organisms with increasing complexity in their architecture over time. Like the plant, its phenotypes also change throughout its life cycle. This motivates the application of advanced time series models for understanding the temporal behavior patterns of the phenotypes aligned with the plant's development phases.

The high-throughput plant phenotyping methods generate high-dimensional data that requires efficient data management infrastructure, as well as proper analytical techniques for meaningful interpretation of phenotypes to leverage the phenotype–genotype gap (Rahaman et al. 2015). The different parts of plants grow non-uniformly over space and time, and this non-uniformity in growth is also affected by genotype and environmental stress factors, e.g., drought, salinity, and pests. Thus, the variation of phenotypes as a function of time may provide greater insight into understanding which genes are expressed by a certain phenotype under a specific environmental condition. This chapter provides a foundational study, using a time series clustering technique, followed by purity analysis, to demonstrate how many plants with the same genotype share the same temporal behavior in the formation of the clusters. In addition, the chapter provides a discussion on an eigenvalue-based analysis of the phenotypes, extracted from plant image sequences to demonstrate the temporal variation of phenotypes regulated by genotypes. Time series prediction, a dynamic research field which has attracted the attention of researchers over the past few decades, can be defined as predicting the future based on analyzing the past (Raicharoen et al. 2003). This chapter also provides discussion on time series prediction in the context of image-based plant phenotyping analysis.

The rest of the chapter is organized as follows. Section 8.2 provides descriptions of the phenotypes used in this study. Next, in Section 8.3, basic concepts of time series are described, and a time series analysis algorithm, based on clustering and angular histograms for application in plant phenotyping, is discussed in Section 8.4. Section 8.5 provides a discussion on time series prediction, using neural networks, along with the most widely used prediction performance measures. Section 8.6 introduces time series analysis based on eigenvalues. The experimental analyses are presented in Section 8.7 on a real phenotypic dataset. Finally, Section 8.8 concludes the chapter.

8.2 STRUCTURAL PHENOTYPES

Image-based plant phenotypes are broadly categorized as either structural or physiological (Das Choudhury et al. 2018). The structural phenotypes characterize the morphological attributes of the plants, whereas physiological phenotypes refer to the traits related to functional processes in plants, that regulate growth and metabolism.

The structural and physiological phenotypes can be computed by considering the plant as either a single object (holistic phenotypes) or as individual parts of the plant, e.g., stem, leaf, fruit, and flower (component phenotypes) (Das Choudhury et al. 2018). Structural holistic phenotypes are usually the measurements of the geometric shape attributes of the plant, e.g., the height of the bounding rectangle characterizes plant height, the width of the minimum enclosing bounding circle at the top view characterizes plant width, and the area of the convex-hull provides information about the spread of the plant. The plant's overall temperature and chlorophyll content are two important examples of holistic physiological phenotypes.

Recent image-based component plant phenotyping methods have mainly considered Arabidopsis (*Arabidopsis thaliana*), tobacco (*Nicotiana tabacum*), or economically important cereal crops, e.g., rice, maize, sorghum, as model plants for breeding studies aimed at developing higher-yielding or stress-tolerant cultivars. Common image-based analytical methods include leaf segmentation using 3-dimensional histogram cubes and superpixels (Scharr et al. 2016), time series analysis of stem angle to account for stem lodging rate (Das Choudhury et al. 2017), automated rice panicle counting using artificial neural networks (Duan et al. 2015), tracking the growth of the ear and silks of maize plants (Brichet et al. 2017), computation of morphological characteristics of tassels, e.g., tassel weight, tassel length, spike length, branch number, curvature, compactness, fractal dimension, and skeleton length (Gage et al. 2017), and fruit quality determination using hyperspectral image analysis (Lu et al. 2017).

In this chapter, we have used four structural phenotypes, namely stem angle, bi-angular convex-hull area ratio, plant aspect ratio, and plant areal density for experimental analysis. The definitions of the four phenotypes and their significance to plant science are also discussed. The last three holistic phenotypes are the ratios of two parameters with the same units, and hence, are scale-invariant.

8.2.1 Stem Angle

We define the stem axis as the straight line formed by the linear regression curve fitting of all junctions and the base of a stem. The slope of the stem axis is computed by

$$m = \frac{y_1 - y_2}{x_1 - x_2},$$

where (x_1,y_1) and (x_2,y_2) are the coordinates of two points, which are the perpendicular projections of the two adjacent junctions on the stem axis. The stem angle (φ) is defined as the angle between the stem axis and the vertical axis using

$$\varphi = \tan^{-1}(m).$$

Stem angle is a measure of a plant's susceptibility to lodging, i.e., the displacement of the stem axis from the vertical, and is primarily caused by the presence of excessive moisture or nitrogen in the soil, high planting density, or certain fungal diseases (Nielsen and Colville 1986).

8.2.2 Bi-Angular Convex-Hull Area Ratio

The bi-angular convex-hull area ratio ($BA_{CH}R$) is defined as

$$BA_{CH}R = \frac{Area_{CH} \text{ at side view } 0^\circ}{Area_{CH} \text{ at side view } 90^\circ},$$

where $Area_{CH}$ is the area of the convex-hull.

The bi-angular convex-hull area ratio provides information on the change in phyllotaxy, i.e., the arrangement of leaves around the stem, to optimize light interception.

8.2.3 Plant Aspect Ratio

The plant aspect ratio (PAR) is defined as

$$PAR = \frac{Height_{BR} \text{ in side view}}{Diameter_{MEC} \text{ in top view}}$$

where $Height_{BR}$ and $Diameter_{MEC}$ denote the height of the bounding rectangle (BR) of the plant in side view at 0° and the diameter of the minimum enclosing circle (MEC) of the plant in the top view, respectively. PAR is useful to distinguish between genotypes with narrow *versus* wide leaf extent when plant height is controlled. It also provides information on canopy architecture.

8.2.4 Plant Areal Density

The plant areal density *(PAD)* is defined as

$$PAD = \frac{Plant_{Tpx} \text{ at a side view}}{Area_{CH} \text{ at the same view}}$$

where $Plant_{Tpx}$ denotes the total number of plant pixels. PAD is a measure of plant biomass and is also used as a guide for planting density.

Figure 8.1 shows the image-processing pipeline to compute these phenotypes. Each original image of the input sequence is segmented to extract the foreground, i.e., the plant. The extracted foreground is then binarized. The binary image of the

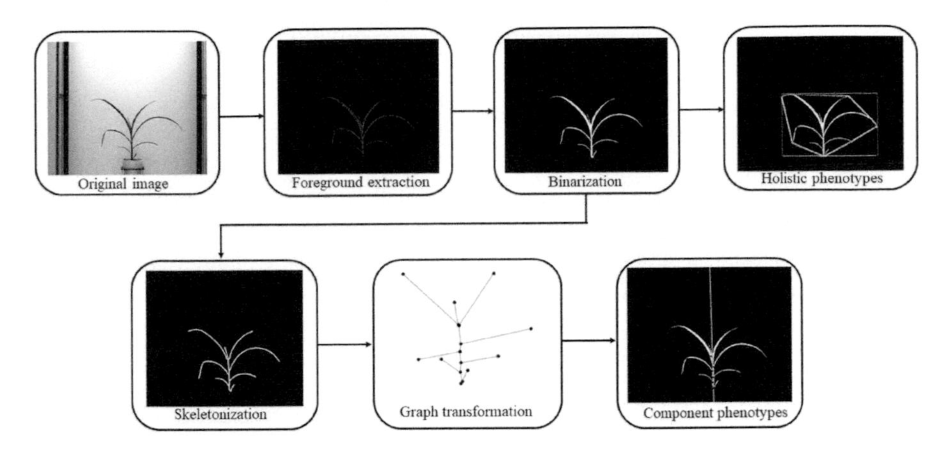

FIGURE 8.1 The image-processing pipeline for computing holistic and component phenotypes.

plant is then characterized by the geometrical shapes, i.e., bounding rectangle, convex-hull, to compute the holistic phenotypes. The computation of component phenotypes essentially requires the separation of individual components of the plants, i.e., leaves and stem. The binary image of the plant is first skeletonized and then transformed into a graphical representation to separate its components for computing the stem angle used in the study (Das Choudhury et al. 2017; 2018). Please refer to Chapter 5 for detailed discussion of this image-processing pipeline.

8.3 BASIC CONCEPTS OF TIME SERIES MODELING

In this section, we discuss some basic concepts of time series modeling as applicable in image-based high-throughput plant phenotyping analysis.

8.3.1 Definition of a Time Series

A time series is an ordered sequence of values of a variable, typically measured over successive times. It is mathematically defined by a vector $v(t)$, where $t=0,1,2,\ldots$ represents different timestamps. The variable $v(t)$ is treated as a random variable, and its measurements are arranged in chronological order. A time series containing records of a single variable is called univariate, whereas, if records of multiple variables are considered, it is termed a multivariate time series.

8.3.2 Types of a Time Series

Based on the values in the time series, the variables can be classified as being either continuous or discrete. In the case of a continuous time series, observations are measured continuously over time, e.g., readings of temperature, the flow of a river, etc. On the other hand, a discrete time series is characterized by the recordings at typically equally spaced time intervals, e.g., daily, weekly, or yearly. Examples of discrete time series include measurement of the population of a particular city and the net production of a business.

In high-throughput plant phenotyping systems, images of plants are captured by cameras in different imaging modalities, e.g., visible, infrared, near infrared, fluorescent, and hyperspectral light, at regular intervals (daily or every alternate day) during a plant's life cycle. Thus, the phenotypes computed, based on analyzing the image sequence of the plant, can be treated as a discrete time series. Figure 8.2 shows an example of a sequence of images of a sunflower plant, captured every alternate day in the Lemnatec Scanalyzer 3-D high-throughput plant phenotyping facility at the University of Nebraska-Lincoln (UNL), using the visible light camera.

8.3.3 Characteristic Features of a Phenotypic Time Series

A phenotypic time series is generally affected by four main components, namely *trend*, *recovery*, *seasonal* and *irregular*. Each is briefly described below.

- *Trend*: The tendency of a phenotypic time series to increase, decrease, or stagnate over time is referred to as a *trend*. For example, the total leaf area of the plant increases with time under normal growth conditions; however, it often starts to decrease as the leaves experience curling or shedding due to the application of any kind of stress, e.g., drought, heat, or salinity. Note that, for many cereal crops, e.g., maize, the height increases monotonically with time before it reaches a stagnant condition on completion of the vegetative stage.
- *Recovery*: In high-throughput plant phenotyping, the speed of recovery from a stressed condition (if the stress is below a threshold) is often studied,

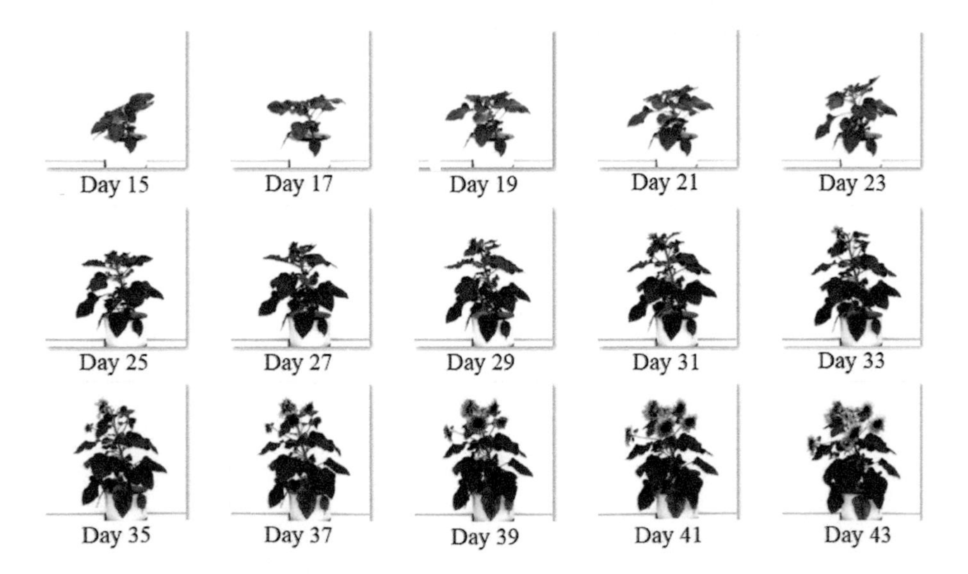

FIGURE 8.2 A sequence of images of a sunflower plant captured at regular time intervals.

e.g., re-watering of a drought-stressed plant. The recovery rate may be used as an important phenotype.

- *Seasonal*: Plants can sense the change of seasons, and respond to seasonal changes in various ways, e.g., leaves changing colors, leaves being shed, flowers blooming, and new leaves being produced.
- *Irregular*: This type of variation in a phenotypic time series is caused by unexpected incidents (e.g., earthquake, storm, or flood) that do not follow any particular pattern but which do impact on plant growth.

A time series is usually represented by a graph to visualize the underlying pattern of the data, where the observations are plotted against the time at which they are recorded. Figure 8.3 shows the time series graphs for the height of the bounding rectangle enclosing the plant (a potential measure of the plant's height) and the diameter of the minimum enclosing circle (a potential measure of the plant's width), computed by analyzing side view and top view images, respectively. The images correspond to a sample maize plant from the Panicoid-Phenomap-1 dataset, starting from Day 4 to Day 26 (both days inclusive). Since the height and the width of the plant increase monotonically with time, these variables are examples of time series with *trend* characteristics. The figure shows that the width of the plant has a faster growth rate than the plant's height after certain days.

Therefore, the variation of phenotypes as a function of time may encode significant genetic influences, along with the responses to environmental conditions. In this chapter, we present a foundation of several types of time series analyses of holistic and component phenotypes, using stem angle, $BA_{CH}R$, PAR, and PAD as references.

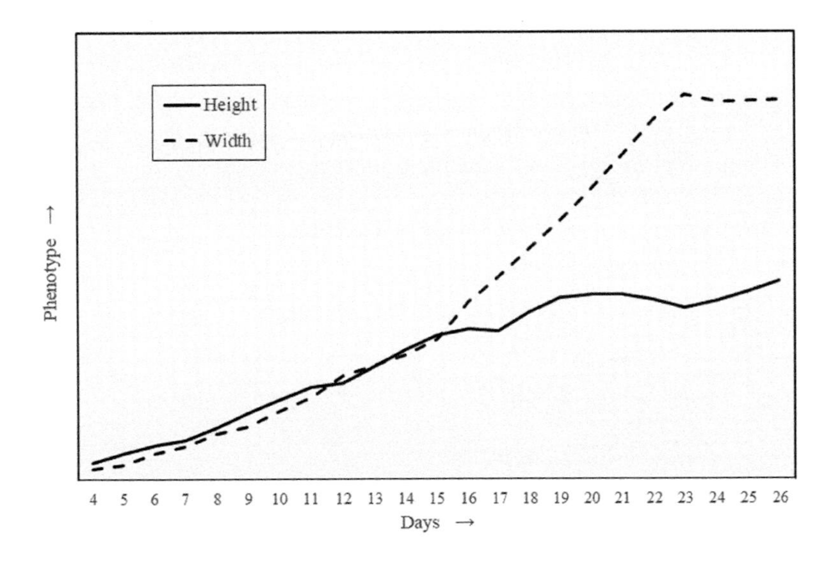

FIGURE 8.3 The height and width of a sample maize plant from Panicoid Phenomap-1 dataset measured on consecutive days.

8.4 TIME SERIES ANALYSIS FOR PHENOTYPE–GENOTYPE MAPPING

Time series analysis has the potential to link phenotype to genotype expression. An algorithm, based on time series analysis (TSA) for the understanding of genetic regulation of phenotypes under a specific environmental condition, is provided in Algorithm 1. The algorithm accepts image sequences of plants grouped into different genotypes as the input. The phenotypes are computed, based on analyzing the image sequences. These phenotypes are used for the computation of time series clustering, cluster purity, and angular histogram. Silhouette width (Rousseeuw 1987) is used to determine the optimal number of clusters, and the Pearson correlation coefficient is used as the similarity measure (Bolshakova and Azuaje 2003). Purity is an external validity measure of cluster quality (Tan et al. 2018), which we have extended here to measure genotype homogeneity. Purity has a value ranging from 0.0 to 1.0. A value of 1.0 implies that all the plants of a particular genotype display a similar temporal pattern of the phenotype. The behavioral patterns of clusters are further investigated, using angular histograms of representative plants of each cluster. The detailed explanation of these concepts, with graphical illustrations, is provided in Section 8.5.

ALGORITHM 1 TSA: TIME SERIES ANALYSIS

Input: A set of plants $P = \{P_1, P_2, \ldots, P_l\}$, where l is the total number of plants. where $P_i = \{p_{i,1}, p_{i,2}, \ldots, p_{i,m}\}$ for $1 \le i \le l$ and m is the number of images per plant.
A set of genotypes $G = \{g_1, g_2, \ldots, g_n\}$, where $n \le l$ is the total number of genotypes.
A set of labels $B = \{b_1, b_2, \ldots, b_l\}$, where $b_i \in G$ is the genotype for P_i.

Output: A set of plant clusters, cluster purity values, and angular histograms.

//Compute phenotype for the plants from their images
for each $P_i \in P$ **do**//each plant
  for each $p_{i,j} \in P_i$ **do**//each image in a sequence
    $t_{i,j} = computePhenotype(p_{i,j})$
  end for
end for

$$T = \{t_{i,j}\}, \quad 1 \le i \le l, \ 1 \le j \le m$$

//Time series cluster analysis
  $C = cluster(P, T)$ where $C = \{C_1, C_2, \ldots, C_k\}$ and k=optimal number of clusters
//Cluster purity analysis

for $C_i \in C$ **do**

$cp_i = computeClusterPurity\left(C_i, T, G, B\right)$

end for

$$CP = \{cp_i\},\quad 1 \le i \le k$$

//Angular histogram analysis

for $P_i \in P$ **do**

$ah_i = computeAngularHistogram\left(P_i, T\right)$

end for

$$AH = \{ah_i\},\quad 1 \le i \le l$$

return ([*C*,*CP*,*AH*])

8.5 TIME SERIES ANALYSIS FOR PREDICTING PHENOTYPES

One of the objectives of time series analysis is to develop a model to describe the inherent structure of the data, based on analyzing the measurements for applications in numerous practical fields, such as business, economics, finance, science, and engineering, and then use the model to obtain predictions. In the context of plant phenotyping, time series analysis can be used to predict the phenotypes for the missing imaging days as well as for a future time and to predict the genotype of a plant based on analyzing its phenotypic time series. The mechanical breakdown of high-throughput plant phenotyping systems might cause missing imaging days, or the intervals of imaging might be increased to allow imaging time slot for all plants in the greenhouse with a capacity for hosting a large number of plants. It is obvious that the accuracy of prediction depends on an appropriate model fitting to the underlying time series. Many different techniques for time series-based predictions have been proposed in the literature, which can be grouped into three main categories, namely stochastic (e.g., autoregressive integrated moving average (Zhang 2003)), support vector machines (SVM; e.g., least-square SVM, dynamic least-square SVM (Fan et al. 2006; Suykens and Vandewalle, 2000)), and artificial neural networks (ANN).

Here, we briefly describe the working principles of two popular ANN models, e.g., non-linear autoregressive neural network and time-lagged neural networks with prediction scenarios in the context of plant phenotyping analysis. ANN models are well-suited for phenomic prediction due to their ability to recognize the inherent structure of the input data, learn from past experiences, and then generalize results. In addition, ANN models are purely data-driven and self-adaptive in nature and do not rely on any assumption about the statistical distribution of the observations. Note that it is important to evaluate the performance of different techniques for time series prediction for a specific application. The commonly used measures to evaluate the performance of time series predictions are mean squared error, the sum of squared error, mean forecast error, mean absolute error, mean absolute percentage error, root mean squared error, normalized mean squared error, and Theil's U-statistic (Adhikari and Agrawal 2013).

8.5.1 Non-linear Autoregressive Neural Network (NARNN)

A NARNN is used to predict a value of a time series y at time t, i.e., $y(t)$, using d past values of the series, and can be represented as follows:

$$y(t) = f\left(y(t-1), y(t-2), \ldots, y(t-d)\right) + \mathcal{E}(t).$$

The most commonly used training algorithm for a NARNN is the Levenberg-Marquardt backpropagation (Hagan and Menhaj 1994). The aim of NARNNs is to approximate the function $f(\cdot)$ by optimizing network weights and neuron bias. The term $\mathcal{E}(t)$ refers to the approximation of the error. The number of hidden layers, and the number of neurons per layer are flexible. Whereas an increase in the number of neurons may add to the complexity, fewer neurons may restrict the generalization capability of the network. Figure 8.4 shows a typical architecture for a NARNN. NARNNs have the potential to be very useful in image-based high-throughput plant phenotyping analyses. They can be used to predict a phenotype for a day, based on analyzing the phenotypes in the past.

8.5.2 Time-Lagged Neural Networks (TLNN)

The input nodes of a feedforward neural network are the successive observations of the time series $y(t)$, and are represented as $y(t) = f\left(y(t-i)\right)$, $i = 1,2,\ldots p$, where p is the number of input nodes. A TLNN is a widely used variant of the feedforward neural network, in which the input nodes are the time series values at a fixed time lag. Figure 8.5 shows an example of a TLNN to work with a time series, with a seasonal

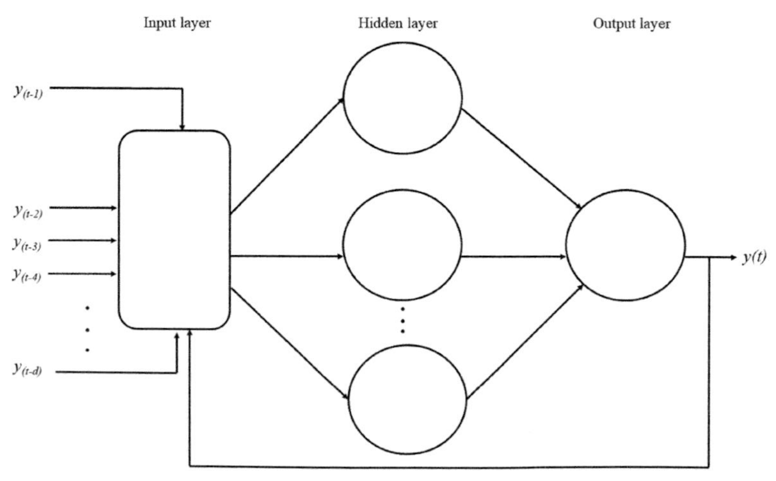

FIGURE 8.4 An architecture for a non-linear autoregressive neural network.

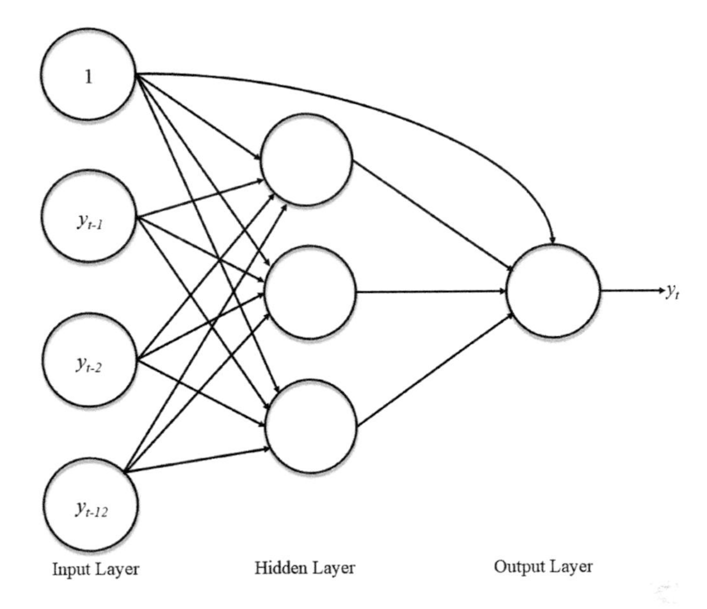

FIGURE 8.5 An architecture for a time-lagged neural network.

period of 12, containing input nodes taking the lagged values at time t–1, t–2, and t–12. The value at time t is to be predicted, using the values at lags 1, 2, and 12.

8.6 EIGENVALUE ANALYSIS

In the domains of machine learning, computer vision, and pattern recognition, the data are commonly represented in the form of a matrix. Eigenvalue analysis provides a fundamental basis with which we can analyze the variability structure of the data. The condition number is a computationally inexpensive measure by which to examine the eigenvalues and their relative strength.

The condition number is calculated as $\sqrt{\frac{\lambda_p}{\lambda_1}}$, where λ_p and λ_l represent the highest and lowest eigenvalues, respectively (Todeschini 1997). A higher condition number indicates a greater similarity, whereas a lower condition number indicates a lower similarity. The correlation matrix for each genotype for each of the three phenotypes is computed. The eigenvalues of the correlation matrix are computed, and the condition number is derived from them. Algorithm 2 summarizes the eigenvalue analysis and the computation of condition numbers.

ALGORITHM 2 EVA: EIGENVALUE ANALYSIS

Input: A set of plants $P = \{P_1, P_2, \ldots, P_l\}$, where l is the number of plants.
A set of genotypes $G = \{g_1, g_2, \ldots, g_n\}$, where $n \leq l$ is the total number of genotypes.

A set of labels $B = \{b_1, b_2, \ldots, b_l\}$, where $b_i \in G$ is the genotype for P_i.
A set of clusters $C = \{C_1, C_2, \ldots, C_k\}$ and k = number of clusters from Algorithm 1

Output: A set of condition numbers of the genotypes, CN

for $g \in G$ **do**//for each genotype
$P_g = \{P_i : P_i \in P \wedge b_i = g\}$// The set of plants in P with genotype g
$n_g = |P_g|$//The number of plants in P with genotype g
for $i = 1, \ldots, h$ **do**// h is the number of phenotypes
$Q_{g,i} = computeMatrix(P_g, T)$//$n_g \times m$ matrix
//Computation of the correlation matrix

$$CM_{g,i} = correlationMatrix(Q_{g,i}\}$$

//Computation of eigenvalues and the condition number

$$EV_{g,i} = computeEigenValues(CM_{g,i})$$

$$CN_{g,i} = computeConditionNumber(EV_{g,i})$$

end for
end for
return (CN)

8.7 EXPERIMENTAL ANALYSIS

Experimental analyses are performed on the Panicoid Phenomap-1 dataset* (Das Choudhury et al. 2016). The images of the dataset are captured by the visible light camera of the Lemnatec Scanalyzer 3-D high-throughput plant phenotyping facility at the University of Nebraska-Lincoln, USA, once daily for 27 days for two side views 90° apart (side view 0° and side view 90°), and a top view, starting at two days after emergence. Panicoid Phenomap-1 consists of images of 40 genotypes of a total number of 176 plants, including at least one representative accession from five panicoid grain crops: maize, sorghum, pearl millet, proso millet, and foxtail millet. Thus, the dataset contains $26 \times 3 = 78$ images for three views (side view 0°, side view 90°, and top view) per plant, totaling $78 \times 176 = 13728$ images. Out of 40 genotypes, the dataset contains 32 genotypes of maize plants. Table 8.1 shows the genotype names corresponding to genotype IDs used in the dataset. Detailed descriptions of the data-capturing process and dataset organization can be found in Das Choudhury et al. (2016).

8.7.1 Time Series Analysis

The genetic regulation of ontogenetical phenotypes is experimentally demonstrated, using cluster analysis, cluster purity analysis, and angular histogram analysis.

* The dataset can be freely downloaded from http://plantvision.unl.edu/.

TABLE 8.1
The Genotype Names Corresponding to the Genotype IDs Used in the Panicoid Phenomap-1 Dataset

G_{ID}	G_{name}	G_{ID}	G_{name}	G_{ID}	G_{name}	G_{ID}	G_{name}	G_{ID}	G_{name}
1	740	9	C103	17	LH82	25	PHG83	33	Yugu1
2	2369	10	CM105	18	Mo17	26	PHJ40	34	PI614815
3	A619	11	LH123HT	19	DKPB80	27	PHH82	35	PI583800
4	A632	12	LH145	20	PH207	28	PHV63	36	Purple Majesty
5	A634	13	LH162	21	DHB47	29	PHW52	37	BTx623
6	B14	14	LH195	22	PHG35	30	PHZ51	38	PI535796
7	B37	15	LH198	23	PHG39	31	W117HT	39	PI463255
8	B73	16	LH74	24	PHG47	32	Wf9	40	PI578074

A phenotype prediction analysis, based on a non-linear autoregressive neural network model, is also presented.

8.7.1.1 Cluster Analysis

We perform cluster analysis, using the stem angles of the plants for 20 days, as explained in Section 8.4. The variation of the average of the stem angles for each cluster over increasing time (days) is shown in Figure 8.6. The optimal number of

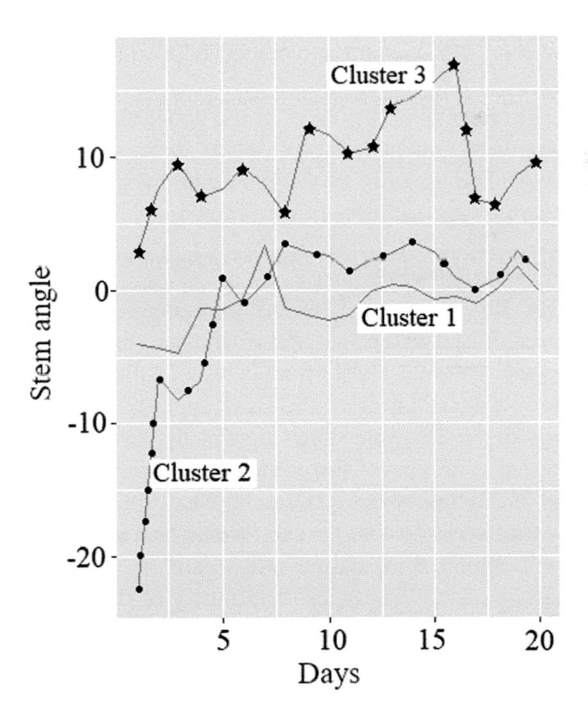

FIGURE 8.6 Time series cluster analysis of stem angles.

TABLE 8.2
Semantic Analysis of the Genotypes

Cluster	Cluster Semantics	Genotype
1	Stem angle reducing over time	{2,14,15,20,23,27,29}
2	Stem angle shows a range bound movement	{3,12,16,17,22,25,26}
3	Stem angle shows an uptrend	{1,5,6,7,10,11,13,21,24,28,30}
Unpredictable	No clear majority trend	{4,8,9,18,19,32}

clusters was determined to be 3. It is evident from the figure that the stem angle of the plants in Cluster 1 and Cluster 2 remained more-or-less range-bound, with an upward trend, although a significant difference in the values for stem angles between Cluster 1 and Cluster 2 was noted in the early days. The average value of the stem angles in Cluster 3 was much higher, compared with the other two clusters. Cluster 3 exhibited the greatest variation in the stem angle.

Table 8.2 shows the distribution of genotypes in the major clusters. The following observations are made from this table:

- *Cluster 1*: The majority of the plants of the seven genotypes included in this cluster show a trend of increasing stem angle.
- *Cluster 2*: The majority of the plants of the seven genotypes included in this cluster show a gradual upward movement, with values of stem angle lower than of the plants in Cluster 1.
- *Cluster 3*: The majority of the plants of the genotypes included in this cluster show the greatest fluctuation of the stem angles over time.
- *Unpredictable*: The majority of the plants of six genotypes in this group show that there is no conclusive trend of stem angle variation over time.

8.7.1.2 Cluster Purity Analysis

All clustering algorithms need some validity measures to evaluate the goodness of the clusters produced. Based on the availability of labels, the measures can be broadly classified as internal or external. Purity is an external validity measure of clustering.

The time series cluster analysis of the stem angle of the plants measured over the vegetative stage of the life cycle shows significant variability between the plants belonging to different genotypes. Cluster analysis reveals that there are three fundamental patterns of the time series movement. Next, the influence of the genotype on the stem angle time series is investigated. It may be assumed that plants from the same genotype will exhibit similar time series patterns. This assumption is tested with a purity analysis of the genotypes. The purity of the ith genotype, i.e., g_i, is measured as

$$p_i(g_i) = \frac{\max(n_{i,j})}{\sum n_{i,j}}$$

TABLE 8.3
Genotype-Wise Purity Analysis

Purity	Genotype
1.0	{21,22,26}
0.8	{10,11,13,14,16,24,28}
0.6	{1,2,3,5,6,7,12,15,17,20,23,25,27,29,30}
0.4–0.5	{4,8,9,18,19,32}

where p_i is the purity of the ith genotype, and $n_{i,j}$ represents the number of plants of genotype i present in cluster j. The results of purity analysis for different genotypes are presented in Table 8.3. This is compared against the purity obtained through a random assignment of the clusters. The results of the analysis are shown in Figure 8.7, using a box-plot. The figure indicates a strong genotypic influence on the time series of the stem angles.

The time series clustering analysis has the potential to provide significant information to the plant scientists. The higher the value of the purity for a particular genotype, the greater is the degree of homogeneity among the plants. It is observed that, for 25 out of the 31 genotypes, the behavior of the stem angle time series of the plants in that genotype is homogeneous. For Genotypes 21, 22, and 26, the temporal

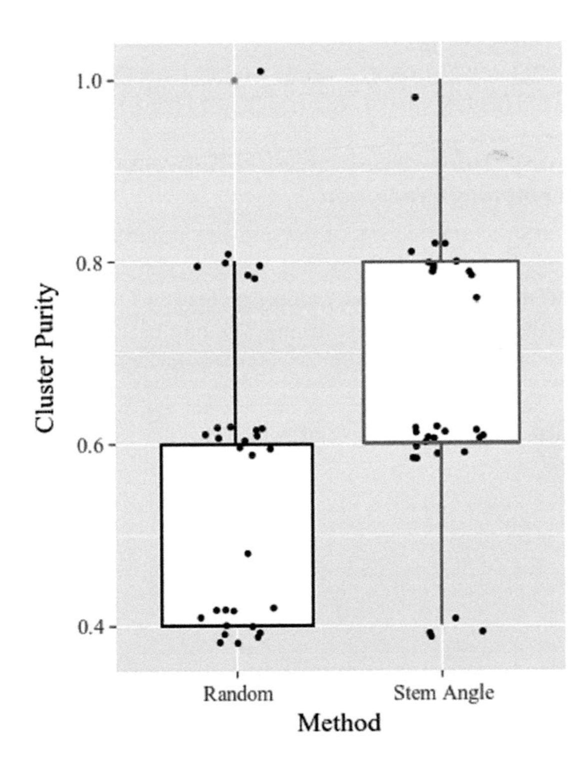

FIGURE 8.7 Analysis of genotypic influence on stem angle, using a box plot.

behavior of the stem angle is completely homogeneous, while, for other genotypes, the majority of the plants for a particular genotype show similar temporal patterns.

8.7.1.3 Angular Histogram Analysis

An angular histogram is a well-known frequency-based data visualization tool. It is useful to analyze the clusters, linear correlations, and outliers in large high-dimensional datasets, while overcoming the issues of over-plotting and clutter associated with traditional parallel coordinates (Geng et al. 2011). We use angular histograms to show the distribution of stem angles, grouped according to numeric range in equally-spaced angular bins for further analysis.

Figure 8.8 shows the angular histograms of stem angle distributions of three representative plants, one from each of the three clusters, i.e., Plant 185 from Cluster 1, Plant 74 from Cluster 2, and Plant 102 from Cluster 3. The following observations are made from the histograms:

Cluster 1 (see Figure 8.8 (left)) and Cluster 3 (see Figure 8.8 (right)) showed more balanced distributions of positive and negative stem angles, compared with Cluster 2 (see Figure 8.8 (middle)).

- In Cluster 1, the positive stem angles are proportionately higher than the negative stem angles (see Figure 8.8 (left)).
- In Cluster 2, most of the stem angles are negative with a small percentage being positive (see Figure 8.8 (middle)).
- In Cluster 3, an opposite outcome of Cluster 1 is noted. In this case, the negative stem angles are proportionately higher than the positive stem angles (see Figure 8.8 (right)).

8.7.1.4 Non-Linear Autoregressive Neural Networks for Phenotyping Prediction

Figure 8.9 illustrates the application of the non-linear autoregressive neural network analysis to predict phenotypes for missing imaging days. The heights of a sample plant from the Panicoid Phenomap-1 dataset measured for 20 consecutive days are used as the input to the network for training. The number of hidden neurons and the

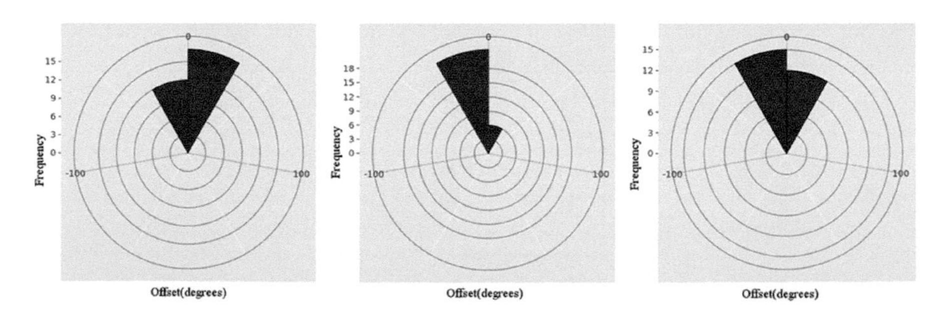

FIGURE 8.8 Angular histograms of stem angles for three representative plants, one from each cluster: (left) Plant 074 from Cluster 1; (middle) Plant 102 from Cluster 2; and (right) Plant 185 from Cluster 3.

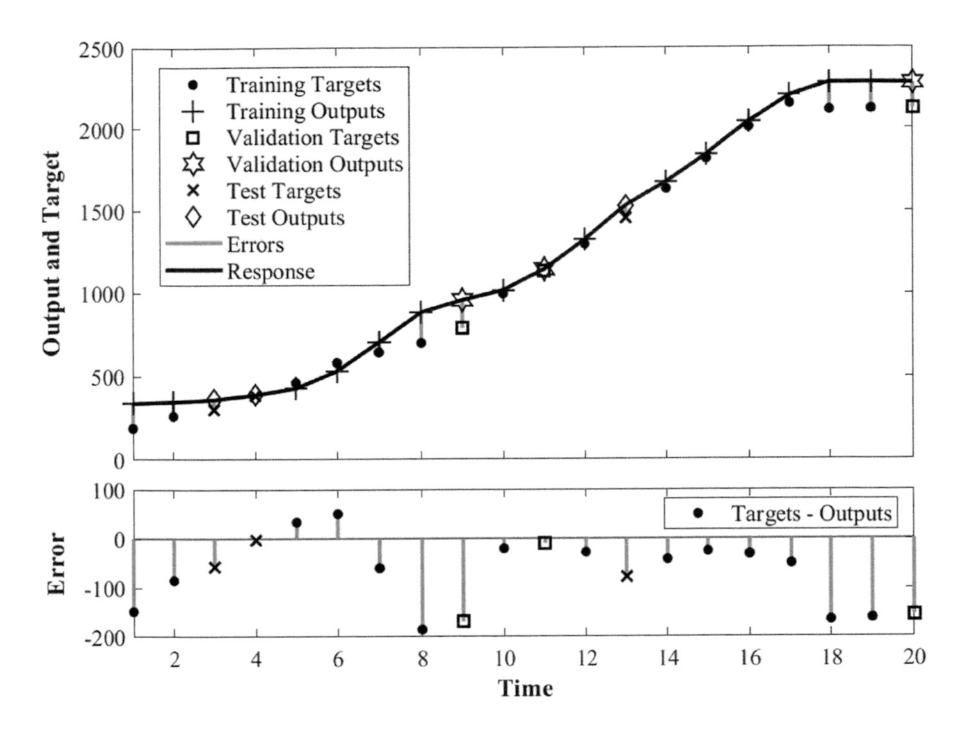

FIGURE 8.9 Illustration of the non-linear autoregressive neural network time series modeling for phenotyping prediction.

number of delays used in this example are 10 and 2, respectively. The network uses the Levenberg-Marquardt algorithm (Levenberg 1944; Marquardt 1963) for training, with 70% of the data for training, 15% for validation, and 15% for testing. In this example, the measurements used for test targets are considered to be missing imaging days, and the test outputs refer to the predicted phenotypes for those days. The figure also shows the errors computed by subtracting the outputs from targets for training and testing as well as for validation.

8.7.2 Condition Number Analysis

An analysis of the condition number is performed for all three phenotypes, namely bi-angular convex-hull area ratio, plant aspect ratio, and plant areal density, using the temporal sequence of the phenotypes. For each of the genotypes, the ratio of the maximum and minimum eigenvalues is computed for each of the phenotypes. As explained before, a higher condition number indicates greater similarity and *vice versa*. Condition numbers of all the three phenotypes are computed and represented, using a univariate box plot. The results of the condition number analysis are shown in Figure 8.10.

It can be observed that bi-angular convex-hull area ratio ($BA_{CH}R$) displays the greatest homogeneity, followed by plant areal density (PAD), and plant aspect ratio (PAR). Except for an outlier, for plant aspect ratio, the phenotypes are generally consistent for each of the genotypes.

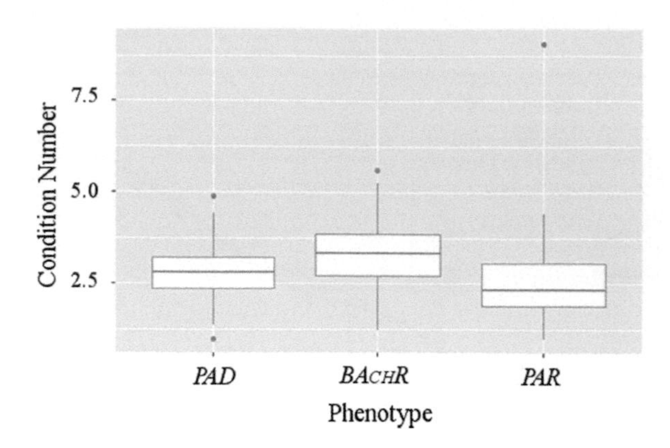

FIGURE 8.10 Condition number analysis of the holistic structural phenotypes.

8.8 CONCLUSION

This chapter provides a description of the basic concepts of time series modeling, covering definition of a time series, identification of types, and its characteristic features. It introduces two novel algorithms to study the genotypic influence on ontogenetical phenotypes, i.e., time series-based clustering and angular histogram analysis, and condition number analysis. The techniques are demonstrated using a component phenotype, namely stem angle, and three holistic phenotypes, namely bi-angular convex-hull area ratio, plant aspect ratio, and plant areal density. The chapter also provides an experimental demonstration of predicting phenotypes for missing imaging days, using height of a plant as an example. The time-series analysis presented in this chapter, can be extended to other holistic and component phenotypes to investigate their temporal variations regulated by genotypes and also to predict phenotypes for a time in future based on analyzing past observations.

The proposed time series-based clustering technique can be used to derive a set of clusters. Semantics may be attached to the derived clusters, in terms of both movement of the stem angles as well as the analysis of the angular histograms. Condition number analysis, based on eigenvalues for all the genotypes, is demonstrated using the holistic phenotypes through a univariate box plot and a scatter plot. The analyses help to compare the phenotypes in terms of genotypic similarity. For example, it is observed that the bi-angular convex-hull area ratio displays the greatest homogeneity among the phenotypes. Such analytical techniques, coupled with visualization, can lead to meaningful insights for a diverse community of researchers in the domain of plant phenotyping.

REFERENCES

Adhikari, R., and R. K. Agrawal. 2013. An introductory study on time series modeling and forecasting. arXiv preprint arXiv:1302.6613.

Bolshakova, N., and F. Azuaje. 2003. Cluster validation techniques for genome expression data. *Signal Processing* 83(4):825–833.

Brichet, N., C. Fournier, O. Turc, O. Strauss, S. Artzet, C. Pradal, C. Welcker, F. Tardieu, and L. Cabrera-Bosquet. 2017. A robot-assisted imaging pipeline for tracking the growths of maize ear and silks in a high-throughput phenotyping platform. *Plant Methods* 13:96.

Das Choudhury, S., V. Stoerger, A. Samal, J. C. Schnable, Z. Liang, and J.-G. Yu. 2016. Automated vegetative stage phenotyping analysis of maize plants using visible light images. In *KDD Workshop on Data Science for Food, Security and Water*, San Francisco, CA.

Das Choudhury, S., S. Goswami, S. Bashyam, A. Samal, and T. Awada. 2017. Automated stem angle determination for temporal plant phenotyping analysis. In *ICCV workshop on Computer Vision Problems in Plant Phenotyping* 41–50. Venice, Italy.

Das Choudhury, S., S. Bashyam, Y. Qiu, A. Samal, and T. Awada. 2018. Holistic and component plant phenotyping using temporal image sequence. *Plant Methods* 14:35.

Duan, L., C. Huang, G. Chen, L. Xiong, Q. Liu, and W. Yang. 2015. Determination of rice panicle numbers during heading by multi-angle imaging. *The Crop Journal* 3(3):211–219.

Fan, Y., P. Li, and Z. Song. 2006. Dynamic least square support vector machine. In *Proceedings of the 6th World Congress on Intelligent Control and Automation* 4886–4889. Dalian, China..

Gage, J. L., N. D. Miller, E. P. Spalding, S. M. Kaeppler, and N. de Leon. 2017. Tips: A system for automated image-based phenotyping of maize tassels. *Plant Methods* 13:21.

Geng, Z., Z. Peng, R. S. Laramee, J. C. Roberts, and R. Walker. 2011. Angular histograms: Frequency-based visualizations for large, high dimensional data. *IEEE Transactions on Visualization and Computer Graphics* 17(12):2572–2580.

Hagan, M. T., and M. B. Menhaj. 1994. Training feedforward networks with the Marquardt algorithm. *IEEE Transactions on Neural Networks* 5(6):989–993.

Levenberg, K. 1944. A method for the solution of certain non-linear problems in least squares. *Quarterly of Applied Mathematics* 2(2):164–168. doi: 10.1090/qam/10666.

Lu, Y., Y. Huang, and R. Lu. 2017. Innovative hyperspectral imaging-based techniques for quality evaluation of fruits and vegetables: A review. *Applied Sciences* 7(2):189.

Marquardt, D. W. 1963. An algorithm for least-squares estimation of nonlinear parameters. *Journal of the Society for Industrial and Applied Mathematics* 11(2):431–441. doi: 10.1137/0111030.hdl:10338.dmlcz/104299.

Montgomery, D. C., C. L. Jennings, and M. Kulahci. 2015. *Introduction to Time Series Analysis and Forecasting*. Wiley Series in Probability and Statistics. Wiley. Hoboken, NJ.

Nielsen, B., and D. Colville. 1986. *Stalk Lodging in Corn: Guidelines for Preventive Management*. AY-Purdue University Cooperative Extension Service, West Lafayette, IN.

Rahaman, M. M., D. Chen, Z. Gillani, C. Klukas, and M. Chen. 2015. Advanced phenotyping and phenotype data analysis for the study of plant growth and development. *Frontiers in Plant Science* 6:619.

Raicharoen, T., C. Lursinsap, and P. Sanguanbhoki. 2003. Application of critical support vector machine to time series prediction. In *Proceedings of the International Symposium on Circuits and Systems* 5:V-741–V-744. Bangkok, Thailand.

Rousseeuw, P. J. 1987. Silhouettes: A graphical aid to the interpretation and validation of cluster analysis. *Journal of Computational and Applied Mathematics* 20:53–65.

Scharr, H., M. Minervini, A. P. French et al. 2016. Leaf segmentation in plant phenotyping: A collation study. *Machine Vision and Applications* 27(4):585–606.

Suykens, J. A. K., and J. Vandewalle. 2000. Recurrent least squares support vector machines. *IEEE Transactions on Circuits and Systems-Part I: Fundamentals Theory and Applications* 47(7):1109–1114.

Tan, P.-N., M. Steinbach, A. Karpatne, and V. Kumar. 2018. *Introduction to Data Mining*. Pearson. Boston, MA.

Todeschini, R. 1997. Data correlation, number of significant principal components and shape of molecules. The K correlation index. *Analytica Chimica Acta* 348(1–3):419–430.

Zhang, G. P. 2003. Time series forecasting using a hybrid ARIMA and neural network model. *Neurocomputing* 50:159–175.

9 Data-Driven Techniques for Plant Phenotyping Using Hyperspectral Imagery

Suraj Gampa and Rubi Quiñones

CONTENTS

9.1 BACKGROUND

A high-throughput plant phenotyping system usually consists of a variety of cameras in different modalities, that are used to image the plants for computing various phenotypes. Most of them, e.g., visible light, fluorescent, and near-infrared, capture the reflected light at wide bands. For example, red-green-blue (RGB) cameras consider the visible part of the spectrum (380 –700 nm) in only three bands (red, green, and blue). Whereas this is adequate for computing many structural and appearance-based phenotypes, this coarse spectral resolution is not sufficient for more subtle phenotypes that may be accessible only at narrow wavelength ranges. Hyperspectral imaging (HSI) addresses this problem by capturing a broad range of wavelengths at very narrow intervals. Hyperspectral cameras can typically capture a scene in hundreds of bands where each band has a resolution of a few nanometers. Figure 9.1 shows

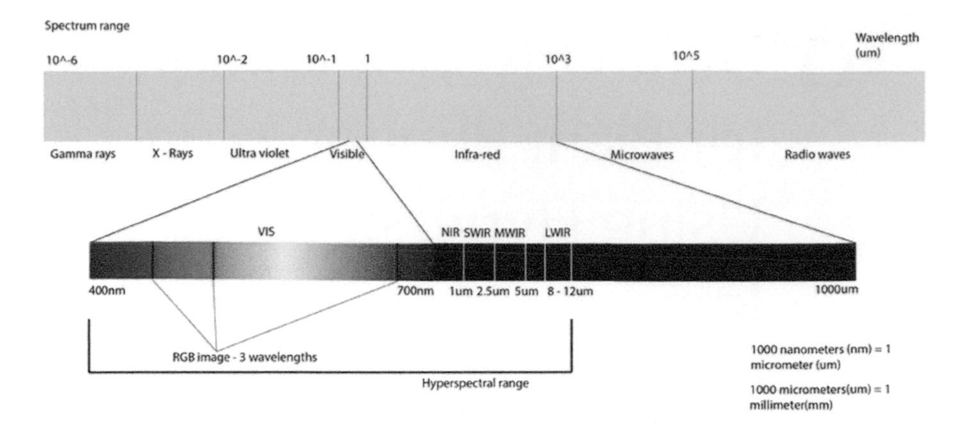

FIGURE 9.1 Overview of the different wavelengths captured with different types of cameras. (Reproduced from Lowe et al. 2017 under Creative Commons Attribution 4.0 International License (http://creativecommons.org/licenses/by/4.0/)).

the wavelengths captured by cameras in different modalities for image-processing applications. It is evident from Figure 9.1 that hyperspectral cameras have the highest coverage of the electromagnetic spectrum (Lowe et al. 2017). A multispectral camera also captures images in narrow wavelength ranges. However, it has only a small number of bands (typically <20), in contrast to those in hyperspectral images.

A hyperspectral image can be represented as a three-dimensional array of intensities, $H(x,y,z)$, where (x,y) represents the location of a pixel at the wavelength λ. This is often called a hyperspectral cube (Figure 9.2 (left)). Each wavelength is represented as a two-dimensional image, and intensity information at a specific location is represented by a spectral reflectance curve (Figure 9.2 (right)).

HSI has been used to detect abiotic (Romer et al. 2012) and biotic stresses (Mahlein et al. 2012) in plants. Measurements from HSI can be used to detect properties such as chlorophyll content, canopy senescence, and water content. State-of-the-art methods have shown that HSI has the potential to determine fruit firmness (Lu 2004; Peng and Lu 2005), detect yellow rust in wheat (Bravo et al. 2003), identify citrus canker in grapefruit (Qin et al. 2008) and detect "early blight" disease in tomato plants (Apan et al. 2005). Mishra et al. (2007) identified the most critical

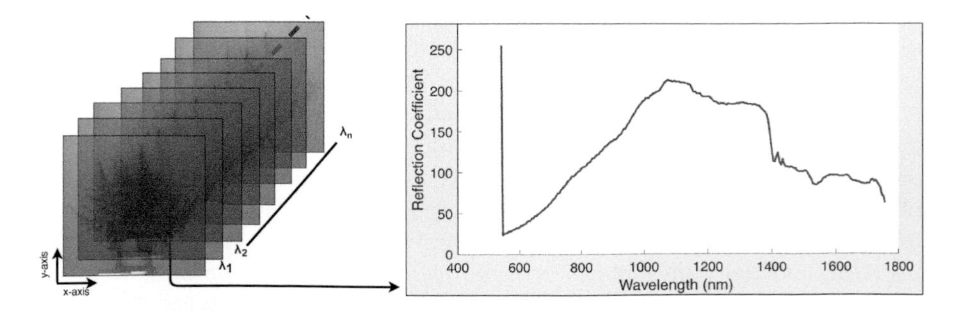

FIGURE 9.2 A hyperspectral cube and a spectral reflectance curve.

wavelengths, using spectral derivative analysis and spectral ratio analysis, to detect citrus greening disease (Huanglongbing) in citrus fruits. Ge et al. (2016) used HSI to analyze the temporal growth, water-use efficiency, and leaf water content of maize plants, whereas Rock et al. (1988) found that the more a leaf is stressed, the more the reflectance curve shifts down toward the red region. This phenomenon is referred to as the "blue shift" of the red edge. In summary, the state-of-the-art research strongly supports the observation that HSI has excellent potential in plant phenotyping in detecting abiotic or biotic stresses in the plants. Efficient algorithms have been developed to compute different types of indices from specific wavelengths of hyperspectral images, that can serve as effective phenotypes, e.g., stress indices, emergence indicators, water content, and biochemical composition (Yendrek et al. 2017).

Despite the advantages, HSI poses challenges to plant phenotyping applications, especially in high-throughput systems. The time required to capture hyperspectral images is significantly higher than that of traditional RGB and fluorescent images. The use of HSI cameras in closed chambers to detect plant diseases or other biotic stresses has not been widely explored due to contamination concerns. Hyperspectral cameras are also more expensive than the more widely used RGB cameras. HSI datasets are significantly larger than the datasets consisting of images in other modalities. For example, a single hyperspectral image with 1000×1000 spatial resolution, 200 bands with 8 bits per pixel results in an image size of 200 Mbytes. Therefore, an experiment with 100 plants imaged daily for 30 days will result in a dataset that is over half a terabyte. Storage, access, and processing data of this magnitude require significant computational resources. Besides the cost of the HSI systems, the lack of standard datasets that can be used for algorithm development exacerbates the problem.

In this chapter, we provide an in-depth discussion of the data-driven techniques that are useful to analyze hyperspectral image sequences for plant phenotyping. The rest of the chapter is organized as follows. Section 9.2 introduces a graphical user interface (GUI)-based HypeRPheno toolbox to assist information extraction from HSI, using machine learning techniques. Section 9.3 describes a dataset that consists of hyperspectral image sequences of tobacco plants. This dataset is used to illustrate three core data-driven techniques widely applicable for plant phenotyping, namely dimensionality reduction, clustering, and classification; these techniques are described in Sections 9.4, 9.5, and 9.6, respectively. Finally, Section 9.7 concludes the chapter and provides directions for future work.

9.2 HYPERPHENO TOOLBOX

A toolbox called HypeRpheno[*] was created to analyze hyperspectral images by using a collection of algorithms that uses the parallel computation abilities in MATLAB to handle high-dimensional computation. The HypeRpheno Toolbox has a GUI and provides many functionalities for image exploration, information extraction, and

[*] HypeRpheno has been developed as a part of the Plant Vision Initiative at the University of Nebraska-Lincoln (http://plantvision.unl.edu) that seeks to provide public, dynamic, and reliable image datasets and software tools.

supervised and unsupervised machine learning to support image-based plant phenotyping. The GUI and the main features of HypeRpheno are briefly described next.

- *Top-level menu*: In the main menu, the user can select the HSI image dataset by providing its location in the file system. It is assumed that the hypercube is stored in a directory organized into a separate image for each band (wavelength range). Once the images are loaded, the user can dynamically display images for specific bands or wavelengths. The toolbox allows multiple hyperspectral images to be loaded and analyzed simultaneously.
- *Spectral reflectance coefficient analysis*: HypeRpheno has the capability of displaying the reflectance values at all wavelengths (i.e., spectral curve) at a single pixel in the image or at multiple pixels simultaneously. It can also display the average spectral curve for a collection of pixels in the image selected by the user.
- *Clustering*: HypeRpheno provides unsupervised learning in the form of clustering of the spectral curves. The user can select from a variety of clustering algorithms including *k*-means, hierarchical, and density-based clustering. Users are prompted to enter the parameters for the chosen algorithm using the GUI. HypeRpheno can then display the results of clustering on top of the image. HypeRpheno can also perform clustering validations, using the silhouette measure or the elbow method.
- *Classification*: HypeRpheno allows the user to choose from a set of widely used classification algorithms, including support vector machines and decision trees, to conduct a more extensive analysis of the hyperspectral image sequences of the plants.

Figure 9.3 shows a preview of some of the core functions of HypeRpheno GUI. It shows the image with band 122, with the wavelength of around 1150 nm, and the average spectral curve of a window (shown in the image) at a selected point. Figure 9.4 shows the three clusters generated by *k*-means clustering.

9.3 DATASET

A dataset consisting of hyperspectral images of two tobacco plants is used for the development and illustration of the algorithms described in this section. The images are captured by the hyperspectral camera of the LemnaTec Scanalyzer 3D High-Throughput Plant Phenotyping System at the University of Nebraska-Lincoln (UNL). Both the plants were imaged for 33 days after emergence. Drought stress was induced in one plant 15 days after emergence, and the other plant underwent normal growth conditions, with adequate water supply. The hyperspectral image consists of 244 bands with a wavelength range of 545 nm to 1760 nm. The spatial resolution of the camera is 320×595. Figure 9.5 shows the time sequence of images (at 545 nm) of an unstressed plant, and Figure 9.6 shows the image sequence of a stressed plant. Figure 9.6 demonstrates the impact of drought stress on a plant over time.

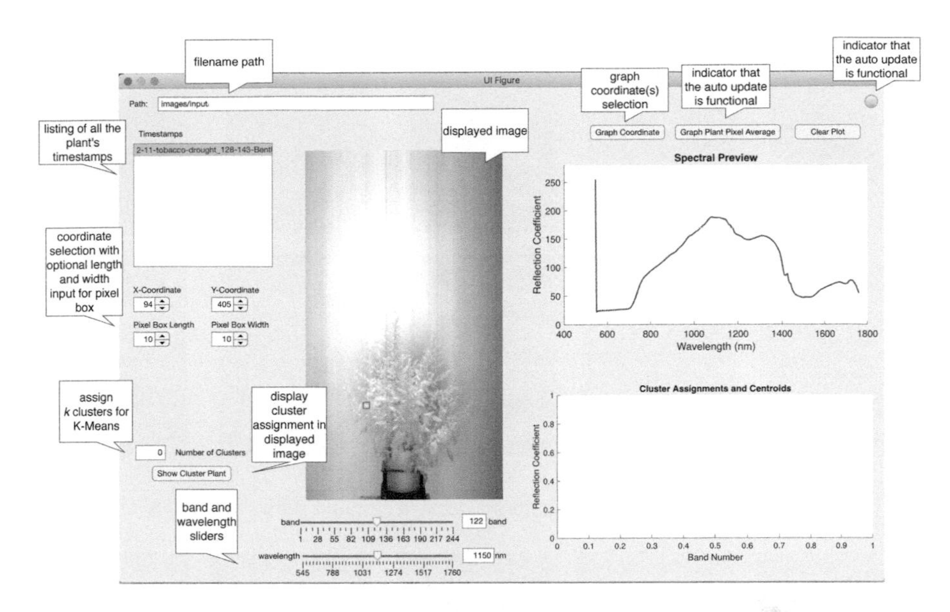

FIGURE 9.3 Sample HypeRpheno window demonstrating some basic I/O (Input/Output) and image exploration functions.

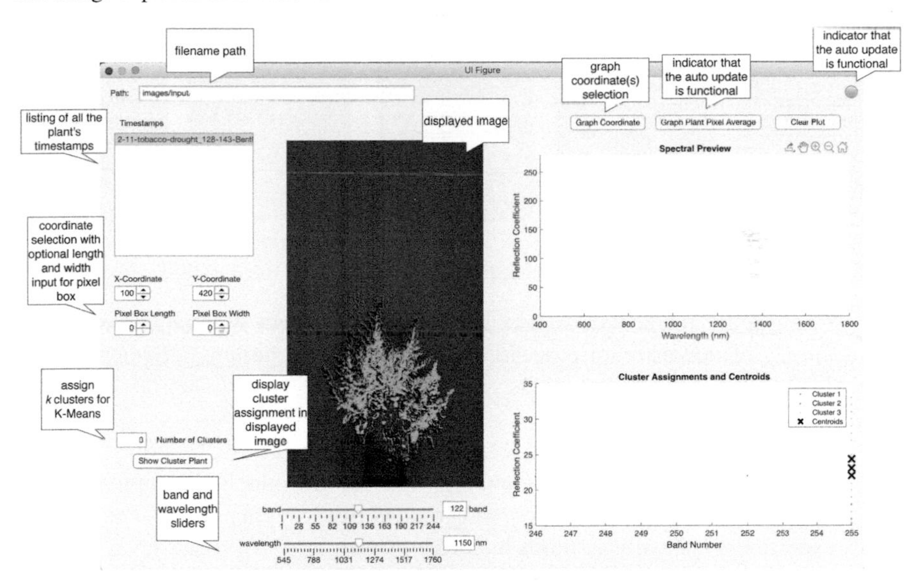

FIGURE 9.4 Sample HypeRpheno window demonstrating the clustering function.

9.4 DIMENSIONALITY REDUCTION TECHNIQUES

Dimensionality reduction is the process of decreasing the number of variables used in downstream analysis. It is critical in hyperspectral image analysis because the information is captured in narrow wavelengths in hundreds of bands. Furthermore, since adjacent bands also have similar wavelengths, the corresponding images are

FIGURE 9.5 Images of the plant under normal growth conditions for the duration of the study.

FIGURE 9.6 Images of the plant under drought conditions for the duration of the study.

likely to be highly correlated. Reducing the number of bands without compromising the accuracy of downstream processing is a significant challenge. Dimensionality reduction has two important benefits:

- *Data compression*: Reducing the number of dimensions helps reduce memory requirements during computing, as well as reducing computational costs. This is even more important in plant phenotyping analysis since an experiment typically contains hundreds of images.
- *Data visualization*: Visual exploration of high-dimensional data is a challenging task, and reducing the number of dimensions makes this task easier. Visualization helps in improving the understanding of data, which is important for developing efficient algorithms for phenotyping applications.

9.4.1 Principal Component Analysis

Principal component analysis (PCA) is the most commonly used dimensionality reduction technique in HSI-based phenotyping analysis. The problem of high dimensionality can be mitigated using techniques that can convert a high-dimensional dataset into a low-dimensional one with minimal information loss. PCA is one such data transformation technique that calculates a new set of features called principal components.

Principal components are new variables that are linear combinations of the original variables. The new variables, i.e., principal components, are computed in a manner where they are uncorrelated, and each principal component carries a certain amount of variance present in the dataset (Craig and Shan 2002). The first principal component typically carries a large amount of the variance in the data, and the variance progressively decreases for the subsequent principal components.

In HSI analysis, PCA is a useful preprocessing step for a variety of tasks, including clustering and classification. The coefficients of the variables in the principal components provide information about their significance. They can be used to identify the spectral bands that are more significant for a given phenotyping application. PCA has been shown to reduce the processing time for classification to one-fifth of the original time, while maintaining a classification rate of 80% or higher (Craig and Shan 2002).

A scatter plot of the first and second principal components visually illustrates the general orientation of the data points and their linear separability. Figure 9.7 (right) shows the scatter plot of the first two principal components of the spectral curves for the stressed plant shown in Figure 9.7 (left). It is clear that the first two components explain most of the variability in the spectral curves. Thus, the 244-dimensional vector of attributes for each pixel (note that there are 244 bands in our HSI dataset) can be replaced by a two-dimensional vector, consisting of only the first two principal components, without significant loss of information.

9.5 CLUSTERING

Clustering is an unsupervised classification technique for grouping similar data points. In HSI image analysis, a data point is a spectral curve at a given pixel of

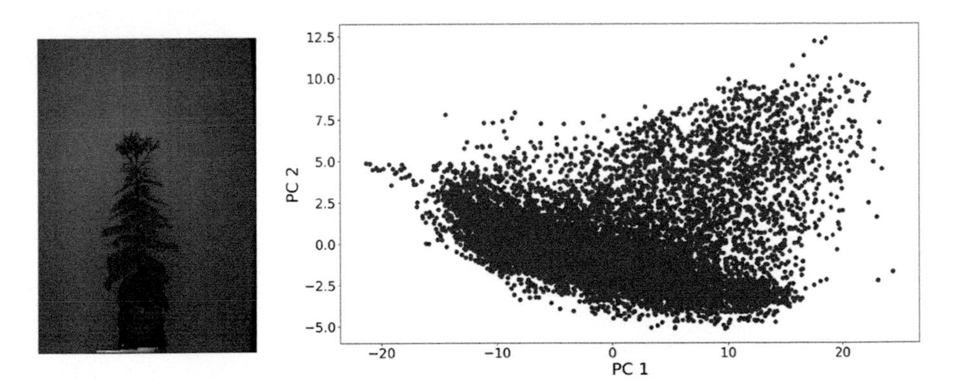

FIGURE 9.7 Illustration of Principal Component Analysis of spectral curves: (left) a sample hyperspectral image of a stressed tobacco plant at a specific wavelength; and (right) scatter plot of the first two principal components of the spectral curves for the plant.

a plant image. In close-range plant phenotyping analysis based on HSI, clustering plays an essential role in identifying differential responses of a plant to biotic and abiotic stresses. Upon clustering, differential responses will probably manifest as different clusters. Some potential applications of clustering include the identification of parts of a plant affected by a particular disease, quantification of the impact of nutrient deficiencies at different severity levels, and detection of parts of a plant affected by different kinds of stress. In this section, we will discuss the most widely used clustering algorithms with experimental demonstrations. We will also examine techniques that will help interpret, validate, and improve cluster analysis results.

9.5.1 *K*-Means Clustering

This is a clustering technique that is used to partition a set of data points into a prespecified number (k) of clusters that minimizes the sum of distances from the data points to their cluster center. A value of k is based on domain knowledge or is determined empirically. In hyperspectral image classification, k-means has been used for feature extraction and data labeling. The steps of the k-means algorithm are given below:

> Select k points as the initial cluster centers
> *repeat*
>   Revise the k clusters by reassigning all points to the closest cluster center
>   Recompute the cluster center for each cluster
> *until* the convergence criterion is satisfied

The k-means algorithm has been widely used in many applications. The main limitation of this algorithm is that its distance function results in the generation of spherical (hyperspherical in higher dimensions) clusters. Figure 9.8 (left) shows the

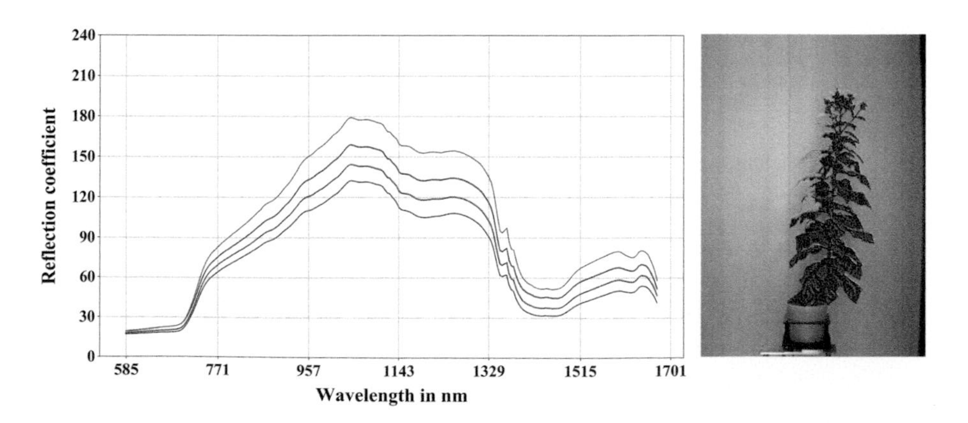

FIGURE 9.8 Mean spectral curves of the clusters generated using the k-means algorithm of the unstressed plant (left) and the constituent pixels of the clusters shown on the plant (right).

mean spectral curves of the clusters and Figure 9.8 (right) shows the constituent pixels of the clusters on the plant for visualization of the plant regions with different physiological characteristics. The number of clusters was determined by the elbow method (Ketchen and Shook 1996). Figure 9.8 shows that most of the leaves in the top part of the plant are in two distinct clusters, represented by the top two mean spectral curves. The other two clusters (represented by the bottom two curves) consist of stem and most of the leaves towards the bottom of the plant.

9.5.2 Hierarchical Agglomerative Clustering (HAC)

Hierarchical clustering builds a hierarchy of clusters: it starts either by treating each data point as a single cluster and sequentially merging similar clusters (*agglomerative* clustering) or initially grouping all the data points into one cluster and successively splitting them into disjoint clusters (*divisive* clustering). Divisive clustering is less commonly used in practice, hence we only focus here on agglomerative clustering. The HAC algorithm is outlined below.

```
Assign each point to be an initial cluster
repeat
        Identify the two clusters that are closest to each other
        Merge the two clusters
until all points are merged into a single cluster
```

The two closest clusters are identified, based on a distance function. Many distance functions, including Euclidean, Manhattan, and Cosine, have been proposed in the literature. These functions define the distance between two data points. Several techniques to determine the closest clusters have been proposed and are briefly summarized below.

- *Complete linkage*: The distance between the farthest points in a pair of clusters, d_{max}, is taken as a measure to merge the clusters. Complete linkage generates compact clusters but does not consider all points in the cluster during the merger process. The distance between two clusters C_i and C_j is given by:

$$d_{\max}\left(C_i, C_j\right) = \underset{p_i \in C_i, p_j \in C_j}{\text{Max}} \text{distance}\left(p_i, p_j\right).$$

- *Single linkage*: The distance between the nearest points in a pair of clusters, d_{min}, is taken as a measure to merge the clusters. This approach can generate non-elliptically shaped clusters but is sensitive to noise and outliers. The distance between two clusters, C_i and C_j, is given by:

$$d_{\min}\left(C_i, C_j\right) = \underset{p_i \in C_i, p_j \in C_j}{\text{Min}} \text{distance}\left(p_i, p_j\right).$$

- *Average linkage*: In this approach, the average of the distances of all pairs of points between two clusters is used to choose the two clusters to merge. The distance between two clusters, C_i and C_j, is given by:

$$d_{\text{avg}}\left(C_i, C_j\right) = \frac{\sum_{p_i \in C_i, p_j \in C_j} \text{distance}\left(p_i, p_j\right)}{\left|C_i\right| \times \left|C_j\right|}$$

where $|C_i|$ and $|C_j|$ are the size of clusters C_i and C_j, respectively. This technique is less susceptible to noise but is biased towards spherically shaped clusters.
- *Ward's Method*: This technique selects the clusters which merge to produce the smallest increase in the sum of the squared distances (Ward's distance) to the cluster centers. Ward's distance between two clusters, C_i and C_j, is given by:

$$d_{\text{Ward}}\left(C_i, C_j\right) = \sum_{p_i \in C_i} \left(p_i - r_i\right)^2 + \sum_{p_j \in C_j} \left(p_j - r_j\right)^2 - \sum_{p_{ij} \in C_{ij}} \left(p_{ij} - r_{ij}\right)^2$$

where r_i, r_j, and r_{ij} are the centroids of C_i, C_j, and C_{ij}, respectively. Ward's clustering algorithm is computationally expensive, which makes it difficult to operate on large, high-dimensional data sets (Pakhira 2014).

Figure 9.9 (left) shows the mean spectral curves of the clusters and Figure 9.9 (right) shows the constituent pixels of each cluster (denoted by the same color used to represent the spectral curves) on the plant obtained using a Ward hierarchical clustering approach. A similar observation of *k*-means clustering is made, i.e., most of the leaves in the upper part of the plant are in two distinct clusters, represented by the top two mean spectral curves, whereas the bottom two curves mostly represent the stalk and the leaves towards the lower part of the plant.

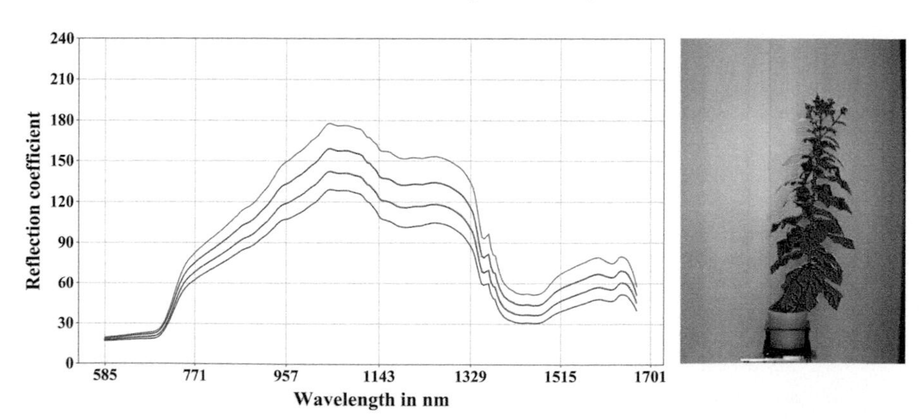

FIGURE 9.9 Mean spectral curves of the clusters generated using the Ward hierarchical clustering approach of an unstressed tobacco plant (left) and the constituent pixels of the clusters shown on the plant (right).

9.5.3 Cluster Validation

Clustering algorithms, like k-means, partition the data into k clusters, irrespective of whether they represent the actual underlying data structure. It is important to measure the quality of clustering, especially when conducting exploratory and data-driven analyses. Some widely used techniques for cluster validation are mentioned below.

Silhouette measure: It is a measure of compactness and distinctness of the clusters. It may be used either for validating or improving the cluster analysis results. Silhouette value, $s(p_i)$, of a data point, p_i, measures how similar it is to the points in the same cluster in comparison with the points in other clusters (Rousseeuw 1987), and is given by*:

$$s(p_i) = \frac{b(p_i) - a(p_i)}{\min\{a(p_i), b(p_i)\}}$$

where $a(p_i)$ is the mean distance between the point p_i and all other points in the same cluster (C_i), and $b(p_i)$ is the mean distance between the point p_i and all points not in the same cluster. The value of $s(p_i)$ ranges between -1 and 1 i.e. $-1 \le s(p_i) \le 1$. A value close to 1 implies that p_i is very well clustered and has been assigned to an appropriate cluster (Rousseeuw 1987). Conversely, a value of -1 means it better fits a neighboring cluster. A value close to 0 implies that it is on the border of two neighboring clusters. The silhouette coefficient for the overall cluster clustering (C) of a set of points (P), also known as overall average silhouette width (Rousseeuw 1987), can be calculated as follows;

$$\bar{s}(C) = \frac{1}{|P|} \sum_{p_i \in P} s(p_i)$$

Elbow Method: This is a heuristic method that examines the total error sums of squares (SSE) as a function of the number of clusters. The SSE of any cluster, C_i, can be defined as the sum of the squares of differences between each data point in C_i and the mean of the points in the cluster (c_i). It can be computed as:

$$SSE(C_i) = \sum_{p_i \in C_i} d^2(p_i - c_i)$$

where d is a distance function. Total SSE of the cluster, C, is the sum of the SSEs of individual clusters.

$$TSSE(C) = \sum_{C_i \in C} SSE(C_i)$$

* If a point p_i is in a cluster of size 1, then $s(p_i) = 0$.

Typically, as the number of clusters increases, the distance between the points to their respective cluster centers decreases, resulting in lower *SSE* values. The decrease is uniform and more pronounced when the number of clusters is small. The point when the rate of decrease significantly flattens out is considered the "elbow" point, and, at that point, the corresponding number of clusters is considered optimal. One of the problems with this approach is that sometimes the elbow point is hard to determine. Figure 9.10 illustrates the elbow method for *k*-means clustering, using the spectral curves for the normal plants for 18 days of the dataset. The figure shows that the elbow occurs when the number of clusters is four.

Dendrogram: This is a tree diagram that visually illustrates the hierarchical relationship between clusters. It is useful in identifying inter-cluster and intra-cluster relationships, using the pairwise distance metric. A dendrogram can be treated as a memory of a hierarchical clustering algorithm, i.e., it is formally created by following every single step of hierarchical clustering for a given application and can provide information on the optimal number of clusters. The working principle of the dendrogram is as follows. It starts with each data point being treated as a cluster and placed along the horizontal axis. The vertical axis represents the Euclidean distance between the clusters. The clusters closest to each other are iteratively connected by horizontal bars, where the height of the bar represents the dissimilarity between them. The greater the distance between two clusters, the greater is the dissimilarity between them. To determine the optimal number of clusters, we find the longest vertical line that does not get intersected by any extended horizontal bar across the dendrogram. A threshold height is chosen, such that a horizontal line drawn at it crosses the longest vertical line. The total number of vertical lines the horizontal threshold

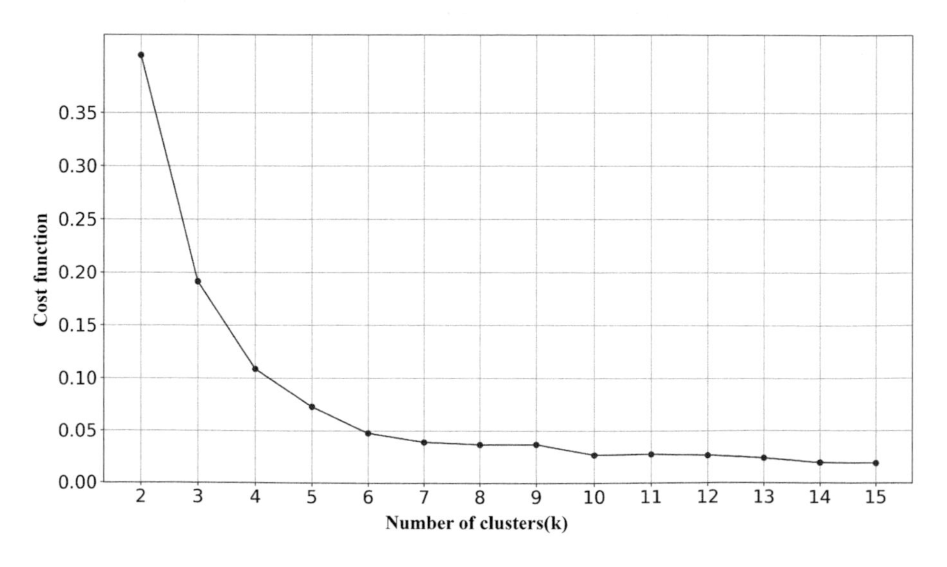

FIGURE 9.10 Illustration of the elbow method for clustering.

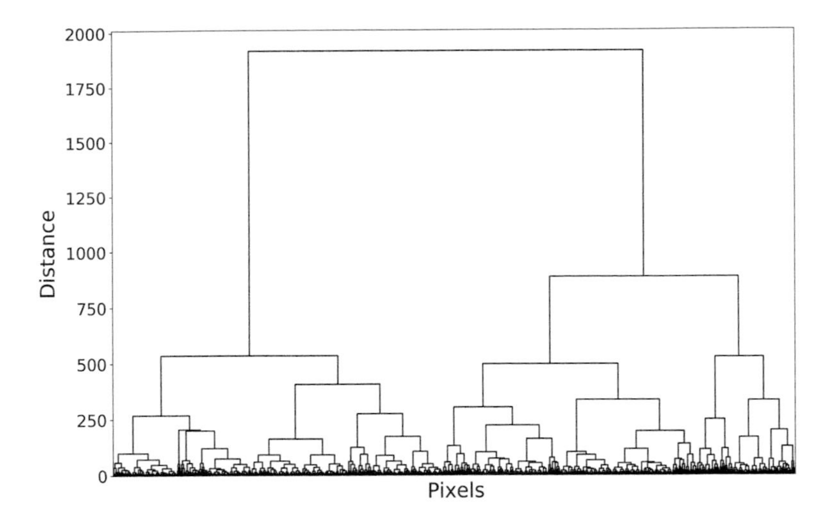

FIGURE 9.11 Dendrogram showing cluster hierarchy for normal plants.

crosses is the optimum number of clusters. Figure 9.11 shows a dendrogram generated using Ward clustering on spectral curves of a plant under normal conditions for 18 days of the experiment. The optimal number of clusters, in this case, is two.

9.6 CLASSIFICATION

Classification is the problem of assigning a discrete label or category to an observation or data point. It is a supervised learning technique where the classifier learns for a set of observations (training data) and their corresponding class labels. It is widely used in many applications, including plant phenotyping. In hyperspectral image analysis, the spectral reflectance curves are used as observations. Many materials have unique reflectance patterns, called spectral signatures. Researchers in remote sensing have developed a number of spectral libraries for various applications including landcover classification for wetland management (Zomer et al. 2009), investigation of spectral characteristics of urban surface materials (Herold et al. 2004), and characterization of soil properties (Shepherd and Walsh 2002). The classification task in HSI, therefore, is to identify the category for a spectral reflectance pattern. The classes will depend on the task at hand, e.g., fruit, flower, and vegetation for organs, or young, mature, and old leaves for age. In this section, we describe classification techniques that are applicable and used in HSI- based plant phenotyping analysis.

9.6.1 Artificial Neural Networks

Artificial neural networks (ANNs) are a class of supervised and unsupervised machine learning algorithms that are inspired by the biological nervous system. ANNs use multiple layers of processing units to extract more abstract features that can achieve higher levels of accuracy in classification (Chen et al. 2014). Training datasets in HSI are high-dimensional, have a high degree of spectral-spatial diversity, and are often very large. ANNs have the potential to automate the process of feature construction by building high-level features from raw spectral reflectance patterns.

An ANN is a collection of artificial neurons (also called perceptrons) structured with an input layer, one or more hidden layers, and an output layer. Each layer can have any number of nodes and is guided by the structure of the input data and the expected output. Each perceptron in a hidden layer takes a weighted sum of all inputs from the training dataset (or the previous hidden layer), passes it through an activation function, and fires an output. Each perceptron in the output layer, in turn, takes a weighted sum of outputs from each perceptron in the last hidden layer, passes it through an activation function, and produces the final output. The output is compared with the actual output to estimate the error in classification. Each observation is fed to the input layer and is passed through the network until the output is generated and compared with the ground-truth. This is often called the forward-propagation. Some widely used activation functions are softmax, sigmoid, tanh, and rectified linear unit (ReLU). Due to smoothness properties, ReLU is currently favored in practice.

The network can improve accuracy by adjusting the weights in the network. This is achieved using the back-propagation algorithm (Rojas 1996), where the classification error is propagated back into the network in reverse order (from the output layer). During back-propagation, the weights of the network are tuned, using the gradient descent algorithm (Rojas 1996). This process of fine-tuning the weights is carried out iteratively until a terminating criterion is reached (e.g., the number of iterations through the dataset, the accuracy of the output, or the gradient of the weights). To design a neural network for a given classification task, we need to define the hyperparameters for the model. Some important hyperparameters include the number of hidden layers, the activation function, optimizer, loss function, dropout scheme, and batch size (Diaz et al. 2017). A correct combination of these parameters can help increase the accuracy of a neural network, avoid overfitting, and achieve optimal generalization. However, tuning hyperparameters is one of the most computationally intensive tasks. A basic hyperparameter optimization technique that is widely used is the grid search approach, where the user defines a finite set of values for each hyperparameter. The grid search generates all possible permutations of the hyperparameters, and the model is evaluated on each of these sets. Although this approach gives a holistic view of the optimization process, it suffers from the problem of the curse of dimensionality (Feurer and Hutter 2019). This problem may be mitigated by using better optimization techniques like random search, Bayesian optimization, etc. (Feurer and Hutter 2019).

The advantage of an artificial neural network in the context of HSI lies in its ability to automatically extract high-level features from a multidimensional, spectral-spatial dataset to perform tasks like classification, regression, segmentation, and dimensionality reduction. However, due to the high dimensionality of the dataset, a proportionally large training set is needed to avoid overfitting (Signoroni et al. 2019). This challenge can be overcome by adopting unsupervised learning techniques to generate labeled data, using data-driven approaches.

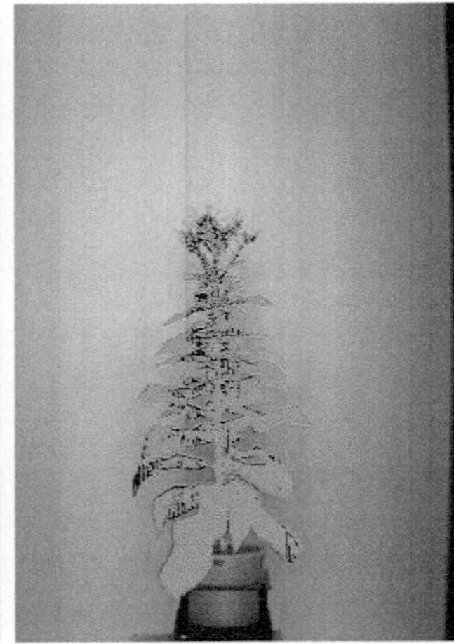

FIGURE 9.12 Illustration of classification using spectral signatures using an ANN. Stressed regions are shown in red for the normal plant (left) and drought-stressed plant (right).

Figure 9.12 illustrates the classification into normal and stressed pixels, using ANN, to identify drought stress in the tobacco plant. Spectral signatures for normal and stressed classes are generated using a data-driven approach as explained by Gampa (2019). An ANN-based classifier is trained using these data. The plant under normal growth conditions (Figure 9.12 (left)) does not show any sign of stress, but there are scattered regions in the drought-stressed plant classified as stressed (shown in red in Figure 9.12 (right)).

9.6.2 Support Vector Machines

Support vector machines (SVMs) (Corinna and Vladimir 1995) are widely applied to classification problems and nonlinear regressions. Unlike neural networks, that perform classification based on statistical estimation, SVMs use a geometrical margin as the basis for classification (Mercier and Lennon 2003). In HSI classification, the SVMs have been applied to significantly reduce the complexity and to improve the accuracy in many real-life applications, e.g., landcover classification (Moughal 2013). SVMs have proved to be an effective alternative to conventional pattern recognition approaches that employ a combination of feature reduction followed by classification to categorize hyperspectral remote sensing data (Melgani and Bruzzone 2004).

The goal of an SVM is to find the best separating hyperplane between classes, i.e., the one with the greatest distance from the data points. It determines the separating hyperplane from two parallel imaginary hyperplanes that separate the observations between the classes with the maximum margin. The wider the margin, the greater is the generalization ability of the classifier. For binary classification, one hyperplane is employed to divide the dataset. For multiclass classification, additional hyperplanes are needed to separate classes.

If the data are not linearly separable in the original input space, they can be transformed into a higher-dimensional feature space where the classes are linearly separable. This is achieved by applying an appropriate kernel function on the input data space, prior to calculating the separation hyperplane. Radial basis, polynomial, and sigmoid functions are widely used nonlinear kernel functions for SVM classifiers (Patle and Chouhan 2013).

9.6.3 Decision Trees

A decision tree classifier builds a model in a top-down approach, while making incremental decisions. The tree structure consists of decision nodes and leaf nodes. The decision node represents a test on an attribute of the dataset, whereas the leaf node is the decision or classification result. Based on the result of the test, one of the edges is followed to either another decision node or a leaf node. This is repeated until a leaf node is reached, which decides the final classification of an observation. Unlike an artificial neural network, where results are difficult to interpret, a decision tree classifier makes it possible to see the hierarchical decision-making process. This transparency makes it a white box model, where every condition can be explained by Boolean logic (Pedregosa et al. 2011). In HSI classification, decision trees can help identify the spectral bands (or the wavelengths) that contribute significantly to classification and hence can be deemed salient. One of the drawbacks of decision trees is that they can create complex models for large datasets, which may not generalize the data and may result in overfitting. This can be mitigated by enforcing minimum samples per leaf node and limiting the depth of the tree.

Numerous decision tree algorithms have been proposed for classification, e.g., ID3 (Quinlan 1986), and Random Forest (Ho 1995). The key idea of ID3 is to select the attribute (for the decision node) that is most likely to divide the observations into homogeneous sets (based on the class label). This is usually determined by some information theory measures, e.g., entropy and information gain (Quinlan 1986). Random Forest is a widely used classification technique based on ensemble learning. It randomly creates subspaces of smaller dimensions than the original training set and associates each subspace with a decision tree. It uses a majority voting policy to combine the classification rates of the classifiers associated with each random subspace, and thus, it reduces the effects of the weak classifiers for improved final classification. Random Forest is characterized by two key parameters: (a) the number of subspaces (decision trees), and (b) the dimension of the subspace, which play a significant role in the classification accuracy. The Random Forest is an ensemble classifier that exploits high dimensionality of the feature space to avoid overfitting.

9.7 CONCLUSION AND FUTURE WORK

HSI has tremendous potential in plant phenotyping applications. The primary advantage of HSI is its ability to capture information at a higher spectral resolution, thus providing the opportunity to create unique profiles for different organs at different stages of growth in a plant and identifying stress response patterns in a plant. Therefore, it has been adopted in agriculture and related fields to estimate vitality and chlorophyll content, which directly relate to nitrogen demand, water content, dry matter, or leaf area.

Although HSI technology has great potential in phenotyping applications, its widespread adoption has been slow due to the high cost of imaging hardware and associated computing cost. There is also a scarcity of tools and techniques to use them in phenotyping applications. This chapter introduced a GUI-based hyperspectral image analysis tool for plant phenotyping (HypeRpheno) that allows efficient and exploratory data-driven analysis of hyperspectral image sequences of a plant, using supervised and unsupervised machine learning algorithms. This chapter presented a detailed overview of several core algorithms that are used in hyperspectral image analysis: dimensionality reduction, clustering, and classification.

While progress has been made, there are many unsolved computer vision problems, both for high-throughput and field-based plant phenotyping systems. Introducing machine learning and deep learning techniques can help eliminate visual error and improve accuracy. Identification of the right combination of bands, that best represents the task-relevant salient information in the large HSI space, remains a challenge. Algorithms to detect critical timing information of important events in a plant's life cycle, that is not visible in RGB images, also remain underdeveloped.

REFERENCES

Apan, A., B. Datt, and R. Kelly. 2005. Detection of pests and diseases in vegetable crops using hyperspectral sensing: A comparison of reflectance data for different sets of symptoms. In *Proceedings of the 2005 Spatial Sciences Institute Biennial Conference: Spatial Intelligence, Innovation and Praxis* 10–18. Melbourne, Australia.

Bravo, C., D. Moshou, J. West, A. McCartney, and H. Ramon. 2003. Early disease detection in wheat fields using spectral reflectance. *Biosystems Engineering* 84(2):137–145. doi: 10.1016/s1537-5110(02)00269-6.

Chen, Y., Z. Lin, X. Zhao, G. Wang, and Y. Gu. 2014. Deep learning-based classification of hyperspectral data. *IEEE Journal of Selected Topics in Applied Earth Observations and Remote Sensing* 7(6):2094–2107. doi: 10.1109/JSTARS.2014.2329330.

Corinna, C., and V. Vladimir. 1995. Support-vector networks. *Machine Learning* 20:273–297. doi: 10.1023/A:1022627411411.

Craig, R., and J. Shan. 2002. Principal component analysis for hyperspectral image classification. *Surveying and Land Information Systems* 62:115–123.

Diaz, G. I., A. Fokoue-Nkoutche, G. Nannicini, and H. Samulowitz. 2017. An effective algorithm for hyperparameter optimization of neural networks. *IBM Journal of Research and Development* 61:9:1–9:11. doi: 10.1147/JRD.2017.2709578.

Feurer, M., and F. Hutter. 2019. Hyperparameter optimization. In *Automated Machine Learning*, eds. F. Hutter, L. Kotthoff, and J. Vanschoren, 3–33. Springer, Cham. doi: 10.1007/978-3-030-05318-5_1.

Gampa, S. 2019. A data-driven approach for detecting stress in plants using hyperspectral imagery. MS Thesis, University of Nebraska-Lincoln, Lincoln, NE.

Ge, Y., G. Bai, V. Stoerger, and J. C. Schnable. 2016. Temporal dynamics of maize plant growth, water use, and leaf water content using automated high throughput RGB and hyperspectral imaging. *Computers and Electronics in Agriculture* 127:625–632.

Herold, M., D. A. Roberts, M. E. Gardner, and P. E. Dennison. 2004. Spectrometry for urban area remote sensing – Development and analysis of a spectral library from 350 to 2400 nm. *Remote Sensing of the Environment* 91(3–4):304–319.

Ho, T. K. 1995. Random decision forests. In *Proceedings of the 3rd International Conference on Document Analysis and Recognition* 1:278–282. Montreal, Canada.

Ketchen, D. J., and C. L. Shook. 1996. The application of cluster analysis in strategic management research: An analysis and critique. *Strategic Management Journal* 17(6):441–458.

Lowe, A., N. Harrison, and A. P. French. 2017. Hyperspectral image analysis techniques for the detection and classification of the early onset of plant disease and stress. *Plant Methods* 13:80. doi: 10.1186/s13007-017-0233-z.

Lu, R. 2004. Multispectral imaging for predicting firmness and soluble solids content of apple fruit. *Postharvest Biology and Technology* 31(2):147–157. doi: 10.1016/j.postharvbio.2003.08.006.

Mahlein, A.-K., E.-C. Oerke, U. Steiner, and H.-W. Dehne. 2012. Recent advances in sensing plant diseases for precision crop protection. *European Journal of Plant Pathology* 133(1):197–209.

Melgani, F., and L. Bruzzone. 2004. Classification of hyperspectral remote sensing images with support vector machines. *IEEE Transactions on Geoscience and Remote Sensing* 42(8):1778–1790.

Mercier, G., and M. Lennon. 2003. Support vector machines for hyperspectral image classification with spectral-based kernels. In *Proceedings of the IEEE International Geoscience and Remote Sensing Symposium* 1:288–290. Toulouse, France.

Mishra, A., R. Ehsani, G. Albrigo, and W. S. Lee. 2007. Spectral characteristics of citrus greening (huanglongbing). In *Proceedings of the ASABE Annual Meeting*, Paper number 073056, St. Joseph, MI. doi: 10.13031/2013.24163.

Moughal, T. A. 2013. Hyperspectral image classification using support vector machine. In *Proceedings of the 6th Vacuum and Surface Sciences Conference of Asia and Australia.* doi: 10.1088/1742-6596/439/1/012042. Islamabad, Pakistan.

Pakhira, M. K. 2014. A linear time-complexity k-means algorithm using cluster shifting. In *Proceedings of the International Conference on Computational Intelligence and Communication Networks* 1047–1051. doi: 10.1109/CICN.2014.220.

Patle, A., and D. S. Chouhan. 2013. SVM kernel functions for classification. In *Proceedings of the International Conference on Advances in Technology and Engineering* 1:9. doi: 10.1109/icadte.2013.6524743. Mumbai, India.

Pedregosa, F., G. Varoquaux, A. Gramfort, et al. 2011. Scikit-learn: Machine learning in Python. *Journal of Machine Learning Research* 12:2825–2830.

Peng, Y., and R. Lu. 2005. Modeling multispectral scattering profiles for prediction of apple fruit firmness. *Transactions of the ASAE* 48:235–242.

Qin, J., T. F. Burks, M. S. Kim, K. Chao, and M. A. Ritenour. 2008. Citrus canker detection using hyperspectral reflectance imaging and PCA-based image classification method. *Sensing and Instrumentation for Food Quality and Safety* 2(3):168–177. doi: 10.1007/s11694-008-9043-3.

Quinlan, J. R. 1986. Induction of decision trees. *Machine Learning* 1:81–106. Kluwer Academic Publishers, Boston. doi: 10.1007/BF00116251.

Rock, B. N., T. Hoshizaki, and J. R. Miller. 1988. Comparison of in situ and airborne spectral measurements of the blue shift associated with forest decline. *Remote Sensing of Environment* 24(1):109–127. doi: 10.1016/0034-4257(88)90008-9.

Rojas, R. 1996. *Neural Networks – A Systematic Introduction* 151–184. Springer-Verlag, Berlin.

Romer, C., M. Wahabzada, A. Ballvora et al. 2012. Early drought stress detection in cereals: Simplex volume maximization for hyperspectral image analysis. *Functional Plant Biology* 39(11):878–890. doi: 10.1071/FP12060.

Rousseeuw, P. J. 1987. Silhouettes: A graphical aid to the interpretation and validation of cluster analysis. *Journal of Computational and Applied Mathematics* 20:53–65.

Shepherd, K. D., and M. G. Walsh. 2002. Development of reflectance spectral libraries for characterization of soil properties. *Soil Science Society of America Journal* 66(3):988–998.

Signoroni, A., M. Savardi, A. Baronio, and S. Benini. 2019. Deep learning meets hyperspectral image analysis: A multidisciplinary review. *Journal of Imaging* 5(5):52.

Yendrek, C. R., T. Tomaz, C. M. Montes, Y. Cao, A. M. Morse, P. J. Brown, L. M. McIntyre, A. D. Leakey, and E. A. Ainsworth. 2017. High-throughput phenotyping of maize leaf physiological and biochemical traits using hyperspectral reflectance. *Plant Physiology* 173(1):614–626. doi: 10.1104/pp.16.01447.

Zomer, R. J., A. Trabucco, and S. L. Ustin. 2009. Building spectral libraries for wetlands land cover classification and hyperspectral remote sensing. *Journal of Environmental Management* 90(7):2170–2177.

10 Machine Learning and Statistical Approaches for Plant Phenotyping

Zheng Xu and Cong Wu

CONTENTS

10.1 INTRODUCTION

10.1.1 Phenotype and Its Determinants

Phenotypes are a plant's traits, such as its morphological and developmental observable characteristics, physiological traits, behaviors, and biochemical properties. Examples of plant phenotypes include biomass, root morphology (Flavel et al. 2012; Kumar et al. 2014), leaf characteristics (Jansen et al. 2009), fruit morphology (Brewer et al. 2006), photosynthetic efficiency (Bauriegel et al. 2011), water content (Penuelas et al. 1997), pigment concentration (Blackburn 2007), and metabolite concentrations (Curran et al. 2001). Plant phenotypes are influenced by three sets of factors (genotype, inheritable epigenetic factors, and non-inheritable environmental factors) and their interactions. The terms phenotype and genotype were coined by the Danish botanist and geneticist Wilhelm Johannsen in 1903. A genotype is the genetic makeup of the organism. It refers to the DNA sequence of the organism. Epigenetic factors are heritable characteristics that can impact and change phenotypes but do not involve changes in DNA sequences (Dupont et al. 2009). Examples of epigenetic mechanisms are methylation and histone modification. Environmental factors are non-heritable factors exogenous to the plant. Examples of environmental factors are precipitation, temperature, soil nutrition, fertilizer concentrations, and light levels. Researchers often conduct integrated analyses, including multiple types of data, to understand the interactions between the various factors that impact phenotypes. Advances in imaging technology, analysis techniques, and in machine learning and statistical approaches can significantly advance the inquiry into phenotyping research.

10.1.2 Changes in Plant Phenotyping

Genotyping refers to the process of measuring genotypes. With recent developments in next-generation sequencing, genotyping techniques have improved significantly. Similarly, phenotyping refers to the process of determining and predicting phenotypes. Phenotyping techniques, using image analysis for high-throughput applications, are gaining popularity in order to increase the research scope compared with traditional phenotyping, using manual human labor.

Traditional phenotyping is low-throughput since it is a manual process. For example, in the *shovelomics* standard for root excavation and phenotyping (Trachsel et al. 2011), researchers excavate a root system at a radius of 20 cm around the hypocotyl and 20 cm below the soil surface. After excavation, the shoot is separated from the root system 20 cm above the soil level, and the root system is washed, using water containing mild detergent, to remove soil. Then the washed root system is placed on a phenotyping board, consisting of a large protractor, to measure root angles at different depth ranges and score the length and density classes of lateral roots. Observed traits derived in this process include root angle, number, density, and diameter. Root traits are measured manually in the shovelomics phenotyping method, which limits its applicability for high-throughput analysis. Recent measurements of phenotypic traits, such as biomass and height of plants are usually image-based, in comparison with traditional phenotyping, that involves manual measurement of the plant's biomass weight and height.

Image-based phenotyping is non-destructive, and therefore, allows for tracking of the dynamics of phenotypes during a plant's life cycle. The scope of traits in plant phenotyping has also expanded over time. New complex traits, such as persistent homology (Li et al. 2018), have been identified and taken account of by researchers. Furthermore, the traits extracted from images are not restricted to phenotypes only. Plant phenotyping is generally regarded as the extracting of *traits* or *features* from images. Since images encode significant information about the plant's structure, approaches from fields such as machine learning, statistics, and computer vision may be applied to plant phenotyping to extract more complex features that may prove significant in the understanding of plant growth. For example, image-based approaches to identify the plant's cultivar and genotype or growth stages are also regarded as plant phenotyping problems, although these characteristics are *traits* and *features* but, strictly speaking, not *phenotypes*. We use the term *feature* and *trait* interchangeably in this chapter.

10.1.3 Digital Images

Many different types of images (and data formats) are used in plant phenotyping research and practice. Widely used imaging modalities in plant phenotyping include Red Green Blue (RGB), hyperspectral, fluorescent, and thermal infrared; magnetic resonance imaging (MRI), positron emission tomography (PET), and computed tomography (CT) are also used. These images are used for different phenotyping purposes. For example, RGB images are typically used to infer features related to the morphology and colors of the plants, such as leaf area, growth dynamics, fruit number and ripeness, and size-based projected biomass. Thermal infrared images can be used to determine the plant's temperature; they indirectly contribute to our understanding of the water and transpiration status in the plant. Hyperspectral images are used to monitor biochemical concentrations and reactions in the plant, as well as in identifying different parts of the plants based on the difference in their reflectance spectra. MRI and CT images are used for underground phenotyping, i.e., root phenotyping. In this chapter, to demonstrate and illustrate machine learning and statistical approaches, we focus on two-dimensional (2D) RGB and hyperspectral images, and three-dimensional (3D) CT images. The methods discussed may be potentially applied to other types of images after appropriate modification.

A digital image may be regarded as a discrete representation of spatial and intensity information in a scene. A 2D image is represented as a two-dimensional array, I, where $I(r,c)\left(r=1,2,\ldots,R;\, c=1,2,\ldots,C\right)$ represents the intensity response of a sensor at fixed locations of 2D Cartesian coordinates. Each location, $I(r,c)$, is termed as a pixel in the image. The indices r and c designate the row and column indices for the pixel, respectively. For a 3D image (r,c,h) $\left(r=1,2,\ldots,R;\, c=1,2,\ldots,C;\, h=1,2,\ldots,H\right)$, it is the response at 3D Cartesian coordinates. Each location is termed a voxel in the image. The indices r, c, and h designate the row, column, and height of the voxel, respectively. At each pixel or voxel, there is intensity information, representing the response to the sensor.

The intensity information can be represented by a single value or multiple values, depending on the type of image. For example, for a 2D RGB image, the intensity information is represented by three values: reflectance at red (R), green (G), and blue (B) wavelengths, respectively, at each pixel. For a 2D hyperspectral image, on the other hand, each pixel stores the intensity information which is represented by multiple reflectance values, each value corresponding to a narrow wavelength range. For a 3D CT image, at each voxel, the intensity information is represented by one value, typically a transformation of the CT number, which is expressed in terms of Hounsfield units, and used to quantify the attenuation of the X-ray beam in the material, relative to water (Ginat and Gupta 2014; Kalender 2000; Larobina and Murino 2014). A dataset generated from imaging is typically comprised of intensity values, which are saved in standard image format with 8 bits, 16 bits, or 32 bits for each pixel or voxel (Bryant et al. 2012). Although less common, image data may also be real numbers, which are defined in the IEEE-754 format to use the single precision 32-bit and the double precision 64-bit (Bryant et al. 2012).

Therefore, a 2D RGB image with R rows and C columns of pixels is saved as an array (also named as a tensor in statistics) with the dimensions $R \times C \times 3$. A 2D grayscale or binary image is saved as an array with the dimension $R \times C \times 1$. A 2D hyperspectral image with W bands sampled is saved as an array with the dimension $R \times C \times W$. A 3D CT image with R rows and C columns and height H of voxels is saved as an array with the dimension $R \times C \times H$.

Image data are often high dimensional because the numbers of rows and columns can be in the order of thousands for high-resolution RGB images. A hyperspectral image typically has hundreds of bands. Therefore, many machine learning and statistical approaches are applicable to the analysis of image data in plant phenotyping. From the perspective of machine learning and statistics, plant phenotyping may be viewed as either a feature extraction problem, i.e., identifying important features in the image, or a prediction problem, i.e., estimating the output, given the input image.

10.1.4 UNL-CPPD for Illustration

In discussing machine learning and statistical approaches to plant phenotyping, we used the University of Nebraska-Lincoln (UNL) Component Plant Phenotyping Dataset (CPPD) for illustration. UNL-CPPD was released by the UNL Plant Vision Initiative, which is committed to developing innovative algorithms to analyze images of plants, using computer vision-based techniques for plant phenotyping, including leaf recognition and leaf tracking. UNL-CPPD consists of images of 13 maize plants from two side views (0° and 90°). Plants were imaged once per day from 2 days after emergence between 28 and 31 days, using the visible light camera of the UNL Lemnatec Scanalyzer 3D High-Throughput Phenotyping Facility. One advantage of the dataset is that, in addition to original images, manually annotated images and ground-truth are included to allow performance evaluation. This facilitates algorithm development for both holistic and component analysis of plant images (Das Choudhury et al. 2018). UNL-CPPD is publicly available at the webpage https://plantvision.unl.edu/dataset.

The rest of this chapter discusses and illustrates different machine learning and statistical approaches used in plant phenotyping in the following order: (1)

unsupervised dimensional reduction approaches, (2) supervised machine learning approaches for continuous traits, (3) supervised machine learning approaches for binary traits, (4) image segmentation approaches, and (5) approaches combining multiple tracks of information or using temporary traits.

10.2 DIMENSIONAL REDUCTION APPROACHES IN PLANT PHENOTYPING

One characteristic of image data is its high dimensionality. To visualize the data in low-dimensional space and to predict traits from images, dimensional reduction approaches have been used.

10.2.1 MULTIDIMENSIONAL SCALING (MDS)

Multidimensional scaling (MDS) is a method to transfer points in high-dimensional space into low-dimensional space. Suppose we have N images, each with p values, where $p = R \times C \times 3$, $R \times C \times W$, and $R \times C \times H$ for RGB, hyperspectral, and CT images, respectively. The value of p is likely to be large; for example, an RGB image with $R = 200$ rows and $C = 300$ columns of pixels will have $p = 180{,}000$. These N images are regarded as N points in the p-dimensional space.

In the p-dimensional space, we can calculate the distance d_{ij} between any two points i and j. The Euclidean distance is widely used, although other types of distance may also be used. MDS can place these images in a low-dimensional space (usually two or three) such that the distance relationships in the original (high-dimensional) space are maintained in the transformed (low-dimensional) space. Thus, MDS can be viewed as a projection of N points from original space to a 2D or 3D space which preserves the pairwise distances between any two points by as much as possible (Alpaydin 2010).

MDS first calculates the pairwise distance between the N images. Then, it generates the coordinates of N transformed points in 2D or 3D with the distances as close as possible to the pairwise distances between the N images. MDS method is an unsupervised machine learning method in that it does not use the information on any response. MDS also allows for easy visualization of the level of similarity of individual images in the dataset.

The coordinates inferred from MDS only specify the location where the points are located in the low-dimensional space to preserve pairwise distance. Dimensions in MDS are not as interpretable as in Principal Component Analysis (PCA) or in Factor Analysis (FA), where the number of principal components or factors have real meaning. Dimensions in MDS means that we project observations in this low-dimensional space to preserve the distance.

An illustration of MDS for images in UNL-CPPD is described next. We overlaid the original images onto the scatter points to show the locations representing images in the 2D space, using the MDS method. We labeled each image with the number of leaves in the plant (leaf number) or the number of days since it was planted (growing days). Figure 10.1 shows the MDS map of UNL-CPPD images labeled with leaf number, and Figure 10.2 shows the MDS map of the images labeled with growing

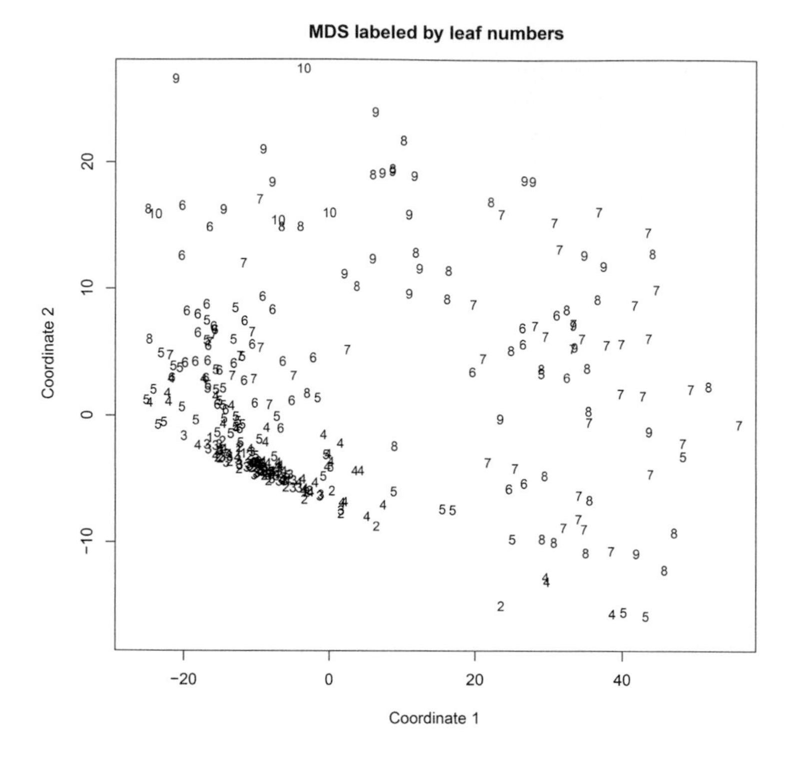

FIGURE 10.1 Map of UNL-CPPD images labeled with leaf number, derived using MDS. Pairwise Euclidean distances between these images are used by MDS to place them in a two-dimensional space to preserve these distances. Leaf numbers (between 1 and 10) are labeled. We divided the plants into three groups according to leaf numbers: 1–3 leaves (Group 1), 4–6 leaves (Group 2), and 7–10 leaves (Group 3).

days. We divided the plants into three groups, according to leaf numbers: 1–3 leaves (Group 1), 4–6 leaves (Group 2), and 7–10 leaves (Group 3). It was observed that plants in Group 1 are all concentrated together. The three groups are generally compact and are well separated from each other. We also divided plants according to growing days: 1–9 growing days (Group 1), 10–18 growing days (Group 2), and 19–28 growing days (Group 3). We observed similar results in that the plants in Group 1 are all concentrated together, and the three groups, according to growing day number, are each compactly clustered, but far away from each other.

10.2.2 Principal Component Analysis (PCA)

Principal Component Analysis (PCA) is another widely used method to transform points in a high-dimensional space into a low-dimensional space. Like the MDS, PCA also transforms a set of N images (with p values/dimensions) to k-dimensional space, where $k<p$, often $k \ll p$ with *minimum* loss of information. PCA is a statistical approach that uses orthogonal transformation to convert a set of observations with high-dimensional, possibly correlated variables into a set of linearly

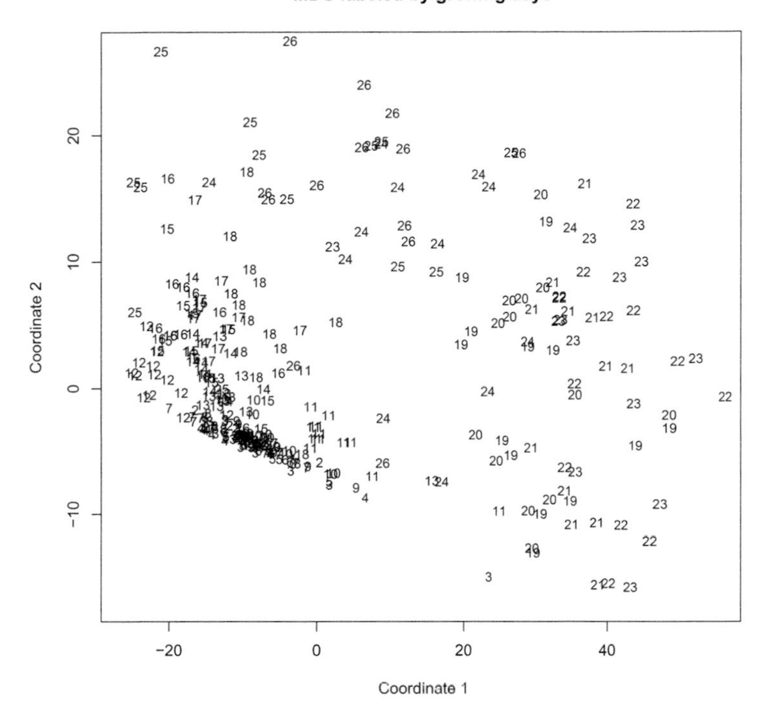

FIGURE 10.2 Map of UNL-CPPD images labeled with growing days, derived using MDS. Pairwise Euclidean distances between these images are used by MDS to place them in a two-dimensional space to preserve these distances. Growing days (between 1 and 28) are labeled. We divided the plants into three groups according to growing days: 1–9 growing days (Group 1), 10–18 growing days (Group 2), and 19–28 growing days (Group 3).

uncorrelated variables with the maximum preservation of information in terms of data variance. The mutually uncorrelated variables after transformation are called principal components (PCs). A high degree of correlation among the data leads to a large reduction in dimensions, using PCA. This is particularly applicable for images since the different dimensions (pixels) are likely to be highly correlated.

10.2.2.1 Extraction of PCs

The linear transformation in PCA is optimized such that the first PC has the greatest possible variance, i.e., accounts for as much of the information/variance in data as possible. Then, each successive component is determined to have the next-highest variance possible under the constraint that it is orthogonal to the preceding components. The resulting components are uncorrelated and explain the maximum proportion of variance in data, given a fixed number of projected variables. The procedure is stopped after a predetermined number of PCs are determined (for example, 10 or 20) or until the cumulative proportion of variance explained by the completed set of components exceeds a threshold value (for example, 80% or 90% of the overall data variance).

PCA is an unsupervised method as it does not require any response information. The objective in determining the projection is to maximize the variance of the original data captured by the PCs. The extracted PCs $z = [PC_1, PC_2, \ldots, PC_k]$ will maximize the variance and are orthogonal to each other (Hastie et al. 2009).

Images, in general, and plant images captured using a high-throughput phenotyping platform, in particular, are high-dimensional, and their pixel intensities are highly correlated. For PCA, each image is converted from an array into a vector of p values. PCA is conducted for the N vectors to extract the top PCs, based on the termination criterion described before. PCA identifies new features in a dataset, the PCs, which are linear combinations of the original features. PCA can infer important traits, reduce the data size to explore the relationship between traits, and explain their variations (Rahaman et al. 2015). For example, the plant phenotyping software LeafProcessor conducted PCA for leaf shape and size, including contour bending energy (Backhaus et al. 2010). We illustrate the use of PCA for images, based on the UNL-CPPD, below.

Figure 10.3 shows the cumulative proportion of variance in the data explained by the number of PCs. As Figure 10.3 shows, the top six PCs account for 80% of the variance in the images. The addition of new PCs beyond the top six causes only a small increase in the proportion of variance explained by the PCs. Overall, the top 18 PCs explain 85% of the variance.

10.2.2.2 Relationship between Top PCs and Traits

The top PCs are expected to capture most of the variance in the data. They can be viewed as new traits themselves, and the top PCs may also be used to extract other traits interesting to researchers. Figure 10.4 plots the absolute correlations of two

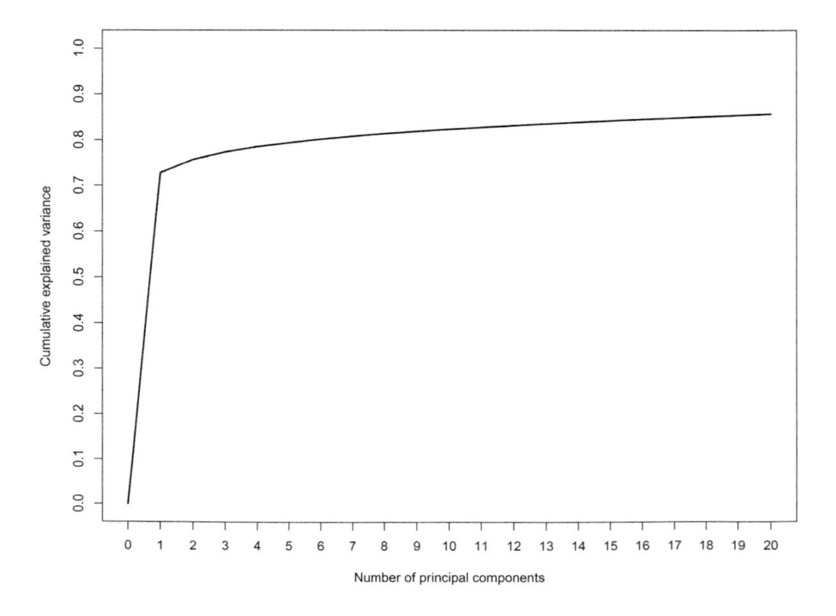

FIGURE 10.3 Cumulative variance explained by PCs as a function of number of principal components.

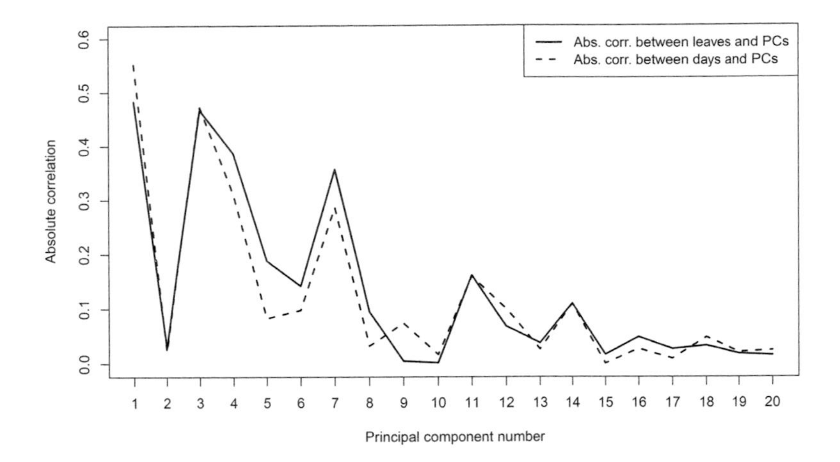

FIGURE 10.4 Absolute correlations between phenotypes and principal components.

plant traits (leaf number and growing days) with the top PCs. It was found that PC1 and PC3 were correlated with both leaf number and growing days. Absolute correlations between PCs and leaf numbers show a similar pattern as the correlation between PCs and growing days. This is due to the high correlation between leaf number and growing days, i.e., correlation coefficient, r, (between leaf number and growing days) =0.927. Thus, the top PCs may be used to predict leaf numbers or growing days. The graph also indicates that the set of top PCs is more likely to be associated with traits than are the other PCs. Researchers in plant phenotyping have applied PC-extracted features to predict complex traits in their studies (Chen et al. 2014; Ge et al. 2016; Pandey et al. 2017; Rahaman et al. 2015).

To further illustrate the relationship between the top PCs and the two traits (leaf number and growing days), scatter plots of PC1 and PC2, PC3 and PC4, labeled by leaf number, are shown in Figures 10.5 and 10.6, respectively. We divided the plants into three groups, according to leaf numbers: 1–3 leaves (Group 1), 4–6 leaves (Group 2), and 7–10 leaves (Group 3). We observed that plants in group 1 were all concentrated together in the scatter plots, and that the three groups were well separated, with the plants within a group being clustered in the scatter plots. Similar results were observed when we divided the plants into three groups on the basis of growing days.

10.2.2.3 Eigenimages and Image Reconstruction

PCA infers an orthogonal basis in projecting UNL-CPPD corn data from p-dimensional space to low-dimensional space. Individual corn images can be reconstructed using the PCs, because they are linear combinations of the orthogonal basis. PCs require *significantly less storage* and are *low-dimensional*, compared with the original images, while still preserving the maximum amount of information possible in the linear mapping process. PCs can reveal the structure of image data set in the low-dimensional space.

Researchers have named the orthogonal basis in PCA for a set of images as *eigenimages* because they are images represented by the *eigenvectors* in PCA analysis. Each original image is reconstructed as the *linear combination* of *eigenimages* with

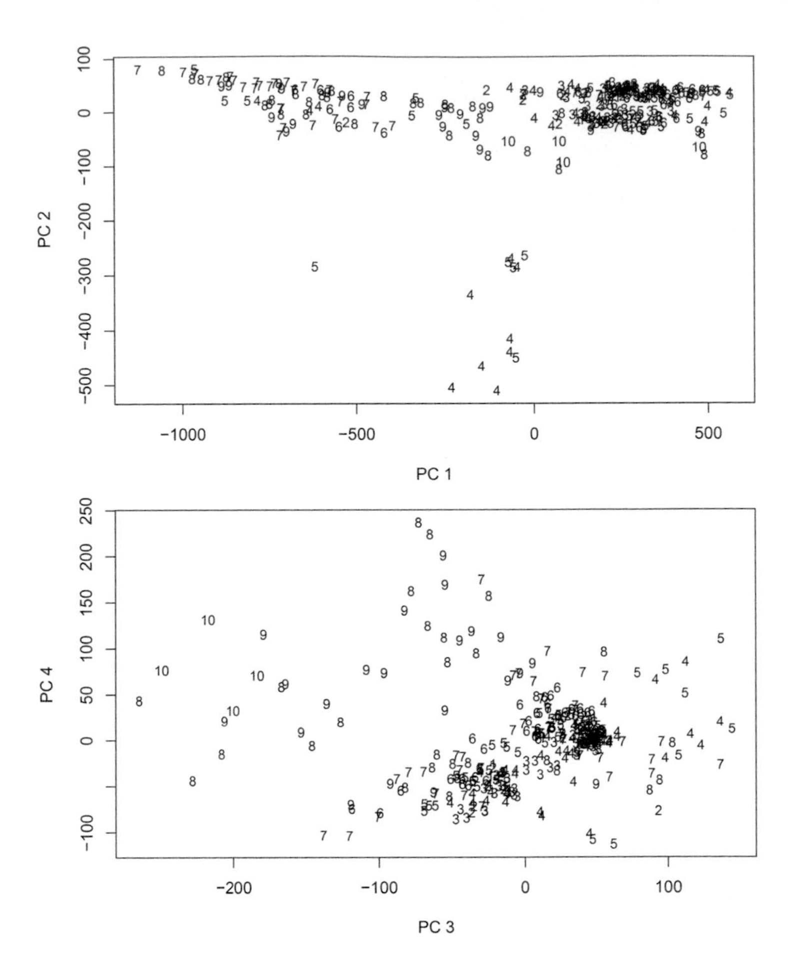

FIGURE 10.5 Scatter plots of PC1 versus PC2 (upper panel) and PC3 versus PC4 (lower panel), labeled by leaf numbers. Leaf numbers (between 1 and 10) are labeled. The plants are divided into three groups according to leaf numbers. Group 1,2, and 3 represent plants with 1–3 leaves, 4–6 leaves, or 7–10 leaves, respectively.

the *PC* values serving as the weights. In the application to face recognition, the eigenimages are called the *eigenfaces* and face reconstruction can be conducted based on top *K* eigenfaces and top *k* PCs, with a pre-specified *k* (Turk and Pentland 1991).

Eigenimage and image reconstruction are illustrated here, using the UNL-CPPD. We use the term *eigencorn* and corn (image) reconstruction because they are inferred from images of corn plants. The PCs and the orthogonal basis inferred will depend on the dataset used to conduct PCA. Thus, the images of the inferred eigencorns are an orthogonal basis for the UNL-CPPD image data, which include corn plants with different leaf numbers and growing days. Eigencorns form a more homogeneous dataset, that has corn plants of the same variety with the same growing day value, which will have a different set of eigencorns.

Figure 10.7 illustrates the image reconstruction of corn plant images. The original corn image is plotted with the reconstructed images, using the top 10, 50 or

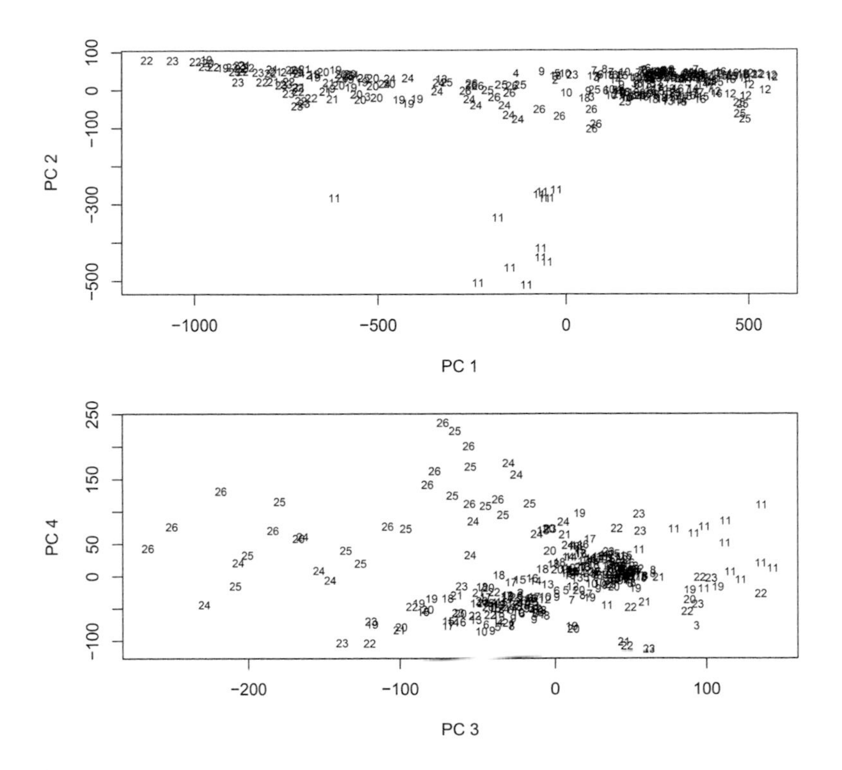

FIGURE 10.6 Scatter plots of PC1 versus PC2 (upper panel) and PC3 versus PC4 (lower panel), labeled by growing days. Growing days (between 1 and 28) are labeled. The plants are divided into three groups according to growing days. Group 1, 2, and 3 represent plants with growing days 1–9, 10–18, or 19–28.

100 PCs. It can clearly be seen that we only need to keep a few eigencorns (and their corresponding values) instead of the original high-dimensional images to generate an accurate reconstruction. In the original space, the number of dimensions (p) is $500 \times 500 \times 3 = 750000$; even this very large number was achieved after scaling down the images to 500×500. The corn image reconstruction demonstrates the benefit of PCA in dimensional reduction. In Figure 10.8, we have plotted the top four eigencorns, corresponding to the top four PCs, to illustrate the orthogonal basis for a better understanding of image reconstruction. Each original image can be reconstructed as the *linear combination* of *eigencorns*, with the weight equal to the PCs.

10.3 SUPERVISED MACHINE LEARNING APPROACHES FOR CONTINUOUS TRAITS

A range of statistical and machine learning approaches to extract traits in plant phenotyping has been proposed. In this section, we discuss three widely used methods (partial least square regression, least absolute shrinkage and selected operator, and random forest regression) which are expected to be useful in the context of image-based plant phenotype research and practice. We have demonstrated the three

FIGURE 10.7 Image reconstruction from PCA. Original image (upper left) and reconstructed images using top 10 PCs (upper right), top 50 PCs (lower left), and top 100 PCs (lower right).

methods in this chapter, using images from the UNL-CPPD. The traits used for the analysis are leaf numbers and growing days. Although they are *integer-valued*, we treat them as *continuous variables* and illustrate the approaches designed for continuous traits.

10.3.1 Partial Least Square (PLS) Regression

Partial least squares (PLS) regression is a statistical method to find a linear regression model by mapping the response Y and predictors X to a new space. Because both X and Y are projected to new spaces, PLS methods are known as bi-linear factor models, which study the relationship between X and Y by finding the multidimensional

FIGURE 10.8 Top four eigencorns for UNL-CPPD image dataset.

direction in the X space that explains the maximum multidimensional variance direction in the Y space. The PLS method is a supervised machine learning method because it uses the information of response Y.

PLS starts by setting $\hat{y}^{(0)} = 0$, $X_j^{(0)} = 0$. Then, in the mth iteration ($m = 1, 2, \ldots, k$), where k is predetermined, it carried out the following steps:

1. Calculate the mth derived input Z_m as the linear combination of variables in X, with the weight equal to their strength to Y. That is, $Z_m = \sum_j \langle X_j^{(m-1)}, Y \rangle X_j^{(m-1)}$, where $\langle a,b \rangle = a^T b$ is the *inner product* between vectors a and b, which have the same dimensions. a^T is the transpose of vector a.

2. Calculate the contribution of the mth derived input Z_m to Y using the formula $\frac{Z_m Z_m, Y}{Z_m, Z_m}$. This formula is understood as regressing Y on Z_m to infer the contribution from Z_m. Then calculate the *cumulative* contribution of the first m derived inputs to Y. Denote the cumulative contribution as $\hat{y}^{(m)}$, which was calculated using the equation $\hat{y}^{(m)} = \hat{y}^{(m-1)} + \frac{Z_m(Z_m, Y)}{(Z_m, Z_m)}$ in this step.
3. Each variable in X is orthogonalized to remove the part associated with Z_m. Orthogonalize each $X_j^{(m-1)}$ by removing the part in Z_m: $X_j^{(m)} = X_j^{(m-1)} - \frac{Z_m Z_m, X_j^{(m-1)}}{Z_m, Z_m}$ in this step.

The procedure of the aforementioned three steps is repeated for k iterations (Hastie et al. 2009).

Partial least squares (PLS) is a latent variable approach to modeling the covariance structure between X and Y. By projecting both X and Y into the new space, it can deal with the situation that original variables in X are highly correlated or X is high dimensional; in such cases, projecting the data into new and lower-dimensional space may lead to significant efficiencies. In each iteration, the algorithm constructs the derived inputs as the linear combination of updated orthogonalized X and puts more weight on the orthogonalized variables, which are more closely associated with Y. The procedure ensures that the top set of derived inputs, i.e., the first few Z_ms, have more explanatory power in Y (Alpaydin 2010; Hastie et al. 2009).

PLS regression is widely used in many areas, including bioinformatics, neuroscience, and plant phenotyping. For image-based plant phenotyping, the PLS method does not directly take the raw image (pixel intensities) as the input. Pixel regression refers to the use of regression methods to infer the response at the pixel level and is used in areas such as image segmentation. To predict traits at the whole-image level, a regression model typically takes derived intermediate traits from images as input variables since it is generally believed that aggregates of pixels, rather than individual pixels, are more salient in extracting traits. Therefore, image-based inference of traits at the whole-image level is considered to be a hierarchical or multistep process. For example, Goodfellow et al. (2016) have shown that the problem of inferring the object's identity from images may involve hierarchical learning. *Input pixels* form *edges*, which, in turn, form *corners* and *contours*, which, in turn, form *object parts*, which are then used to characterize the object's identity. A PLS-based method has been proposed to predict growth, water use, and leaf water content in maize plants (Ge et al. 2016). A PLS-based method to predict chemical concentrations (six macronutrients and six micronutrients, including the three macroelements, N, P, and K, used in fertilizers) based on hyperspectral images has also been developed (Pandey et al. 2017). In the PLS-based regression, aggregate *mean intensity* values *within the plant area* are used instead of original pixel-level intensity (Pandey et al. 2017).

We illustrate the PLS regression, using the UNL-CPPD. We again predict leaf numbers and plant growing days. PCs of images are used as inputs. Performance evaluation of the PLS method is based on six-fold cross validation. Figure 10.9 (top) shows a plot of predicted *versus* actual leaf numbers. Since both predicted traits and actual traits are integers, there are overlapping points in the plot; we use *the number of overlapping points* to label them. Similarly, Figure 10.9 (bottom) shows the predicted *versus* actual growing days, labeled by the number of overlapping points. Figure 10.9 illustrates the validity of using PLS regression to predict continuous traits. The correlation coefficient between predicted values and actual values is 0.934 for leaf number and 0.923 for growing days. The mean absolute deviation is 0.502 leaves and 1.917 days, for leaf numbers and growing days, respectively. As mentioned earlier, in the UNL-CPPD, leaf numbers range from 1 to 10, for up to 28 growing days.

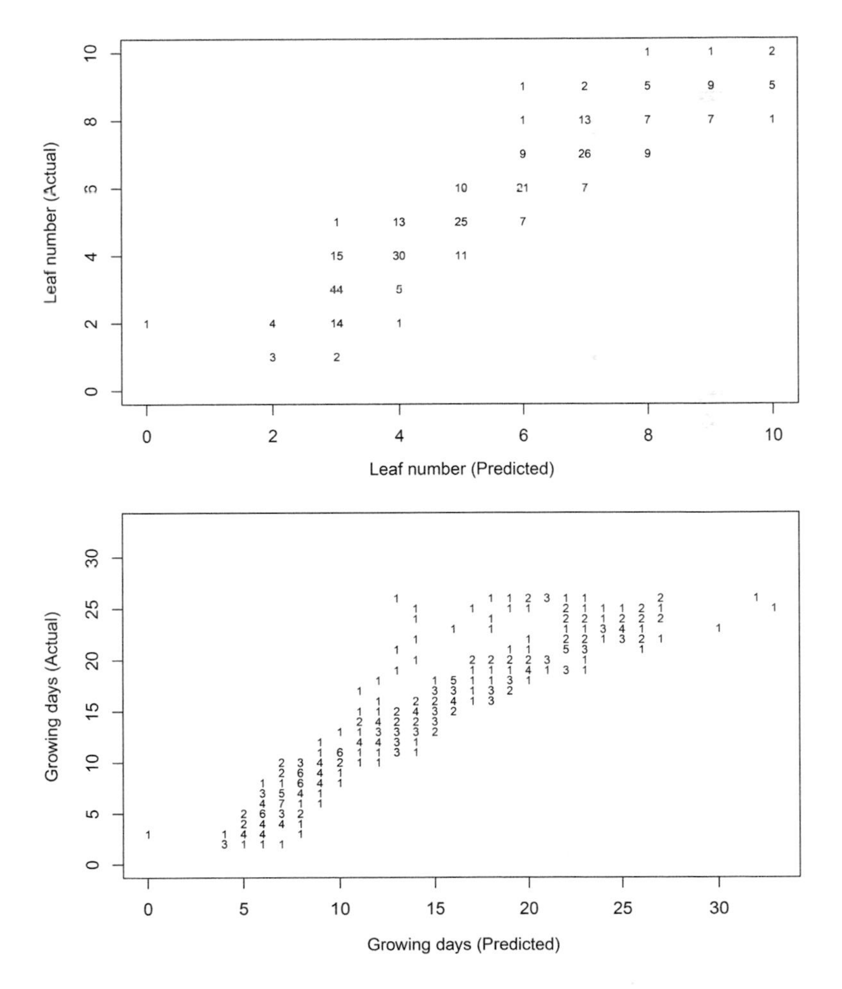

FIGURE 10.9 Predicted traits *versus* actual traits: leaf numbers (top) and growing days (bottom), using PLS labeled with the number of overlapping points.

10.3.2 Least Absolute Shrinkage and Selected Operator (LASSO) for Regression

Least Absolute Shrinkage and Selection Operator (LASSO) for regression is a shrinkage regression method that can simultaneously perform variable selection and regularization to achieve good prediction performance. Prior to LASSO, *ridge regression* was widely used to improve prediction by shrinking large regression coefficients to reduce *model overfitting*. LASSO for regression improves ridge regression in that it can also perform *variable selection* (Tibshirani 1996).

LASSO for regression is especially suitable for high-dimensional regression analysis and variable selection problems. LASSO and its variants have been used extensively to solve problems involving high-dimensional variables or variable selection. Given N observations, the ith observation has p predictors $\left(x_{i1}, x_{i2}, \ldots, x_{ip}\right)$, and the response y_i in the least square regression, the LASSO estimate, is defined by:

$$\hat{\beta}_{LASSO} = \text{argmin}_{\beta}\left\{\frac{1}{2}\sum_{i=1}^{N}\left(y_i - \beta_0 - \sum_{j=1}^{P} x_{ij}\beta_j\right)^2 + \lambda\sum_{j=1}^{P}\left|\beta_j\right|\right\}$$

where $\beta = \left(\beta_0, \beta_1, \ldots, \beta_p\right)$, β_0 is the intercept, $\left(\beta_1, \ldots, \beta_p\right)$ are the slopes of the p predictors, and λ is the weight parameter in balancing the penalty term $\sum_{j=1}^{P}\left|\beta_j\right|$ and the least square term $(1/2)\sum_{i=1}^{N}\left(y_i - \beta_0 - \sum_{j=1}^{P} x_{ij}\beta_j\right)^2$, argmin solves the minimization problem and returns the minimizer of β, i.e. $\hat{\beta}_{LASSO}$, which minimizes the penalized least square function in the curly braces.

We have illustrated LASSO for regression, using UNL-CPPD below. In Figure 10.10, we show a plot of the predicted traits (leaf number and growing days) *versus* the actual traits, labeled by the number of overlapping points. Figure 10.10 illustrates the validity of using LASSO for regression to predict continuous traits. The correlation between predicted values and actual values is 0.903 leaves for leaf numbers and 0.879 days for growing days. The mean absolute deviation is 0.588 leaves and 2.342 days, for leaf numbers and growing days, respectively. In our UNL-CPPD, leaf numbers ranged from 1 to 10, and growing days from 1 to 28.

10.3.3 Random Forests (RF) for Regression

The random forests approach is an ensemble method used widely for both regression and classification. For regression, RF constructs multiple decision trees during training, and outputs the mean prediction value of the individual trees. RF draws a new bootstrap sample from the training data, grows an RF tree to this *bootstrapped sample,* selects m variables at random from the original p variables, picks the *best*

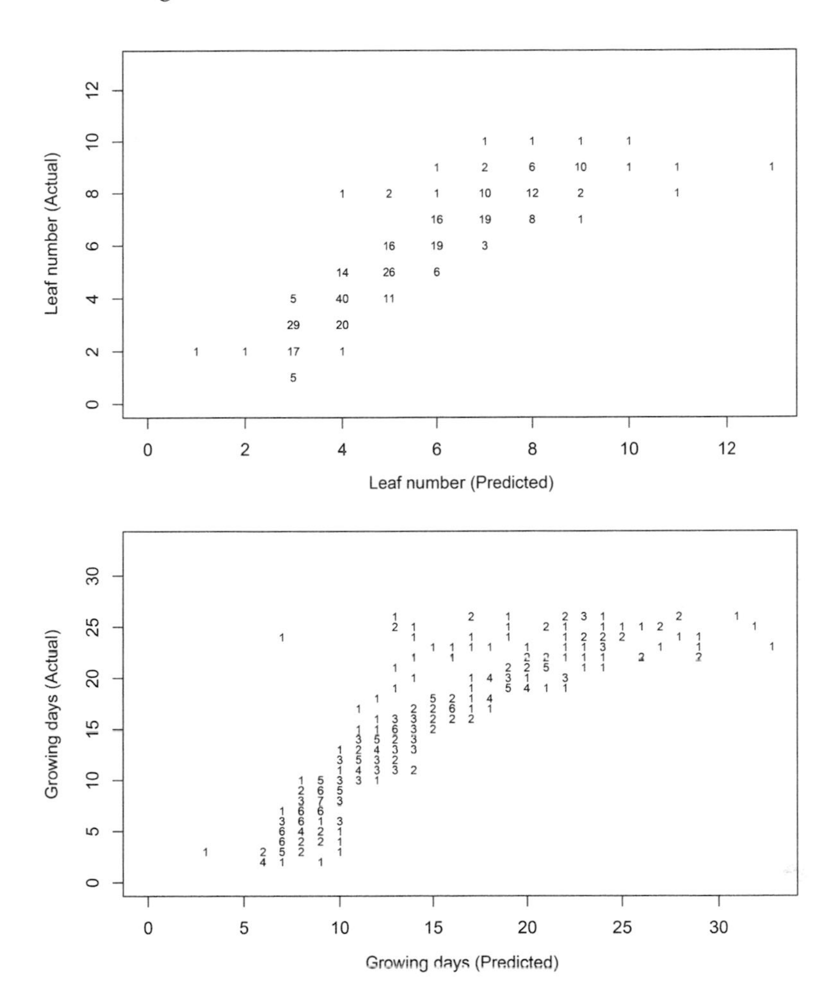

FIGURE 10.10 Predicted traits *versus* actual traits: leaf numbers (top) and growing days (bottom), using LASSO, labeled with the number of overlapping points.

variable/split-point among them, and finally splits the node into two daughter nodes (Hastie et al. 2009).

RFs are widely used in many applications, including plant phenotyping. For example, Singh et al. (2016) proposed several machine learning tools, including RF for high-throughput stress-phenotyping in plants. A recently proposed automated phenotyping system included the use of a super-pixel-based RF algorithm for the task of segmentation in large-scale plant image datasets (Lee et al. 2018).

We illustrate RF for regression, using UNL-CPPD. In Figure 10.11, we have plotted the predicted traits (leaf number and growing days) *versus* the actual traits, labeled by the number of overlapping points. Figure 10.11 illustrates the validity of using RF for regression to predict continuous traits. The correlation between predicted and actual values is 0.913 for leaf numbers and 0.935 for growing days, with mean absolute deviations of 0.645 leaves and 1.815 days, for leaf numbers and

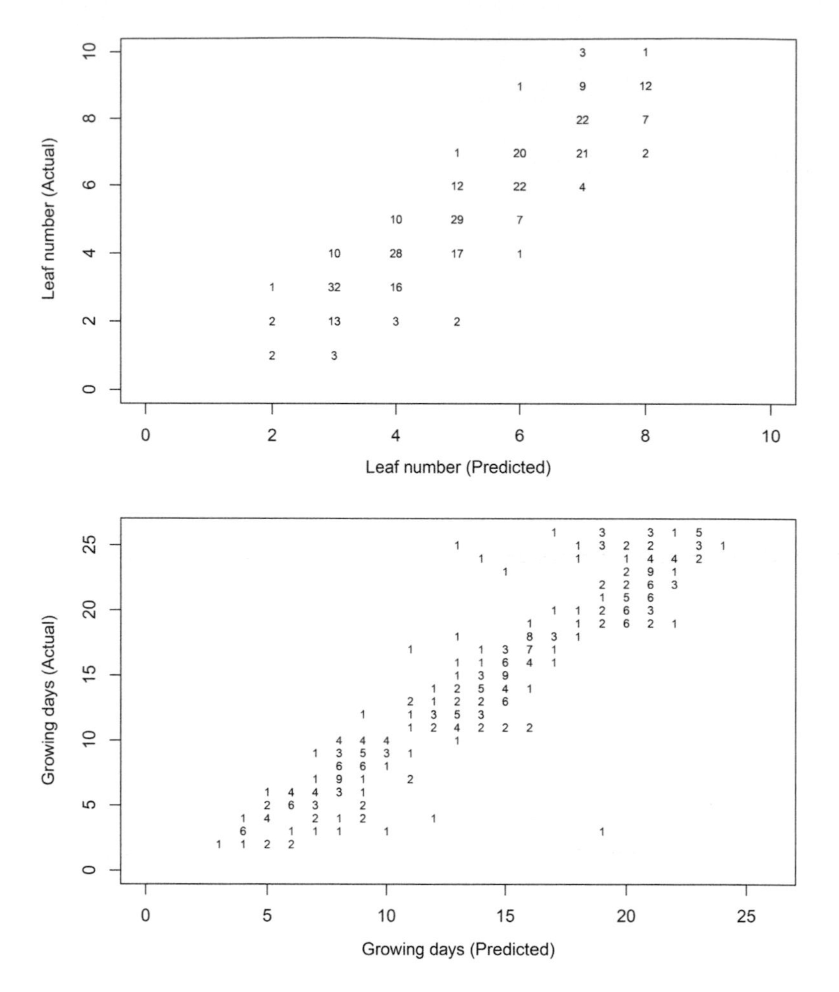

FIGURE 10.11 Predicted traits *versus* actual traits: leaf numbers (top) and growing days (bottom), using random forests, labeled with the number of overlapping points.

growing days, respectively. In UNL-CPPD, leaf numbers ranged from 1 to 10, and growing days from 1 to 28.

Comparing the three methods (PLS, LASSO, and RF), the performances are similar in terms of correlations and mean absolute deviations. The correlations between predicted leaf numbers and actual leaf numbers using PLS, LASSO, and RF are 0.934, 0.903, and 0.913, respectively. The mean absolute deviations between predicted leaf numbers and actual leaf numbers using PLS, LASSO, and RF are 0.502, 0.588, and 0.645, respectively. The correlations between the predicted and actual growing days, using PLS, LASSO, and RF are 0.923, 0.879, and 0.936, respectively. The mean absolute deviation between predicted and actual growing days for the three methods are 1.917, 2.342, and 1.815 days, respectively. Although the performance looks similar in the three methods in terms of correlations and mean absolute deviations, there are differences observed by inspecting the scatter plots of the three methods. RF tends to underestimate high values but overestimates low values, so that the range of predicted values is narrower than the actual range. Similarly, the predicted values using PLS and LASSO methods both have a narrower range than the actual values.

10.4 SUPERVISED MACHINE LEARNING APPROACHES FOR BINARY TRAITS

There is a range of machine learning and statistical methods for classification, including support vector machine (SVM), linear discriminant analysis (LDA), quadratic discriminant analysis (QDA), logistic regression, *k*-means clustering, LASSO for classification, and Bayesian classifiers. Based on RGB and near-infrared images, Guo et al. (2017) studied 37 phenotypic traits and conducted discriminant analysis of plant root zone water status in a greenhouse setting, using three classification models, namely neural networks (NN), support vector machines (SVM), and random forests (RF) for different scenarios and obtained an overall classification accuracy of more than 90% (Guo et al. 2017). Lopatin et al. (2016) used generalized linear models (GLM) and random forests to model vascular plant species richness using LiDAR (Laser + Radar) images (Lopatin et al. 2016).

In this section, we discuss two widely used methods: partial least square discriminant analysis (PLS-DA) and random forest (RF) classification, which both have good explanatory power and provide robust performances. We have illustrated the two methods using the UNL-CPPD. We dichotomized the trait "leaf numbers" into a binary trait LeafyOrNot, such that LeafyOrNot = 1 if the value of leaf numbers is ≥5, which is the median value for the leaf numbers trait in the dataset. Similarly, we dichotomized the trait "growing days" into a binary trait LateStageOrNot, such that LateStageOrNot = 1 if the value of growing days ≥14 days, which is the median number of growing days in the dataset. We use these two *binary traits* for our illustration.

10.4.1 Partial Least Square Discriminant Analysis (PLS-DA)

Partial least square discriminant analysis employs a partial least square (PLS) method to perform discriminant analysis (DA). There are different versions of PLS-DA, based on the PLS method used. PLS-DA involves forming a *regression* model between X and c, where X is the set of attributes and c is a class label; in this case, c is a discrete number with two values for binary classification (Barker and Rayens 2003). PLS-DA provides good explanations for the discriminant analysis *via* the weight and loading (Brereton and Lloyd 2014).

10.4.2 Random Forest for Classification (RF-C)

Random forests for classification (RF-C) is another method widely used for classification, including in plant phenotyping. Similar to random forests for regression, which has already been discussed, RF-C is an ensemble method which draws bootstrapped samples and, for each bootstrapped sample, a random forest tree grows, with the selection of m variables at random from the original p variables; the best variable/split point among them is selected and then the node is split into two daughter nodes. The prediction in RF-C is based on the *majority vote* of these random forest trees, compared with RF-R, which is based on the *mean prediction level* of these random forest trees (Hastie et al. 2009).

We used PLS-DA and RF-C to classify images of UNL-CPPD and evaluated their performance, using six-fold cross-validation. The two binary traits used are LeafyOrNot and LateStageOrNot. Classification accuracy for the binary trait LeafyOrNot was 91.4% and 91.1% for PLS-DA and RF-C methods, respectively. Classification accuracy for the binary trait LateStageOrNot was 86.3% and 94.6% for PLS-DA and RF-C methods, respectively.

10.5 STATISTICAL APPROACHES FOR IMAGE SEGMENTATION

Image segmentation is the process of partitioning an image into different homogeneous regions, i.e., assigning a label to each pixel in an image, such that pixels with the same label have the same meaning or characteristics. Image segmentation is performed at the pixel level. In plant phenotyping, segmentation is useful for isolating individual plants or for identifying the components (leaves, flowers, fruits, etc.) of a plant.

In high-throughput phenotyping, plants are grown under controlled conditions and are imaged individually in customized settings, i.e., background, lighting, camera location, etc. This facilitates the separation of the plant from the background in the image, using segmentation. Once the plant pixels are identified, segmentation can further help to distinguish individual components, e.g., leaves and stems, to facilitate component analysis.

A large number of image segmentation approaches have been proposed in the literature. One of the simplest approaches is thresholding, based on intensity values. Other approaches have been developed for specific applications. General-purpose segmentation algorithms may need to be fine-tuned for plant phenotyping applications, especially in complex situations, e.g., field phenotyping or computation of advanced traits. In this section, we will discuss and illustrate several widely used machine learning and statistical approaches to image segmentation.

10.5.1 *K*-MEANS CLUSTERING

The simplest image segmentation approach is thresholding based on intensities. All pixels above a threshold are assigned to one segment and the remainder to a second segment. Thus, it is used when there are only two classes in the image, e.g., foreground object and the background. However, the thresholding method has the limitation that (1) the threshold value needs to be pre-determined, and (2) thresholding results often need to be further processed (such as morphological operations) for better results. Images with a static background may also be segmented using the *image difference* between the image and the background image. In a greenhouse setting, where plants are often imaged with the same background, this approach may be effective.

In controlled environments, where plants are imaged one at a time, there is a clear difference between the plant and the background pixels. In this case, instead of a threshold-based segmentation, a clustering algorithm may be used. For example, k-means clustering can be used in this case, where the number of clusters (k) can logically be set to 2. The method of k-means clustering has the advantage of not defining the threshold value. The k-means clustering method can be directly applied by simply converting pixel intensities in an image into an input vector. The standard k-means algorithm (Hastie et al. 2009) has been used for plant phenotyping in a controlled environment. For complex image segmentation problems, the variant of the k-means clustering method, which was proposed by Barghout and Sheynin (2013), may be used. It should be noted that the actual image segmentation procedure is typically a multiple-step process and often requires significant post-processing. Thresholding and the k-means methods generate only the initial segmentation in the pipeline.

We have illustrated the thresholding method and the k-means method, using UNL-CPPD. Figure 10.12 shows a plant image (top left) and the results of segmentation,

FIGURE 10.12 Original image (top left), image difference (top right), segmentation based on thresholding (bottom left), and segmentation based on k-means clustering (bottom right).

using the difference image approach (top right), the thresholding method (bottom left), and *k*-means clustering (bottom right).

10.5.2 Approaches for Binary Images

After initial segmentation using either the *k*-means clustering or thresholding approach, further processing is often performed on segmented *binary* images to improve segmentation results. Morphological changes (dilation, erosion, opening, and closing) are commonly used image- processing methods on binary images after initial segmentation (Solomon and Breckon 2011). The method of labeling connected components in a binary image, and then filtering out components by statistical properties (such as by component area and shape statistics in 2D images, or by volume and eigenvalue ratio in 3D binary images) is widely used. Morphological traits such as the plant's height or area can be extracted from binary images. Interesting regions and points (such as plant root and leaf tips) can be extracted by component phenotyping analysis, based on skeletonization and graph-based approaches. We have illustrated the approach to binary images using the UNL-CPPD. Figure 10.13 (top left) shows the results after the initial segmentation by *k*-means clustering, followed by some post-processing. The steps in processing are: (1) morphological closing (dilation followed by erosion), (2) label-connected components, and (3) retaining the largest component (the plant) for phenotype computation. Figure 10.13 (top right) shows the edges (boundaries) of the plant obtained using Frei and Chen's method for edge detection (Frei and Chen 1977). Figure 10.13 (bottom left) shows an illustration of the concept of a convex-hull (the smallest enclosing convex polygon) using a set of points. Figure 10.13 (bottom right) shows the convex-hull of the plant.

10.5.3 Multiple Step State-Of-The-Art Image Segmentation

Approaches to deal with more complex scenarios typically involve multiple steps and have been proposed. To illustrate a multiple-step image-processing segmentation, we present our root-system reconstruction algorithm developed for cassava, based on 3D CT images (Xu et al. 2018). Steps in the algorithm include (1) segmentation based on thresholding, (2) component filtering based on three shape statistics (volume, extent, and principal length ratio), (3) morphological closing by 10 voxels, (4) labeling of connected components, and (5) retaining the smaller components within a distance of 30 voxels from the largest component. Figure 10.14 shows the result of the initial segmentation of the 3D root image (left) and the improved segmentation after the post-processing steps (right).

10.6 APPROACHES COMBINING INFORMATION AND INFERRING TEMPORARY TRAITS

10.6.1 Combining Information

Recent plant phenotyping systems typically include multiple imaging systems for capturing images at regular intervals during a plant's life cycle. Integrating

FIGURE 10.13 Additional processing steps after initial segmentation. Largest component after k-means clustering and closing operation (top left), edge detection results (top right), convex-hull for N points (bottom left), and convex-hull for the plant (bottom right).

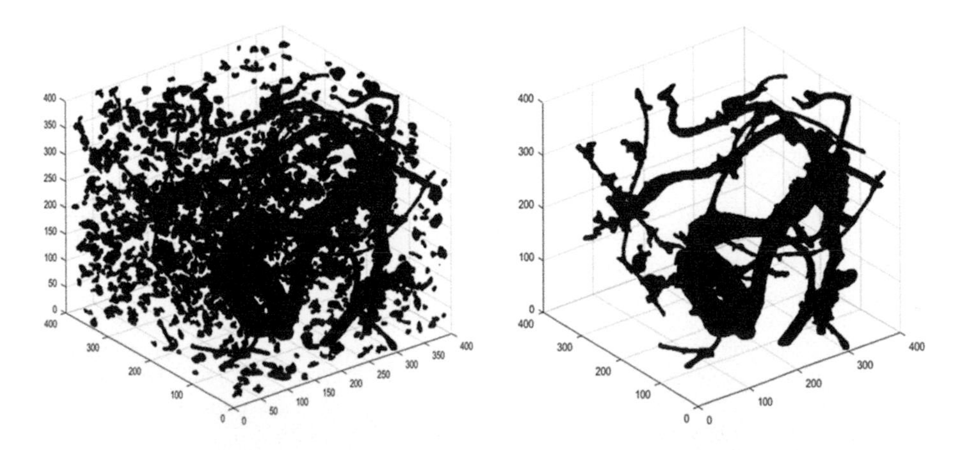

FIGURE 10.14 Cassava root after initial segmentation (left) and further processing (right).

information extracted from multiple images in the same or different modality may improve statistical power and prediction accuracy. For the UNL-CPPD, we found that the addition of growing days as an additional predictor in our machine learning models improved the prediction of leaf number values. This is attributed to the fact that growing days and leaf numbers are highly correlated.

10.6.2 Functional Data Analysis for Temporal Traits

Temporal traits measure the properties of a plant over time. Dynamics in traits can be modeled by curve fitting and functional data analysis. For example, the PhenoCurve system uses a curve-fitting algorithm to identify relationships between phenotypes and environments to study both values and trends in the phenomic data (Yang et al. 2017). Functional data analysis can be used to infer a continuous dynamic curve, typically a growth curve, and make inferences based on it. Temporal traits at a time between two observed time points, when the plants are imaged, can be inferred using nonparametric smoothing methods, whereas plant traits at non-overlapping days can be compared by functional ANOVA (Xu et al. 2018).

10.7 CONCLUSIONS

Plant phenotyping is treated as a feature extraction or trait prediction problem, based on images from the perspective of machine learning and statistics. This chapter discussed and illustrated several machine learning and statistical approaches widely used in image-based plant phenotyping. They included dimensional reduction, supervised learning for continuous traits, supervised learning for binary traits, image segmentation, information pooling, and inferring temporary traits as dynamic curves. The UNL Component Plant Phenotyping Dataset (UNL-CPPD) was used for the illustration of the methods. With rapid progress in high-throughput plant phenotyping, machine learning and statistical approaches are widely and increasingly used in this field and should be explored more in research and practice.

REFERENCES

Alpaydin, E. 2010. *Introduction to Machine Learning.* MIT Press, Cambridge, MA. doi: 10.1007/978-1-62703-748-8_7.

Backhaus, A., A. Kuwabara, M. Bauch, N. Monk, G. Sanguinetti, and A. Fleming. 2010. LEAFPROCESSOR: A new leaf phenotyping tool using contour bending energy and shape cluster analysis. *The New Phytologist* 187(1):251–261. doi: 10.1111/j.1469-8137.2010.03266.x.

Barghout, L., and J. Sheynin. 2013. Real-world scene perception and perceptual organization: Lessons from computer vision. *Journal of Vision* 13(9):709–709. doi: 10.1167/13.9.709.

Barker, M., and W. Rayens. 2003. Partial least squares for discrimination. *Journal of Chemometrics* 17(3):166–173. doi: 10.1002/cem.785.

Bauriegel, E., A. Giebel, and W. B. Herppich. 2011. Hyperspectral and chlorophyll fluorescence imaging to analyse the impact of *Fusarium culmorum* on the photosynthetic integrity of infected wheat ears. *Sensors* 11(4):3765–3779. doi: 10.3390/s110403765.

Blackburn, G. A. 2007. Hyperspectral remote sensing of plant pigments. *Journal of Experimental Botany* 58(4):855–867. doi: 10.1093/jxb/erl123.

Brereton, R. G., and G. R. Lloyd. 2014. Partial least squares discriminant analysis: Taking the magic away. *Journal of Chemometrics* 28(4):213–215. doi: 10.1002/cem.2609.

Brewer, M. T., L. Lang, K. Fujimura, N. Dujmovic, S. Gray, and E. Van Der Knaap. 2006. Development of a controlled vocabulary and software application to analyze fruit shape variation in tomato and other plant species. *Plant Physiology* 28:213–225. doi: 10.1104/pp.106.077867.

Bryant, J. A., N. A. Drage, and S. Richmond. 2012. CT number definition. *Radiation Physics and Chemistry* 81(4):358–361. doi: 10.1016/j.radphyschem.2011.12.026.

Chen, D., K. Neumann, S. Friedel, B. Kilian, M. Chen, T. Altmann, and C. Klukas. 2014. Dissecting the phenotypic components of crop plant growth and drought responses based on high-throughput image analysis. *The Plant Cell* 26(12):4636–4655. doi: 10.1105/tpc.114.129601.

Curran, P. J., J. L. Dungan, and D. L. Peterson. 2001. Estimating the foliar biochemical concentration of leaves with reflectance spectrometry: Testing the Kokaly and Clark methodologies. *Remote Sensing of Environment* 76(3):349–359. doi: 10.1016/S0034-4257(01)00182-1.

Das Choudhury, S., S. Bashyam, Y. Qiu, A. Samal, and T. Awada. 2018. Holistic and component plant phenotyping using temporal image sequence. *Plant Methods* 14:35. doi: 10.1186/s13007-018-0303-x.

Dupont, C., D. R. Armant, and C. A. Brenner. 2009. Epigenetics: Definition, mechanisms and clinical perspective. *Seminars in Reproductive Medicine* 27(5):351–357. doi: 10.1055/s-0029-1237423.

Flavel, R. J., C. N. Guppy, M. Tighe, M. Watt, A. McNeill, and I. M. Young. 2012. Non-destructive quantification of cereal roots in soil using high-resolution X-ray tomography. *Journal of Experimental Botany* 63(7):2503–2511. doi: 10.1093/jxb/err421.

Frei, W., and C. C. Chen. 1977. Fast boundary detection: A generalization and a new algorithm. *IEEE Transactions on Computers* 10:988–998. doi: 10.1109/TC.1977.1674733.

Ge, Y., G. Bai, V. Stoerger, and J. C. Schnable. 2016. Temporal dynamics of maize plant growth, water use, and leaf water content using automated high throughput RGB and hyperspectral imaging. *Computers and Electronics in Agriculture* 127:625–632. doi: 10.1016/j.compag.2016.07.028.

Ginat, D. T., and R. Gupta. 2014. Advances in computed tomography imaging technology. *Annual Review of Biomedical Engineering* 16:431–453. doi: 10.1146/annurev-bioeng-121813-113601.

Goodfellow, I., Y. Bengio, and A. Courville. 2016. *Deep Learning*. MIT Press, Cambridge, MA.

Guo, D., J. Juan, L. Chang, J. Zhang, and D. Huang. 2017. Discrimination of plant root zone water status in greenhouse production based on phenotyping and machine learning techniques. *Scientific Reports* 7(1):1–12. doi: 10.1038/s41598-017-08235-z.

Hastie, T., R. Tibshirani, and J. Friedman. 2009. *The Elements of Statistical Learning: Data Mining, Inference, Prediction*. Springer Verlag, New York. doi: 10.1007/978-0-387-84858-7.

Jansen, M., F. Gilmer, B. Biskup et al. 2009. Simultaneous phenotyping of leaf growth and chlorophyll fluorescence via GROWSCREEN FLUORO allows detection of stress tolerance in *Arabidopsis thaliana* and other rosette plants. *Functional Plant Biology* 36(11):902–914. doi: 10.1071/FP09095.

Kalender, W. 2000. *Computed Tomography*. Publicis MCD Verlag, Oslo, Norway.

Kumar, P., C. Huang, J. Cai, and S. J. Miklavcic. 2014. Root phenotyping by root tip detection and classification through statistical learning. *Plant and Soil* 380(1–2):193–209. doi: 10.1007/s11104-014-2071-3.

Larobina, M., and L. Murino. 2014. Medical image file formats. *Journal of Digital Imaging* 27(2):200–206. doi: 10.1007/s10278-013-9657-9.

Lee, U., S. Chang, G. A. Putra, H. Kim, and D. H. Kim. 2018. An automated, high-throughput plant phenotyping system using machine learning-based plant segmentation and image analysis. *PLOS ONE* 13(4):4. doi: 10.1371/journal.pone.0196615.

Li, M., H. An, R. Angelovici et al. 2018. Topological data analysis as a morphometric method: Using persistent homology to demarcate a leaf morphospace. *Frontiers in Plant Science* 9:553. doi: 10.3389/fpls.2018.00553.

Lopatin, J., K. Dolos, H. J. Hernández, M. Galleguillos, and F. E. Fassnacht. 2016. Comparing generalized linear models and random forest to model vascular plant species richness using LiDAR data in a natural forest in central Chile. *Remote Sensing of Environment* 173:200–210. doi: 10.1016/j.rse.2015.11.029.

Pandey, P., Y. Ge, V. Stoerger, and J. C. Schnable. 2017. High throughput in vivo analysis of plant leaf chemical properties using hyperspectral imaging. *Frontiers in Plant Science* 8:1348. doi: 10.3389/fpls.2017.01348.

Peñuelas, J., J. Pinol, R. Ogaya, and I. Filella. 1997. Estimation of plant water concentration by the reflectance water index WI (R900/R970). *International Journal of Remote Sensing* 18(13):2869–2875. doi: 10.1080/014311697217396.

Rahaman, M. M., D. Chen, Z. Gillani, C. Klukas, and M. Chen. 2015. Advanced phenotyping and phenotype data analysis for the study of plant growth and development. *Frontiers in Plant Science* 6:619. doi: 10.3389/fpls.2015.00619.

Singh, A., B. Ganapathysubramanian, A. K. Singh, and S. Sarkar. 2016. Machine learning for high-throughput stress phenotyping in plants. *Trends in Plant Science* 21(2):110–124. doi: 10.1016/j.tplants.2015.10.015.

Solomon, C., and T. Breckon. 2011. *Fundamentals of Digital Image Processing: A Practical Approach with Examples in MATLAB*. John Wiley & Sons, Hoboken, NJ. doi: 10.1002/9780470689776.

Tibshirani, R. 1996. Regression shrinkage and selection via the Lasso. *Journal of the Royal Statistical Society: Series B (Methodological)* 58(1):267–288. doi: 10.1111/j.1553-2712.2009.0451c.x.

Trachsel, S., S. M. Kaeppler, K. M. Brown, and J. P. Lynch. 2011. Shovelomics: High throughput phenotyping of Maize (*Zea mays* L.) root architecture in the field. *Plant and Soil* 341(1–2):75–87. doi: 10.1007/s11104-010-0623-8.

Turk, M., and A. Pentland. 1991. Eigenfaces for recognition. *Journal of Cognitive Neuroscience* 3(1):71–86. doi: 10.1162/jocn.1991.3.1.71.

Xu, Y., Y. Qiu, and J. C. Schnable. 2018. Functional modeling of plant growth dynamics. *The Plant Phenome Journal* 1(1):1. doi: 10.2135/tppj2017.09.0007.

Yang, Y., L. Xu, Z. Feng, J. A. Cruz, L. J. Savage, D. M. Kramer, and J. Chen. 2017. PhenoCurve: Capturing dynamic phenotype-environment relationships using phenomics data. *Bioinformatics* 33(9):1370–1378. doi: 10.1093/bioinformatics/btw673.

11 A Brief Introduction to Machine Learning and Deep Learning for Computer Vision

Eleanor Quint and Stephen Scott

CONTENTS

11.1 INTRODUCTION

While the concept of machine learning (ML) has been around for decades, the past 25 years have seen significant advances in the theory underlying ML as well as the breadth of applications. This is in large part due to exponential increases of data available for training and computational power to train complex models. We briefly introduce the basic concepts of machine learning in general and deep learning in particular, especially as they relate to computer vision and their applications to plant phenotyping.

11.2 MACHINE LEARNING

11.2.1 What is Learning?

There exist numerous formal definitions of learning, depending on the field in question (psychology, education, neurobiology, etc.). For our purposes, we generally define learning as *generalizing from past experiences.*, e.g., given instances of toys each labeled as "truck" or "not truck," a child can quickly learn to identify new toys as trucks based on their shape, size, wheel configuration, etc. This is a natural task that humans take for granted but is non-trivial for a computer to perform. In contrast, a computer can easily memorize gigabytes of data for immediate retrieval and exact comparison to other data, which would tax the abilities of most humans.

Based on this distinction, we would typically employ machine learning in cases where (Mitchell 1997):

1. Human expertise does not exist (e.g., navigating on Mars),
2. Humans are unable to explain their expertise (e.g., speech recognition; face recognition; driving),
3. An optimal solution changes in time (e.g., routing on a computer network; adapting ads on a webpage based on browsing history; driving), and
4. A solution needs to be adapted to particular cases (e.g., biometrics; speech recognition; spam filtering).

In short, machine learning is often employed when one needs to generalize from experience in a non-obvious way. In contrast, one would typically not use machine learning in more straightforward tasks like calculating payroll, sorting a list of words, serving web pages, word processing, monitoring CPU usage, and querying a database. That is, we would not use machine learning when we can definitively specify how all cases should be handled.

More formally, we turn to Mitchell (1997) for his definition of machine learning:

> A computer program is said to *learn* from experience E with respect to some class of tasks T and performance measure P, if its performance at tasks in T, as measured by P, improves with experience E.

There are many ways for, E, and P to manifest. We explore a sampling of these below.

11.3 EXAMPLE TASKS

Supervised learning: In *supervised learning*, the learner is given a *training set* $X = \{x_1, \ldots, x_n\}$ of *examples* (or *instances*), each x_i with a *label* y_i. Each instance is described as a vector of values for a collection of *features*. For example, a description of a toy vehicle might include features *number-of-wheels* (integer), *height-over-width* (real), and *hauls-cargo* (Boolean). Label *yi* of example x_i could be from the set {*truck*, *non-truck*}. The learning algorithm then searches a set $\mathcal{F}$ of functions for one specific $f \in \mathcal{F}$ that optimizes a performance criterion, say, minimizing prediction error on X. This function f (sometimes called the *hypothesis* or *model*) is used to later predict the labels of new instances $x' \notin X$, i.e., to generalize.

What we describe above is a simple example of a *binary classification problem*. A generalization of binary classification is *k-class classification*, where the possible labels come from a discrete set of size k. If the set of possible labels is instead the real line, the learning problem is called *regression*.

Unsupervised learning: In *unsupervised learning*, the learner is again given a set X of training data, but no labels are provided. One still wants to infer a model of the data, but instead of predicting labels, we want to understand its *structure* or *distribution*. Specific unsupervised learning tasks include *clustering*, *density estimation*, and *feature extraction*.

Two example applications using unsupervised learning are *missing data imputation* and *data cleaning*. In missing data imputation, a model fills in small gaps in collected data with the values that were most likely present. Data cleaning is a process by which partial or noisy observations are transformed to remove to resemble the correct values (Bora et al. 2018). Another application of unsupervised learning is inferring *generative models*, which are models that will generate new instances according to a probability distribution resembling the one that produced the training set. Such models include hidden Markov models (HMMs) and Bayesian networks. Powerful generative models are also represented by deep neural networks.

Semi-supervised learning: In *semi-supervised learning*, labels are available for a (typically relatively small) subset $X' \subset X$ of the training data. For example, if image classification is the learning task, it is easy to gather a very large set X of images from the web, but there is a cost associated with a human labeling those in X'. The learning algorithm could then use the unlabeled instances of $X \backslash X'$ to *pre-train* a model (e.g., to identify what features are relevant in distinguishing examples), and then those in X' can be used to choose the hypothesis f.

A variation of semi-supervised learning is *active learning*, where the learner can select some unlabeled instances from $X \backslash X'$ to additionally label, at some cost. The goal is to learn a good model f while purchasing as few labels as possible.

Reinforcement learning: In *reinforcement learning*, an agent interacts with its environment. At each step, the agent perceives the *state* $s \in S$ of its environment and takes *action* $a \in A$. Choosing action a results in some reward r and changes the state to $s' \in S$. These are the basic steps of a *Markov decision process* (MDP) M. The learner's goal is to learn a *policy* $\pi: S \rightarrow A$ that, when applied to M, maximizes the agent's expected long-term reward. Applications of reinforcement learning include games such as Backgammon, Go, and video games, as well as controllers for robots and self-driving cars.

Reinforcement learning differs from other learning tasks in that the feedback (reward) is typically delayed. It often takes several actions before a reward is received (e.g., no reward is received in Backgammon until the game ends). This leads to the problem of *temporal credit assignment*, in which the learner needs to decide how much each action contributed to the final reward. Reinforcement learning is covered in detail by Sutton and Barto (2018).

11.3.1 Performance Measures

In supervised and semi-supervised learning, performance can be measured in several ways. *Classification error* is the fraction of n labeled test examples* of an independent *test set* that f misclassifies. *Squared error* is:

$$\sum_{i=1}^{n}\left(y_i - f\left(x_i\right)\right)^2.$$

This error measure is commonly used in regression problems. If the labels y_i and the predictions $f(x_i)$ both come from {0,1}, then this evaluates to the same as classification error. Finally, *cross-entropy* is a performance measure that is useful when $f(x_i)$ is a prediction of the *probability* that $y_i = 1$:

$$-\sum_{i=1}^{n} y_i \ln f\left(x_i\right) + \left(1 - y_i\right) \ln\left(1 - f\left(x_i\right)\right).$$

This generalizes to $k > 2$ classes by setting $y_{ij} = 1$ if x_i's class is class j and 0 otherwise, and having $f_j(x_i) > 0$ be the model's predicted probability that x_i is class j:

$$-\sum_{i=1}^{n}\sum_{j=1}^{k} y_{ij} \ln f_j\left(x_i\right).$$

A key advantage of squared loss and cross-entropy is that both functions are continuous and differentiable. This allows the use of gradient-based methods to optimize the choice of f, which is common in neural network-based methods.

In reinforcement learning, the performance of a policy is typically measured by repeatedly using the policy to control an agent in a simulated environment and computing the total reward per episode (e.g., a full run of a game).

Unsupervised models that perform instance generation have their own measures of quality. Because they define a probability distribution, one very straightforward metric is the likelihood of data under the model distribution. This is usually reported in the form of negative log likelihood. Unfortunately, this doesn't necessarily correlate with the quality of output (Wu et al. 2016), so a set of rough guiding metrics have been developed. These include measuring the quality of a classifier trained using the generated data (Salimans et al. 2016), checking that the likelihood of out-of-distribution data under the model is appropriately high (Du and Mordatch 2019),

* A generalization of classification error often used in image classification tasks is *top-k classification error*. In this case, given a labeled test instance x_i, f predicts the k classes that it considers most likely for x_i. The error rate is defined as the fraction of labeled test examples whose label is not in f's top k. For image classification competitions with 100 or more classes, $k = 5$ is common.

and comparing the statistics of real data against those of generated data to ensure sufficient similarity (Heusel et al. 2017). These are meant to ensure that the model produces data that humans judge to "look right", aren't spurious, and captures all types of information about the dataset, respectively.

11.3.2 Overfitting and Underfitting

Selecting the set of candidate model functions $\mathcal{F}$ is a critical step in machine learning. If $\mathcal{F}$ is not expressive enough to represent the important aspects of the data, then we say that our chosen $f \in \mathcal{F}$ *underfits* the data X. For example, say that each label $y_i = 2x_i^2 - 3x_i + \epsilon_i$, where ϵ_i is a Gaussian-distributed noise parameter. Thus, labels in this one-dimensional regression problem are quadratic in the single attribute. However, if $\mathcal{F}$ is limited to linear functions in the attribute, then it underfits since there will not exist any $f \in \mathcal{F}$ that has low error on the training set X, and presumably also will not generalize well.

On the other hand, if $\mathcal{F}$ is *too* expressive, then it can *overfit* X by fitting the noise in the data. For example, if $|X|=n$, and $\mathcal{F}$ consists of all polynomials of degree $n-1$, then there exists a function f that perfectly fits the data, including each noise term ϵ_i. Such a function f is unlikely to generalize well. Figure 11.1 gives a simple example of under- and overfitting for a one-feature regression problem. The data points (circles) can be perfectly fit by a complex model (dashed curve), but that model is unlikely to generalize well. On the other hand, the linear function (dotted line) is too simple to model the data. Finally, the solid parabolic curve nicely balances error minimization with model simplicity.

A popular method to address underfitting and overfitting is to choose $\mathcal{F}$ to be expressive enough to model X, but limit its representational power beyond that. Often

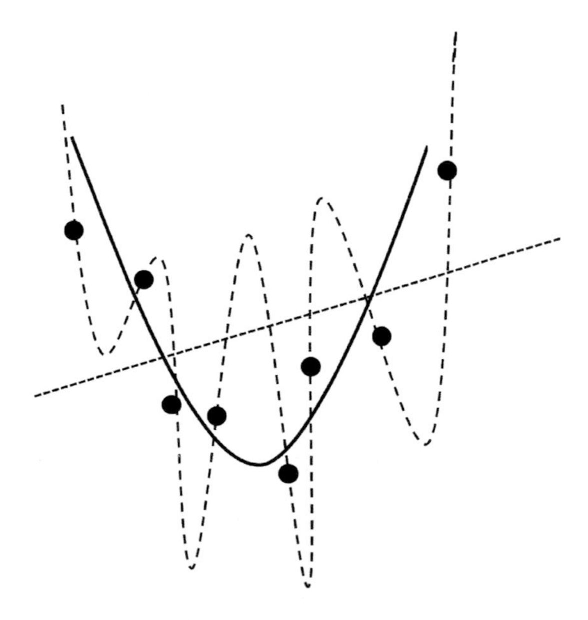

FIGURE 11.1 Examples of under- and overfitting.

this entails optimizing a function that combines training set performance of f (say, squared loss) with a measure of f's complexity, which is called a *regularization* term. For example, if f is a non-linear, real-valued function of the attributes that has d parameters $w_1, \ldots, w_d$, then we can measure f's complexity by the sum of the squares of these parameters (so-called L_2 regularization), and penalize large values:

$$\sum_{i=1}^{n}\left(y_i - f\left(x_i\right)\right)^2 + \alpha \sum_{j=1}^{d} w_j^2,$$

where α weights the complexity term against the loss term. Other forms of regularization include L_1 regularization (replacing w^2 above with $|w|$), and *activation penalization* (for a neural network, replacing w^2 above with the squares of neuron outputs). *Dropout* is another neural network regularization method in which neuron inputs are dropped out (set to zero) at random with probability β. Finally, another popular regularizer is *early stopping*, where after every training iteration (or *epoch*), the current model is evaluated against a *validation set*, which is independent of the training set. So long as performance on the validation set continues to improve, training proceeds. Once validation performance begins to decline, overfitting is likely occurring, so training stops and the final model is the one that was best on the validation set.

Another general mitigator against overfitting is making the training set X as large as possible. If the noise in the labels has zero mean, then it will tend to average out as X grows. If acquisition of new data is difficult or impossible, one can *oversample* the training set, making copies of training instances and then adding some form of zero-mean noise. Such an approach is also helpful in addressing *class imbalance*, where some classes in the training set dominate others in proportion. In this case, it is critical to not change each copied instance in such a way as to change what its label should be. SMOTE (Chawla et al. 2002) satisfies this constraint by taking a minority-class example x along with one of its nearest minority-class neighbors x', and creating a synthetic example from a random point on the line segment connecting x and x'.

11.4 DEEP LEARNING

In ML, there are many possible sets of functions $\mathcal{F}$ that the learning algorithm can choose from, including decision trees, support vector machines, Bayesian networks, and hidden Markov models. Deep learning focuses on a particularly expressive set of functions called *deep neural networks*. A neural network consists of many simple, nonlinear processing units called *neurons*, organized into several *layers*. By feeding the output of one layer into subsequent layers as input, the network's output is the composition of many nonlinear functions. This grants neural networks great expressive power, and the deeper the network, the more expressive it is.

11.4.1 Basic Unit: The Artificial Neuron

An artificial neuron (Figure 11.2), so called because it was loosely inspired by the neurons of the human brain, consists of a set of weights that decide how strongly the neuron should activate when exposed to inputs. Many artificial neurons operating

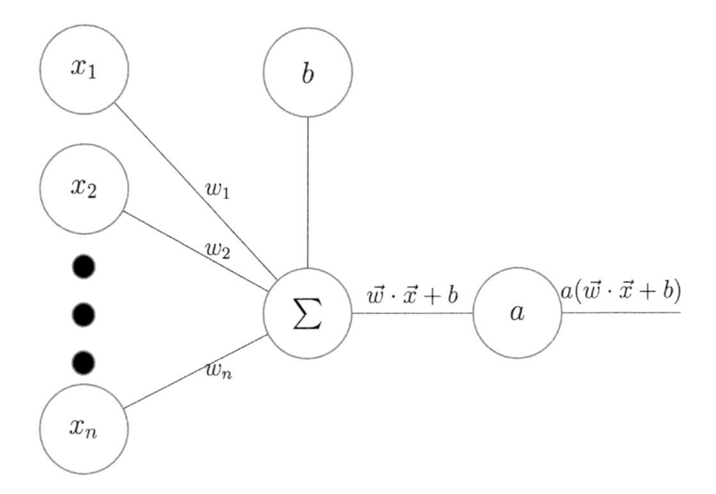

FIGURE 11.2 Artificial neuron with n inputs, n weights, and bias b.

on the same input can be arrayed into a layer so that each can recognize and activate in response to a distinct pattern. Complex functions can be represented by stacking many layers of these transformations so that the output of one layer becomes the input in the following layer. Then, each layer only needs to make a relatively simple transformation so that their composition accomplishes the task. We call this layer a *dense layer* (Figure 11.3) to contrast it with a convolutional layer, which we describe later.

The weight vectors of the nodes in a layer can be combined into a matrix, so applying the layer function to an input is effectively a matrix-vector multiplication. Then, each neuron learns one additional weight called its *bias*, which it adds to its activation. So far, the mathematical operation of applying a layer looks like $\vec{f}\left(\vec{x}\right) = W\vec{x} + \vec{b}$, where $\vec{x}$ is the input vector, W is the layer's weight matrix, and $\vec{b}$ is the bias vector. Then, a network, which is just the composition of these functions, looks like $\vec{f}_k\left(\vec{f}_{k-1}\left(\ldots\vec{f}_1\left(\vec{f}_0\left(\vec{x}\right)\right)\ldots\right)\right)$, where $\vec{f}_i$ is the function returning the vector-valued output of the ith layer.

The astute reader will recognize the layer function as an affine transformation, and further, might remember that the composition of affine transformations is itself an affine function. Thus, to give networks more power, we apply an activation function to the output of each layer. This can be any non-linear function, the most common being variants on a piecewise linear function called the *rectified linear unit*, or ReLU. It is written as* $a(x) = \max(x,0)$. Then, the layer function looks like $\vec{f}\left(\vec{x}\right) = \vec{a}\left(W\vec{x} + \vec{b}\right)$, and networks can express very complex, nonlinear functions.

* There are popular variants of ReLU with different behaviors when $x < 0$. *Leaky ReLU* outputs βx when $x < 0$ for some $\beta < 1$. An *exponential Linear Unit* (ELU) outputs $\alpha(\exp(x) - 1)$ when $x < 0$ for parameter α.

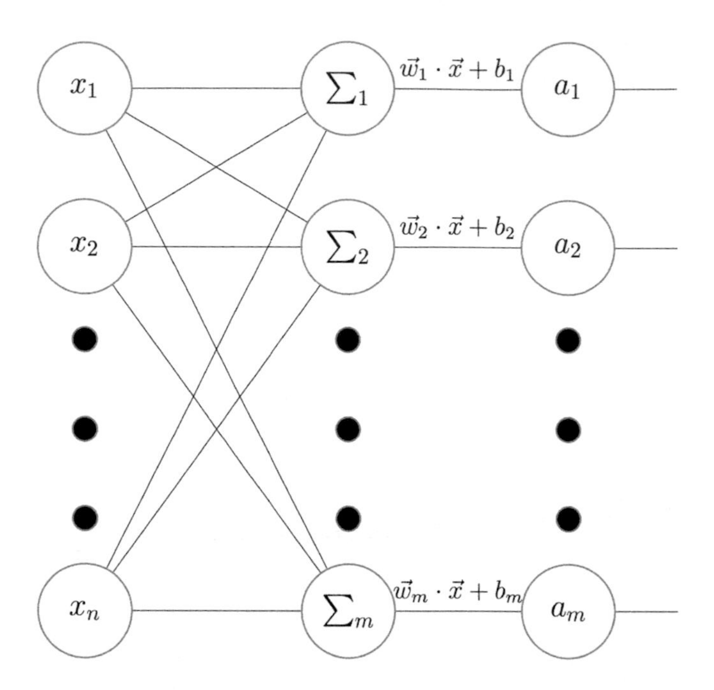

FIGURE 11.3 Dense layer. "$\overrightarrow{w_i}$" denotes vector of weights for node i, "b_i" (omitted for clarity) denotes scalar bias for node i. The layer's output is a vector whose ith component is $a_i\left(\overrightarrow{w_i} \cdot \vec{x} + b_i\right)$.

ReLU activation and its variants work well in the hidden layers (layers $i<k$) of a deep neural network because their output values don't *saturate* like those of sigmoid (s-shaped) functions such as the *logistic function* $1/(1+\exp(-x))$ and *hyperbolic tangent function*, $\tanh(x)$, which used to be common in neural network hidden layers. However, as architectures became deeper, it was discovered that the fact that both of these functions have a derivative that tends to zero as $|x|$ grows makes the hidden layers difficult to train when using them. This is not the case for ReLU-based functions.

On the other hand, for nodes in the output layer $\vec{f}_k$, sigmoid activations are useful when the network's output needs to be constrained to the intervals [0,1] (logistic) or [−1,1] (tanh). Other useful output activation functions include the identity function $a(x)=x$ (called a *linear unit* and used often in regression) and *softmax* for multi-class classification. In softmax, there is one output node per class to be predicted. Let the weighted sum of output node i be z_i. Then the softmax output of node i is

$$\exp(z_i)/\left(\sum_j \exp(z_j)\right).$$

I.e., each weighted sum is exponentiated and normalized so the sum of all the softmax outputs is 1. Since the outputs sum to 1 and each output is positive, this yields a probability distribution over the classes. Training can then be naturally done by

x_{11}	x_{12}	x_{13}	x_{14}
x_{21}	x_{22}	x_{23}	x_{24}
x_{31}	x_{32}	x_{33}	x_{34}
x_{41}	x_{42}	x_{43}	x_{44}

w_{11}	w_{12}
w_{21}	w_{22}

$w_{11}x_{11} + w_{12}x_{12}$ $+w_{21}x_{21} + w_{22}x_{22}$	$w_{11}x_{12} + w_{12}x_{13}$ $+w_{21}x_{22} + w_{22}x_{23}$	$w_{11}x_{13} + w_{12}x_{14}$ $+w_{21}x_{23} + w_{22}x_{24}$
$w_{11}x_{21} + w_{12}x_{22}$ $+w_{21}x_{31} + w_{22}x_{32}$	$w_{11}x_{22} + w_{12}x_{23}$ $+w_{21}x_{32} + w_{22}x_{33}$	$w_{11}x_{23} + w_{12}x_{24}$ $+w_{21}x_{33} + w_{22}x_{34}$
$w_{11}x_{31} + w_{12}x_{32}$ $+w_{21}x_{41} + w_{22}x_{42}$	$w_{11}x_{32} + w_{12}x_{33}$ $+w_{21}x_{42} + w_{22}x_{43}$	$w_{11}x_{33} + w_{12}x_{34}$ $+w_{21}x_{43} + w_{22}x_{44}$

FIGURE 11.4 Convolutional layer example. (Top-left) Two-dimensional input to convolutional layer. (Top-right) 2×2 convolution, defined by 4 parameters. (Bottom) Result of applying convolution to input.

optimizing cross-entropy, and top-k classification error can easily be computed by comparing the k largest outputs with an instance's true label.

11.4.2 Variations

One of the first applications neural networks were specialized to was computer vision. The *convolutional layer* (Figure 11.4) is similar in structure to the dense layer, but rather than learning a weight at each neuron for each input, one learns small filters to transform an input image. A stack of these layers will learn to specialize the filters such that earlier layers recognize smaller, simpler patterns like lines and colors, and later layers will learn to recognize more complicated, compound objects based on the output of the earlier layers. A stack of convolutional layers combined with *pooling layers* (each node taking the maximum or average of its inputs) is often used in image processing to *downsample* an image, reducing its size and memory footprint by identifying and passing forward important features in a translation-invariant way. Such a *convolution stack* is followed by a small stack of dense layers (Figure 11.5), yielding an effective model for recognizing, localizing, and segmenting objects in images.

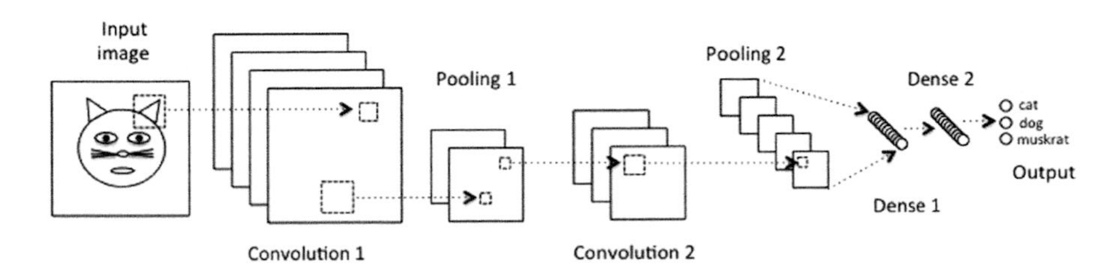

FIGURE 11.5 Convolution and pooling stack feeding a stack of dense layers.

Another variation, used to process time-series data, is called the *recurrent layer*. It is designed to maintain state over time, which allows it to "remember" significant details about recent input data. At the first time step, the recurrent layer is initialized with its default, starting state, and takes in the first step of input data and produces output as well as its updated state. At subsequent time steps, the layer state is propagated from the previous time step. The most common types of recurrent layer are the *long short-term memory* (LSTM) unit and the *gated recurrent unit* (GRU). Recurrent networks are used for language processing, audio recognition and generation, analysis of brain waves, and more.

11.4.3 Training

Generally, a machine learning algorithm searches the space $\mathcal{F}$ of possible hypotheses to find one that optimizes a performance criterion. In the case of artificial neural networks, the candidate hypotheses are the possible assignments of values to the (possibly millions of) weights in the network. The performance criterion is typically the loss function (e.g., square loss or cross-entropy), possibly added to a regularizer such as L_2. When the function L to be optimized is continuously differentiable in its parameters (which is typical for neural networks), then *gradient descent* (GD) is often the preferred method for searching for network weights to optimize the function. Given a training instance, one can compute the gradient of L with respect to the weights. This gradient is an N-dimensional vector (where N is the total number of network weights), and the ith component of the vector is the partial derivative $\partial L/\partial w_i$. I.e., the ith component of the gradient vector quantifies how sensitive L is to the ith weight, and which direction w_i should change in order to reduce L. One can then compute the gradient with respect to each instance in training set X and average the vectors to get an estimate of the expected gradient of L on a randomly-selected instance. Gradient descent then updates each weight w_i by an amount proportional to the ith component of the average gradient, scaled by a *learning rate* η, which controls how large of steps are taken for each update. Values of η vary, but are often in the range 10^{-4}–10^{-1}. A relatively large value of η is useful early in training, since it is likely that the weights at that time are not close to optimal, and so larger, more aggressive steps are permissible (and encouraged, in fact, since it speeds up training). However, if later in training the weights are near a local optimum, then large values of η can cause the GD update to "jump over" the local optimum. On the other hand, since small learning rates can slow training, it is advisable to start with a larger value of η. Thus, one often will initialize η to a relatively large value, then decrease it during training, a process sometimes called *learning rate decay*.

Since typical deep learning data sets consist of training sets of sizes larger than 10^4 and the number of parameters can exceed 10^6, it is typically infeasible to fit the entire training set into memory to perform GD. Thus, one will randomly split the training set into *mini-batches* of size 10^2–10^3 and update the weights one mini-batch at a time. This injects more randomness into the estimate of the gradient, hence its name *stochastic gradient descent* (SGD). Adam (Kingma and Ba 2015) is a commonly used SGD algorithm, which includes the use of a *momentum* term that bases parameter updates not only on the current gradient but also on the direction of prior updates. I.e., the direction of the current update step is a weighted combination of the current gradient and the direction of the previous step's momentum term. This can help "push" an update past a local optimum in the search.

In addition to the automatically learned parameters of the model, a number of values must be fixed prior to training. These are called *hyperparameters* and they include the learning rate η (and possible decay schedule) of the optimizer, early stopping strategy, the training mini-batch size, the maximum number of training *epochs* (passes through the training set), weight of the momentum term, weight of the regularizer term, and more. Often, one will train the same neural network architecture with multiple hyperparameter values, across different subsets of the training set X (called *cross-validation*) to determine the values that work the best. Often combined with a hyperparameter search is an *architecture search*, where one varies the number of layers and number of nodes per layer.

In addition to the numerical considerations, there are some practical ones that can greatly improve the training process. Using a graphics processing unit (GPU) for training a deep neural network typically reduces the training time by at least an order of magnitude, though there do exist other *PU variants that are more specialized for deep learning.* Another speed-up opportunity is using a pre-trained network and doing *transfer learning* to fine-tune the model on a particular dataset. This also enables more powerful models to be trained on smaller datasets. Best practices further include parallelization, checkpointing, data augmentation, batch normalization, training visualization, and good statistical practices for verifying model quality. For more details, refer to Géron (2017).

11.4.4 Technologies

At the time of writing, there are many frameworks designed to help users specify machine learning and deep learning models. Commonly used are Python libraries. *scikit-learn* is the standard for general machine learning, and TensorFlow, Keras, and PyTorch are the most widely used deep learning packages. There exist a variety of pre-trained image classification and segmentation models that are mostly pre-trained, and may be easily adapted to specific tasks.

11.5 APPLICATIONS TO COMPUTER VISION

Deep learning has found significant application in computer vision. Before deep learning, learning from images required the manual design of a pre-processing pipeline that would extract features from the image before feeding the features into a machine learning algorithm. Not only does deep learning obviate the need for such manual labor in images by performing "end-to-end learning," but its automatically-learned features often dramatically outperform human-designed feature extraction.

Early research started by solving simple problems on datasets of small images. Some notable examples include digit recognition using MNIST (a set of 28×28 pixel

* E.g., the Tensor Processing Unit (TPU): https://cloud.google.com/tpu/

handwritten digits) or object detection using CIFAR-10 (a set of 32×32 pixel natural images) (Krizhevsky and Hinton 2009; LeCun et al. 2010). Application of deep learning to computer vision dramatically took off when deep learning models began winning the ImageNet Large Scale Visual Recognition Competition (ILSVRC) (Krizhevsky et al. 2012).

The simplest task for which deep learning is efficient is classification, where the neural network chooses which one of a set of labels best matches an input image (e.g., "cat", "dog", etc.). This can be specialized to *object detection*, where an object's location is indicated in the image by marking a bounding box around its pixels. The You Only Look Once (YOLO) architecture was one of the first popular object detection models. It has the advantage of speed over other region proposal-based models (Redmon et al. 2016). Alternatives to YOLO run many proposed bounding boxes through a convolutional network and check the score of each box, outputting the best. YOLO, on the other hand, simultaneously checks class probabilities and proposes bounding boxes in one forward pass. Since the introduction of the first YOLO model in 2015, many iterations on the basic concept have been proposed (Zhao et al. 2019). Related to the task of object detection is *image segmentation*, where, rather than drawing a bounding box, image pixels are directly labeled, which can be more useful in some applications (Hariharan et al. 2014; Long et al. 2015).

Another application area deep learning has found use in is the reconstruction and enhancement of low-resolution images. This is made possible by probabilistically guessing what some missing information should be, based on the information that is already present. This principle has been applied to the reconstruction of MRIs from measurement data (Cheng et al. 2018), replacing small holes in images (Liu et al. 2018), and completing large portions missing from an image (Oord et al. 2016). Image super-resolution is another problem to which deep learning can be applied (Lai et al. 2017; Ledig et al. 2017).

11.6 APPLICATIONS TO PLANT PHENOTYPING

Applications of deep learning to plant phenotyping are varied, but many share common aspects. First, most employ convolutional and pooling layers, since they have been shown repeatedly to effectively learn to extract features from image and spectral data. Second, many phenotyping applications suffer from a paucity of labeled data, so transfer learning (refining the weights of a pre-trained model rather than training from scratch) and/or data augmentation (enlarging the training set by copying existing labeled instances, each copy with minor modifications) are oft employed. Finally, it is not uncommon to utilize image segmentation to help differentiate plants from the background.

11.6.1 COUNTING

A very popular phenotyping application of deep learning is counting, particularly leaf counting. Dobrescu et al. (2017; 2019) used a neural network for regression based on a pre-trained convolutional stack from ResNet (He et al. 2016), to deal with small training sets. Their experiments used data from the CVPPP 2017 Leaf

Counting Challenge.* Their architecture was able to accommodate images of different sizes and scales and deal with leaves of a variety of shapes. They also analyzed their trained network to determine the most salient parts of the input images to the network's prediction. They found that image backgrounds were largely ignored, and leaf edges were most relevant to counting.

Aich and Stavness (2017) also experimented with data from the CVPPP challenge. Their network took as input not only the image, but also a segmented version (foreground versus background) of the input image. Segmentation was done by the SegNet architecture (Badrinarayanan et al. 2017), and the counting network was based on VGGNet (Simonyan and Zisserman 2015).

Other counting applications of deep learning include fruits and flowers. Rahnemoonfar and Sheppard (2017) built models to count tomatoes. They used synthetic data generated by adding random red circles on a green and brown background. Zabawa et al. (2019) used semantic segmentation, where each pixel is classified as "berry," "edge," and "background," and then they post-processed these pixel-level predictions to count berries. Nellithimaru and Kantor (2019) counted grapes by using three-dimensional representations of images generated from a stereo camera and extracting a variety of features about them. Finally, Xu et al. (2018) counted newly opened cotton flowers by first using a convolutional neural network to classify flowers, constructing point clouds from the CNN's output, registering the detected flowers across images to eliminate double counting, and, finally, counting the flowers in the field.

To predict sorghum biomass, Masjedi et al. (2019) trained on hyperspectral and LiDAR data. They first trained an *autoencoder*, which is a symmetric architecture that is often trained in an unsupervised manner, where the goal is to output a copy of the input. In the case of image inputs, imagine Figure 11.5, but replacing the output layer with a reverse of the architecture (through Dense 2). Once trained, the autoencoder can be used to generate features, by feeding in an image and using the output† of Dense 2 as a low-dimensional representation of the input. Finally, since the data were also temporal in nature, they used a recurrent architecture for the final prediction. They also made predictions via a support vector regression (SVR) approach, an alternative to ANNs.

11.6.2 Other Phenotyping

Besides counting leaves, etc. in images, deep learning has been applied to numerous other plant phenotyping tasks. Singh et al. (2016; 2018) gave detailed overviews of machine learning approaches to identifying stressors. Ubbens and Stavness (2017) trained deep models on a variety of phenotyping tasks, including leaf counting, mutant detection, and age prediction. Pound et al. (2017a) studied root and shoot feature identification and localization. Their models identified if an image contains a root tip, and if an image of a plant shoot contains leaf tips, bases, etc.

---

* https://www.plant-phenotyping.org/CVPPP2017-challenge

† The actual architecture used by Masjedi et al. differs slightly from a symmetric version of Figure 11.5, but the basic principle is the same.

Mardanisamani et al. (2019) trained a binary classifier to identify crop lodging (when stems break or bend). Input images were run through a convolutional stack, whose output was concatenated with additional features extracted from images, e.g., contrast-based features. This concatenation was fed into a connected layer and a 2-node softmax output layer. Training data was augmented by transformations such as horizontal and vertical flipping.

Namin et al. (2018) worked on genotype classification based on top-view images of plants. They used the pre-trained convolutional network AlexNet (Krizhevsky et al. 2012, 2017) due to a small training set. The data they used came from time-lapse recordings. To handle the time-series nature of the data, they used the LSTM recurrent architecture in their model. They also performed data augmentation via rotation of the images.

The model of Pound et al. (2017b) does what is termed "multi-task" crop phenotyping, simultaneously locating spike features and classifying the awned phenotype. Locating spike features was done via pixel-wise classification using an unusual architecture. The architecture consisted of four "hourglass" networks stacked back-to-back, where each hourglass resembles an autoencoder in its symmetry, and utilizing *residual*, or "skip" connections that are known to speed up training (He et al. 2016). They performed data augmentation by choosing random ears and applying random transformations to them.

Deep learning can also be used to estimate measurements. In Vit et al.'s (2019) work on length phenotyping, they analyzed images to estimate the height of banana plants and to estimate the lengths and widths of banana leaves. Images were fed into a convolutional "backbone network," whose output fed into two other networks. The first other network used object detection to specify rectangles around possible leaves, each with a confidence score. Then a separate network took the backbone network's output and identifies "points of interest," which were used in estimating measurements.

11.6.3 Data Augmentation

Throughout this section, we indicated multiple studies where data augmentation was employed. This process typically involves making copies of training instances, each copy modified from its original via basic transformations such as rotation and translation (Mardanisamani et al. 2019; Namin et al. 2018; Pound et al. 2017). Other examples of transformations in data augmentation include the following. Kuznichov et al. (2019) used an augmentation process that preserves the physical appearance of the leaves. They segmented the images to identify the leaves, then applied geometric transformations to these leaves and pasted them randomly to form new images. Ubbens et al. (2018) trained on real data from the PRL dataset (Minervini et al. 2015) and synthetic data based on an L-system model (Lindenmayer 1968). They discovered that, while synthetic data was easier to fit with the network, adding the synthetic data to enlarge the entire training set helped mitigate overfitting.

As an alternative to transforming copies of instances, Giuffrida et al. (2017) built a generative model for training data, based on *deep convolutional generative adversarial networks* (DCGANs) (Radford et al. 2016). Generally, a generative adversarial network (GAN) (Goodfellow et al. 2014) consists of two parts: a *generator G* and a *discriminator D*. The generator *G* is trained to generate synthetic images that resemble the training set, and *D* is trained to distinguish the real images from the fakes. *G* and *D* alternate turns in this game until convergence, at which point *G* is typically capable of generating images similar to the training set. Giuffrida et al. (2015) used this approach to train a generator of images of Arabidopsis plants. They found that their model could generate realistic-looking images after only about 30 training epochs. Further, after adding the generated images to the A4 dataset from the CVPPP 2017 challenge, they saw improved performance in a leaf counting network (Giuffrida et al. 2015).

11.7 CONCLUSIONS AND FURTHER READING

Machine learning, and deep learning in particular, have made enormous strides the past few decades, in terms of advances in technique as well as performance on a wide variety of applications. However, it must be acknowledged that such approaches are not a panacea, and some applications of deep learning require careful tuning of hyperparameters and significant training times.

Although it lacks coverage of recent advances, Mitchell's (1997) textbook is still considered one of the best introductions to machine learning. An excellent more recent textbook is Alpaydin (2014), which covers additional topics beyond Mitchell, including support vector machines and hidden Markov models. A more applied approach to machine learning, including deep learning, is given by Géron (2017). This book presents sample *scikit-learn* implementations of popular ML approaches as well as TensorFlow implementations of neural network architectures. Those seeking a more formal view of deep learning should consult Goodfellow et al. (2016). Those more applied-minded can see Montavon et al. (2012).

REFERENCES

Aich, S., and I. Stavness. 2017. Leaf counting with deep convolutional and deconvolutional networks. In *Proceedings of the IEEE International Conference on Computer Vision Workshop on Computer Vision Problems in Plant Phenotyping* 2080–2089. Venice, Italy.

Alpaydin, E. 2014. *Introduction to Machine Learning*, 3rd ed. MIT Press, Cambridge, MA.

Badrinarayanan, V., A. Kendall, and R. Cipolla. 2017. SegNet: A deep convolutional encoder-decoder architecture for image segmentation. *IEEE Transactions on Pattern Analysis and Machine Intelligence* 39(12):2481–2495.

Bora, A., E. Price, and A. G. Dimakis. 2018. AmbientGAN: Generative models from lossy measurements. In *Proceedings of the 6th International Conference on Learning Representations*. Vancouver, Canada.

Chawla, N. V., K. W. Bowyer, L. O. Hall, and W. P. Kegelmeyer. 2002. SMOTE: Synthetic minority over-sampling technique. *Journal of Artificial Intelligence Research* 16:321–357.

Cheng, J. Y., F. Chen, M. T. Alley, J. M. Pauly, and S. S. Vasanawala. 2018. Highly scalable image reconstruction using deep neural networks with bandpass filtering. arXiv preprint arXiv:1805.03300.

Dobrescu, A., M. Valerio Giuffrida, and S. A. Tsaftaris. 2017. Leveraging multiple datasets for deep leaf counting. In *The IEEE International Conference on Computer Vision Workshops* 2072–2079.Venice, Italy.

Dobrescu, A., M. Valerio Giuffrida, and S. A. Tsaftaris. 2019. Understanding deep neural networks for regression in leaf counting. In *The IEEE Conference on Computer Vision and Pattern Recognition Workshops*. Long Beach, CA.

Du, Y., and I. Mordatch. 2019. Implicit generation and generalization in energy-based models. arXiv preprint arXiv:1903.08689.

Géron, A. 2017. *Hands-On Machine Learning with Scikit-Learn and TensorFlow: Concepts, Tools, and Techniques to Build Intelligent Systems*. O'Reilly Media, Inc. Boston, MA.

Giuffrida, M. V., M. Minervini, and S. Tsaftaris. 2015. Learning to count leaves in rosette plants. In *Proceedings of the Computer Vision Problems in Plant Phenotyping*. Swansea, UK.

Giuffrida, M. V., H. Scharr, and S. A. Tsaftaris. 2017. ARIGAN: Synthetic Arabidopsis plants using generative adversarial network. In *2017 IEEE International Conference on Computer Vision Workshops* 2064–2071. Venice, Italy.

Goodfellow, I. J., J. Pouget-Abadie, M. Mirza, B. Xu, D. Warde-Farley, S. Ozair, A. Courville, and Y. Bengio. 2014. Generative adversarial nets. In *Advances in Neural Information Processing Systems* 2672–2680. Montreal, Canada.

Goodfellow, I.,Y. Bengio, and A. Courville. 2016 *Deep Learning*. MIT Press, Cambridge, MA.

Hariharan, B., P. Arbeláez, R. Girshick, and J. Malik. 2014. Simultaneous detection and segmentation. In *European Conference on Computer Vision* 297–312. Zurich, Switzerland.

He, K., X. Zhang, S. Ren, and J. Sun. 2016. Deep residual learning for image recognition. In *Proceedings of the IEEE Conference on Computer Vision and Pattern Recognition* 770–778. Las Vegas, NV.

Heusel, M., H. Ramsauer, T. Unterthiner, B. Nessler, and S. Hochreiter. 2017. GANs trained by a two time-scale update rule converge to a local Nash equilibrium. In *Advances in Neural Information Processing Systems* 6626–6637. Long Beach, CA.

Kingma, D. P., and J. Ba. 2015. Adam: A method for stochastic optimization. In *3rd International Conference on Learning Representations*. San Diego, CA.

Krizhevsky, A., and G. Hinton. 2009. Learning multiple layers of features from tiny images. Technical Report, University of Toronto.

Krizhevsky, A., I. Sutskever, and G. E. Hinton. 2012. ImageNet classification with deep convolutional neural networks. In *Advances in Neural Information Processing Systems* 1106–1114. Lake Tahoe, NV.

Krizhevsky, A., I. Sutskever, and G. E. Hinton. 2017. ImageNet classification with deep convolutional neural networks. *Communications of the ACM* 60(6):84–90.

Kuznichov, D., A. Zvirin, Y. Honen, and R. Kimmel. 2019. Data augmentation for leaf segmentation and counting tasks in rosette plants. In *Proceedings of the IEEE Conference on Computer Vision and Pattern Recognition Workshops*. Long Beach, CA.

Lai, W.-S., J.-B. Huang, N. Ahuja, and M.-H. Yang. 2017. Deep Laplacian pyramid networks for fast and accurate super-resolution. In *Proceedings of the IEEE Conference on Computer Vision and Pattern Recognition* 624–632. Honolulu, HI.

LeCun, Y., C. Cortes, and C. Burges. 2010. MNIST handwritten digit database. http://yann.lecun.com/exdb/mnist.

Ledig, C., L. Theis, F. Husz´ar, J. Caballero, A. Cunningham, A. Acosta, A. Aitken et al. 2017. Photo-realistic single image super-resolution using a generative adversarial network. In *Proceedings of the IEEE Conference on Computer Vision and Pattern Recognition* 4681–4690. Honolulu, HI.

Lindenmayer, A. 1968. Mathematical models for cellular interaction in development, Parts I and II. *Journal of Theoretical Biology* 18(3):280–315.

Liu, G., F. A. Reda, K. J. Shih, T.-C. Wang, A. Tao, and B. Catanzaro. 2018. Image inpainting for irregular holes using partial convolutions. In *Proceedings of the European Conference on Computer Vision* 85–100. Munich, Germany.

Long, J., E. Shelhamer, and T. Darrell. 2015. Fully convolutional networks for semantic segmentation. In *Proceedings of the IEEE Conference on Computer Vision and Pattern Recognition* 3431–3440. Boston, MA.

Mardanisamani, S., F. Maleki, S. H. Kassani, et al. 2019. Crop lodging prediction from UAV-acquired images of wheat and canola using a DCNN augmented with handcrafted texture features. In *Proceedings of the IEEE Conference on Computer Vision and Pattern Recognition Workshops*. Long Beach, CA.

Masjedi, A., N. R. Carpenter, M. M. Crawford, and M. R. Tuinstra. 2019. Prediction of sorghum biomass using UAV time series data and recurrent neural networks. In *Proceedings of the IEEE Conference on Computer Vision and Pattern Recognition Workshops*. Long Beach, CA.

Minervini, M., A. Fischbach, H. Scharr, and S. A. Tsaftaris. 2015. Finely-grained annotated datasets for image-based plant phenotyping. *Pattern Recognition Letters* 81:80–89.

Mitchell, T. M. 1997. *Machine Learning*. McGraw-Hill. New York, NY.

Montavon, G., G. B. Orr, and K. Müller (Eds.). 2012. *Neural Networks: Tricks of the Trade*. Lecture Notes in Computer Science 7700. Springer. Heidelberg, Germany.

Namin, S. T., M. Esmaeilzadeh, M. Najafi, T. B. Brown, and J. O. Borevitz. 2018. Deep phenotyping: Deep learning for temporal phenotype/genotype classification. *Plant Methods* 14:66.

Nellithimaru, A. K., and G. A. Kantor. 2019. Rols: Robust object-level slam for grape counting. In *Proceedings of the IEEE Conference on Computer Vision and Pattern Recognition Workshops*. Long Beach, CA.

Oord, A. V. D., N. Kalchbrenner, and K. Kavukcuoglu. 2016. Pixel recurrent neural networks. In *Proceedings of the 33rd International Conference on Machine Learning* 1747–1756. New York, NY.

Pound, M. P., J. A. Atkinson, A. J. Townsend et al. 2017a. Deep machine learning provides state-of-the-art performance in image-based plant phenotyping. *GigaScience* 6(10):1–10.

Pound, M. P., J. A. Atkinson, D. M. Wells, T. P. Pridmore, and A. P. French. 2017b. Deep learning for multi-task plant phenotyping. In *Proceedings of the 2017 IEEE International Conference on Computer Vision Workshops* 2055–2063. Venice, Italy.

Radford, A., L. Metz, and S. Chintala. 2016. Unsupervised representation learning with deep convolutional generative adversarial networks. In *4th International Conference on Learning Representations*. San Juan, Puerto Rico.

Rahnemoonfar, M., and C. Sheppard. 2017. Deep count: Fruit counting based on deep simulated learning. *Sensors* 17(4):905.

Redmon, J., S. Divvala, R. Girshick, and A. Farhadi. 2016. You only look once: Unified, real-time object detection. In *Proceedings of the IEEE Conference on Computer Vision and Pattern Recognition* 779–788. Las Vegas, NV.

Salimans, T., I. Goodfellow, W. Zaremba, V. Cheung, A. Radford, and X. Chen. 2016. Improved techniques for training GANs. In *Advances in Neural Information Processing Systems* 2234–2242. Barcelona, Spain.

Simonyan, K., and A. Zisserman. 2015. Very deep convolutional networks for large-scale image recognition. In *3rd International Conference on Learning Representations*. San Diego, CA.

Singh, A., B. Ganapathysubramanian, A. K. Singh, and S. Sarkar. 2016. Machine learning for high-throughput stress phenotyping in plants. *Trends in Plant Science* 21(2):110–124.

Singh, A. K., B. Ganapathysubramanian, S. Sarkar, and A. Singh. 2018. Deep learning for plant stress phenotyping: Trends and future perspectives. *Trends in Plant Science* 23(10):883–898.

Sutton, R. S., and A. G. Barto. 2018 *Reinforcement Learning: An Introduction*. The MIT Press, Cambridge, MA.

Ubbens, J. R., and I. K. Stavness. 2017. Deep plant phenomics: A deep learning platform for complex plant phenotyping tasks. *Frontiers in Plant Science* 8:1190.

Ubbens, J., M. Cieslak, P. Prusinkiewicz, and I. Stavness. 2018. The use of plant models in deep learning: An application to leaf counting in rosette plants. *Plant Methods* 14:6.

Vit, A., G. Shani, and A. Bar-Hillel. 2019. Length phenotyping with interest point detection. In *Proceedings of the IEEE Conference on Computer Vision and Pattern Recognition Workshops*. Salt Lake City, UT.

Wu, Y., Y. Burda, R. Salakhutdinov, and R. Grosse. 2016. On the quantitative analysis of decoder-based generative models. arXiv preprint arXiv:1611.04273.

Xu, R., C. Li, A. H. Paterson, Y. Jiang, S. Sun, and J. S. Robertson. 2018. Aerial images and convolutional neural network for cotton bloom detection. *Frontiers in Plant Science* 8:2235.

Zabawa, L., A. Kicherer, L. Klingbeil, A. Milioto, R. Töpfer, H. Kuhlmann, and R. Roscher. 2019. Detection of single grapevine berries in images using fully convolutional neural networks. In *Proceedings of the IEEE Conference on Computer Vision and Pattern Recognition Workshops*. Long Beach, CA.

Zhao, Z.-Q., P. Zheng, S.-T. Xu, and X. Wu. 2019. Object detection with deep learning: A review. arXiv preprint arXiv: 1807.05511.

Part III

Practice

12 Chlorophyll *a* Fluorescence Analyses to Investigate the Impacts of Genotype, Species, and Stress on Photosynthetic Efficiency and Plant Productivity

Carmela Rosaria Guadagno and Brent E. Ewers

CONTENTS

12.1 INTRODUCTION

Food, energy, and water demands are increasing worldwide due to an ever-growing population, expected to surpass 9 billion by 2050 (IPCC 2014; UN Report 2017). Ongoing environmental changes aggravate global needs, hindering the realization of sustainable development (Cook et al. 2014, 2015). Climate models predict further increases in temperature at all latitudes over the course of the 21st century, with more frequent and severe "drought spells" and changes in rainfall patterns (Choat et al. 2018; Liang et al. 2017; Verduzco et al. 2018). Moreover, elevated atmospheric carbon dioxide and recurring pathogen and insect outbreaks also pose serious threats to natural and managed ecosystems (Breshears et al. 2018; Buras and Menzel 2018; Challinor et al. 2014; Kurz et al. 2008; Lobell and Gourdji 2012; Matesanz et al. 2010; McDowell et al. 2019; Parmesan and Hanley 2015; Rosenzweig et al. 2014; Trumbore et al. 2015).

Although these changes have already been impacting managed and natural ecosystem functions and services, mechanistic models of plant productivity have only provided us with partial information, not yet sufficient to fully predict and improve agricultural and land management practices (Baldocchi et al. 2001; Power 2010). Many plant biophysical responses to the environment are finely regulated by gene expression and the complexity of this interaction limits the predictive capacity of existing approaches (Confalonieri et al. 2016; Pogson et al. 2012; Wallach and Thorburn 2017). There is currently a pressing need for process-based models of productivity to capture both genotypic and phenotypic variation in changing environments to explore the fullest potential of a certain genotype in particular conditions (Pieruschka and Poorter 2012; Ziska and Bunce 2007). While technology and data analysis for the genome have flourished in the last two decades (Jiao and Schneeberger 2017), quantitative analysis for the phenome has been developing at a much slower pace (Bolger et al. 2019; Furbank and Tester 2011). To predictively understand how the environment shapes phenotypes, a multitude of spatial and temporal dimensions, from cells to whole plants to ecosystems must be explored (Houle et al. 2010; Pratap et al. 2019).

New developments in sensor technology and levels of automation have recently increased the throughput for plant phenotyping, i.e., the number of individuals and their traits that can possibly be measured (Fiorani and Schurr 2013; Furbank and Tester 2011; Pratap et al. 2019; Zhao et al. 2019). Modern phenomics, the study of multidimensional phenotypes by high-throughput technologies, is now an emerging field and is likely to close the gap between genome and phenome (Costa et al. 2019). High-resolution sensors, image-based methods, and more robust pipelines for quantitative data analysis have significantly improved our capacity to obtain empirical evidence to test mechanistic understanding of plant responses to a changing environment (Blais 2004; Fahlgren et al. 2015; Marko et al. 2018). High-throughput plant phenotyping platforms (HTPPs) are being utilized and improved around the globe, providing critical phenomic data on enough individuals and their traits, with high enough temporal resolution to be useful in functional genomics methods for gene discovery and regulation (Kumar et al. 2015; Pieruschka and Poorter 2012). Today, phenomic information still lack precise ground-truthing and definitive mechanistic insights (Marko et al. 2018; Pratap et al. 2019), leading to inconclusive results, which limit how these data improve predictive understanding at the temporal and spatial scales relevant to plant controls over ecosystem productivity.

In this chapter, we investigate the use of chlorophyll *a* fluorescence (ChlF) in phenotyping. Easy to measure and monitor, scalable from cells to the ecosystem (Adams and Demming-Adams 2004; Baker 2008; Murchie and Lawson 2013), ChlF may have many of the desirable properties of a high-throughput and mechanistically relevant phenotyping tool. ChlF is directly connected to complex physiological traits of photosynthesis, such as the quantum efficiency of photochemistry and the dissipation of electron energy, and it might help to dissect the complex quantitative genetics of photosynthesis (Hammer et al. 2016; van Bezouw et al. 2019) and their response to stress (Großkinsky et al. 2015; Rungrat et al. 2016). Here, we review the basics of the Pulse Amplitude Method (PAM) and compare low and high-throughput instruments that utilize this ChlF technique. Several applications of the method are also presented to illustrate the sensitivity of ChlF in capturing genotypic difference, species variability, and stress response. We aim to showcase ChlF as an integrative phenomics tool. Such a tool should be able to provide a range of empirical and mechanistic connections to biophysical processes, which are ultimately relevant for yield predictions as well as for the implementation of breeding and agricultural practices to address global climate change and increasing droughts (Pleban et al. 2020).

12.2 MATERIALS AND METHODS

12.2.1 Studied Species

In this chapter, we present physiological data from several species and genotypes under different environmental conditions to highlight the vast applicability of ChlF as a robust tool in plant phenotyping. Specifically, we synthesize data collected on the herbaceous crop *Brassica rapa* (different varieties, crop accessions, and inbred lines), the gymnosperm *Pinus ponderosa*, and two angiosperms, *Populus tremuloides*, a tree, and *Artemisia tridentata*, a shrub. *Brassica rapa* is globally cultivated as a winter annual as well as rapid-cycling fall and spring annual crops. Its wide geographical distribution indicates an evolutionary history of variable growing conditions which makes it an excellent system to study vigor in changing environmental conditions. We use a panel of 12 extreme genotypes, carrying a large variation in the expression of circadian parameters with a consequent difference in morphological and physiological traits (Edwards et al. 2010, 2011, 2012). Results obtained on a Chinese cabbage (CC, CGN06867, pekinensis); a vegetable turnip (VT, CGN10995, D'Auvergne Hative); the oilseed R500 (R500) and Yellow Sarson (YS); the fast-cycling Wisconsin FastPlant self-compatible and its derived Imb211; r46, r301, r171, r213, r215, r325, recombinant inbred lines from the R500xImb211 *BraIRRi* population (Ashraf and Mehmood 1990; Iniguez-Luy et al. 2009; Yarkhunova et al. 2016) are presented. To test the robustness of ChlF and to support its utilization as a sensitive ground-truthing tool for canopy imagery studies in the field, we also examine the response to drought of pine and aspen as two of the highly represented tree species in the world with different susceptibility to soil water limitation. *P. ponderosa* is one of the most drought-tolerant native tree species in the arid western USA, and it has been shown to maintain high water-use efficiency due to superb regulation of osmotic adjustments and the presence of an extensive root system (Daubenmire 1968; Kolb and Robberecht 1996; Kolb et al. 2013). Trembling aspen (*P. tremuloides*) is one of the most abundant broad-leaved species in the world, and it shows spatial heterogeneity in responses to drought and other stressors (Michaelian et al. 2011; Perala 1990). This species significantly reduces leaf area and stem growth to avoid stress with consequent decreased net carbon uptake, eventually leading to mortality (Anderegg et al. 2012; Barr et al. 2007). Finally, we present data collected on a high desert shrub, big sagebrush (*A. tridentata*), a species highly adapted to harsh environments characterized by both high and low temperatures, low rainfalls and high winds (Beverly et al. 2019; Lorenz and Lal 2005). Big sagebrush plays a vital role in the biogeochemical cycles of high-desert ecosystems because of its remarkable water-use efficiency, characterized by hydraulic redistribution of deep ground-water reserves (de Sousa Mendes and de Oliveira 2014) and highly adapted canopy architecture (Knapp et al. 2008; Mencuccini 2003), directly comparable to other species in the genus around the world (Wilske et al. 2010).

12.2.2 Growing Conditions

Data presented in this chapter are the result of several experiments, which took place between 2013 and 2018 in growth chamber or greenhouse conditions. For each experiment, all *Brassica rapa* seeds were planted in pots filled with a soil mix (Miracle-Gro Moisture control Potting Mix (20% v/v), Marysville, OH, and Profile Porous Ceramic (PPC) Greens Grade (80% v/v), Buffalo Grove, IL) amended with 2 ml of Osmocote 18-6-12 fertilizer (Scotts, Marysville, OH) *per* pot, while pine, aspen, and sagebrush were planted as seedlings in native soils using the same amendment. For growth chamber conditions (PGC-9/2 Percival Scientific, Perry, IA), the photoperiod was set at 14 hr/10 hr (day/night) with a photosynthetic photon flux density (PPFD) at the plant height of ~130 mmol photons m^{-2} s^{-1}. Temperature was set to 21°C (±2)/18°C (day/night) cycle with relative humidity maintained between 28–33% (Greenham et al. 2017). For greenhouse experiments, plants were grown at 1800 μmol photons m^{-2} s^{-1} maximum average photosynthetic photon flux density (PPFD), 20.75–25.7°C/20.9–21.3°C day/night, 26.5–73.8% relative humidity with an average of 47.1% (Wang et al. 2019; Guadagno et al. 2017).

12.2.3 Principles of Chlorophyll *a* Fluorescence and the Pulse Amplitude Method (PAM)

In all photosynthetic organisms, the excitation energy absorbed by the light-harvesting complexes is quickly funneled to chlorophyll *a* molecules (reaction centers), which can use it for photochemistry or dissipate it as heat or fluorescence (Butler 1978). This original model for photosystem II (PSII) processes, based on the decrease in fluorescence when electrons are transferred from the reaction centers to the primary quinone acceptor of PSII (Q_A), allowed for the separation of photochemical and non-photochemical (heat) quenching of fluorescence. Before the development of the Butler's model though, the strong correlation between ChlF emission and changes in photosynthetic activity was already known, thanks to observations by Hans Kautsky (Kautsky et al. 1960; Kautsky and Hiersch 1931; Kautsky and Zedlitz 1941). Kautsky et al. showed that the signal of fluorescence changes upon illumination of dark-acclimated leaves and later these changes were mechanistically connected to the leaf photosynthetic activity (Baker 2008; Genty et al. 1989; Govindjee 2004; Kautsky et al. 1960; Maxwell and Johnson 2000).

TABLE 12.1
Chlorophyll *a* Fluorescence Variables Commonly Used in Low and High-Throughput Methods. "Direct" Indicates That the Instrument Provides a Direct Numerical Value for the Measurement

Variable	Measurement	Definition
F_o	Direct	Minimal fluorescence in the dark (open reaction centers)
F_v	Direct	Variable fluorescence signal
F_m	Direct	Maximum fluorescence in the dark (closed reaction centers)
F'_m	Direct	Maximum fluorescence in light (closed reaction centers)
F'_o	Direct	Minimal fluorescence in light (open reaction centers)
F_s	Direct	Steady-state fluorescence
F_v/F_m	Calculated as $(F_m - F_o)/F_m$	Theoretical maximum photochemical efficiency of PSII in dark
F'_v / F'_m	Calculated as $(F'_m - F'_o) / F'_m$	Maximum photochemical efficiency of PSII in light
Φ_{PSII}	Calculated as $(F'_m - F_s) / F'_m$	PSII operating efficiency
NPQ	Calculated as $(F_m \quad F'_m) / F'_m$	Non-photochemical quenching of fluorescence; heat dissipation

* Note: calculation of Φ_{PSII} (Genty et al. 1989) does not require the use of far-red pulse to close all reaction centers, necessary to calculate F'_v / F'_m (Maxwell and Johnson 2000).

Pulse-amplitude modulated (PAM) fluorometry is one of the most common methods to measure ChlF and to quantify the three fates for the excitation energy in the thylakoid membranes (Brooks and Niyogi 2011; Schreiber 2004; Schreiber et al. 1986). The PAM technique uses different light sources to identify changes in ChlF signals. Every PAM fluorometer is equipped with a weak light source to trigger fluorescence (<0.1 μmol photons $m^{-2} s^{-1}$), an actinic light source to drive photosynthesis (<4,000 μmol photons $m^{-2} s^{-1}$) and a high-intensity light to apply saturating (up to 18,000 μmol photons $m^{-2} s^{-1}$) pulses. Accurately switching between these three types of light, fluorometers are able to record ChlF traces over time (Baker 2008; Brooks and Niyogi 2011). Table 12.1 describes the most common direct and indirect variables derived from ChlF traces; extensive reviews of fluorescence nomenclature may be found in Maxwell and Johnson (2000) and Rosenqvist and van Kooten (2003). In dark-acclimated leaves, exposed to a weak pulsing light, Q_A is fully oxidized and ChlF rises to F_o (minimal fluorescence in the dark), representing the 'open state' for the reaction centers of PSII. Upon application of a short (<1 second) saturating pulse, complete reduction of Q_A can be obtained, causing complete 'closure' for the reaction centers and maximum ChlF in the dark, F_m. The difference between F_m and F_o is called variable fluorescence, F_v. Since direct ChlF variables are mostly influenced by leaf optical properties and pigment composition, fluorometry often uses ratios of ChlF values to detect stress and comparing between leaf samples (Baker 2008).

The ratio F_v/F_m represents the maximum (theoretical) quantum efficiency of PSII and the decrease of its value has been widely used as an indicator of stress in several species (Baker 2008; Björkmann and Demming 1987; Butler 1978; Genty et al. 1989; Murchie and Lawson 2013). When leaves are exposed to continuous actinic light, after an initial increase, ChlF is diminished (quenched) by photochemistry and heat dissipation occurring at the PSII level. After stabilization, ChlF reaches F_s, termed steady-state fluorescence. The application of a saturating pulse after acclimation to actinic light discriminates between photochemical and non-photochemical (heat) quenching of the ChlF signal. Upon application of the pulse, the maximum ChlF in light conditions F_m' is recorded. The difference between F_m and F_m' is due to non-photochemical quenching while the difference between F_m' and F_s is equal to the amount of energy utilized in photochemistry (Brooks and Niyogi 2011; Baker, 2008). Non-photochemical quenching mechanisms (NPQ) are energetic safety valves for plants, and they have been defined in their different mechanistic components (Guadagno et al. 2010; Horton and Hague 1988; Lazar 2015). With the application of a very short far-red pulse on light-acclimated samples, Q_A can be immediately oxidized in the dark and another variable can be recorded, F_o', which is used to calculate the maximum efficiency of PSII (F_v' / F_m'). An estimate of the quantum yield of photochemistry within given light conditions (operating efficiency) can also be estimated, avoiding the use of F_o' (Genty et al. 1989) (Table 12.1). The efficiency of PSII estimated from ChlF traces highly correlates to the efficiency of CO_2 assimilation and oxygen evolution, making it a robust proxy, with mechanistic insight, of overall plant photosynthetic status (Baker 2008; Genty et al. 1989, 1992; Harbinson et al. 1990; Horton 1983; Murchie and Lawson 2013; Oxborough and Baker 1997).

12.2.4 Low and High-Throughput Fluorometers

Major challenges in plant phenotyping, such as screening large breeding populations for many relevant traits in a high-throughput manner, increasingly need easy-to-measure and accurate methods to compute them. ChlF has become a routine proxy for photosynthetic efficiency, and demonstrations that this technique is fast, reliable, and non-invasive have accelerated the development of new instruments able to rapidly capture even slight variations in the fluorescence signal (Govindjee 2004). Measurements of ChlF are not in the interest of specialists alone anymore, with several types of PAM fluorometers currently on the market. However, fluorescence data is often misinterpreted which leads to inaccurate conclusions on the photosynthetic performance of the plants (Govindjee 2004; Kalaji et al. 2016; Murchie and Lawson 2013). Moreover, quantitative comparisons between low and high-throughput PAM instrumentations are still scarce (Kuhlgert et al. 2016; Pleban et al. 2020; Tietz et al. 2017) and the interpretation of ChlF information at temporal and spatial scales relevant for ecosystem energetics remains a challenge (Campbell et al. 2019; Lichtenhaler et al. 1986).

In this chapter, we summarize the features of different PAM fluorometers, comparing their mechanistic value and the time utilized for collection (Table 12.2). ChlF can be explained by purely physical phenomena occurring at the cellular level (Butler 1978), yet closely linked to the whole plant physiological status (Genty et al. 1989).

TABLE 12.2
Capability and Characteristics of Low and High-Throughput Instruments for Chlorophyll *a* Fluorescence PAM Measurements Described in This Chapter. The "Mechanistic Value" (Low, Medium, High) Is Assigned Based on the Amount of Biophysically Relevant Information the Instrument Can Provide in the Indicated Time of Collection

Instrument	Manufacturer	Collected Variables (Mechanistic Value)	Time of Collection
FluorPen	Photon System Instrument, Drasov (Czech Republic)	F_v/F_m, F_v'/F_m', F_s *(Low)*	$t < 10$ sec
MultispeQ	PhotosynQ LLC, East Lansing (MI)	F_o, F_m, F_m', F_o', F_s, F_v/F_m, F_v'/F_m', $\Phi PSII$, NPQ *(Highest)*	$t < 30$ sec
FluorCAM	Photon System Instrument, Drasov (Czech Republic)	F_o, F_m, F_m', F_o', F_s, F_v/F_m, F_v'/F_m', $\Phi PSII$, NPQ *(High)*	$t > 3$ min
LI-COR 6400-40 Fluorometer	LI-COR Biosciences, Lincoln (NE)	F_o, F_m, F_m', F_o', F_s, F_v/F_m, F_v'/F_m', $\Phi PSII$, NPQ *(High)*	$t > 5$ min

This mechanistic nature of the fluorescence signal makes it a robust phenotyping tool, sensitive and specific under a wide range of conditions. We assign higher mechanistic value to instruments that in their basic (fastest) configuration allow for collecting the greatest biophysical knowledge to inform on the physiological status of the plant.

- *PAR-FluorPen*: The PAR-FluorPen is a portable PAM fluorometer that enables precise high-throughput measurements of ChlF (https://handheld.psi.cz/products/fluorpen-and-par-fluorpen/). The FluorPen offers straightforward operation at an affordable price, with hundreds of measurements *per* hour taken when utilized in its regular configuration (Table 12.2). Despite its robust results, Fluorpen in its standard configuration does not allow simultaneous measurement and illumination with the ambient light because the sample leaf is inserted between the measuring light source and a clamp. FluorPen has a low mechanistic value because it only records the quantum yield of PSII using a fast protocol, although it offers the option to run light curves and NPQ relaxation dynamics with longer times of collection and more differences from the ambient light status of the leaf.
- *MultispeQ*: The MultispeQ combines the functionality of a handheld fluorometer, a chlorophyll meter, and a bench-top spectrometer into one affordable tool. This instrument is among the most innovative leaf-level devices, allowing for measurements in less than a minute. Absorbance at different

wavelengths, photosynthetic yield, light intensity, etc. are simultaneously taken, making the MultispeQ the instrument currently offered on the market recording the highest amount of mechanistic information in the shortest time. The throughput of this instrument is remarkable, with hundreds of samples possibly measured in a few minutes and immediately shared using the global network, PhotosynQ (https://photosynq.org).

- *FluorCam*: The closed FluorCam FC 800-C represents a highly innovative, robust, and user-friendly system for kinetic fluorescence imaging. The system generates images of ChlF signal and presents them using a false color scale in its basic configuration. This imager allows for analyses of whole leaves to whole plants, trays of seedlings in small pots, and/or in Agar plates. We assign medium mechanistic value to it since full kinetic analysis is only available with a longer recording time (https://fluorcams.psi.cz/products/closed-fluorcam/).
- *LI-COR 6400-40* is a Leaf Chamber PAM Fluorometer that transforms the LI-6400XT infrared gas analyzer into an integrated and powerful phenotyping tool, able to measure ChlF and gas exchange almost simultaneously (https://www.licor.com/env/products/photosynthesis/LI-6400XT/). Based on the principle of extrapolation, this fluorometer uses a multiphase technique that allows for measurements of F_m' in a single flash event. Although the mechanistic value of this instrument is very high because it directly measures ChlF and gas exchange of CO_2 and H_2O simultaneously, it has long times of collection due to the equilibrium state that the leaf has to reach in the measuring head. Now, a new dynamic method for gas exchange has been developed (RACiR), which uses CO_2 ramps instead of discrete values, notably shortening the experimental times in the LI-COR6800 (Stinziano et al. 2019). Although the new instrument equilibrates faster for gas exchange, improving the relationship between ChlF and the other photosynthetic parameters, the LI-COR 6400 is still more present in the literature.

All fluorometers but the FluorPen have complete control over the actinic and saturating lights and measure both direct and calculated variables upon flash application when using more or less articulated protocols (Table 12.2). Despite the medium/high mechanistic value of the FluorCAM and LI-COR instrument, the MultispeQ is currently the most useful PAM fluorometer for its informative and highly mechanistic results obtained within very short times of operation.

Using a variety of species, from annual crops to trees, we compared ChlF data collected with both high and low throughput methods.

Specifically, we compared F_v' / F_m' values collected with a FluorPen (high-throughput) and with the LI-COR fluorometer (low-throughput) (Figure 12.1). Data shown are the result of both growth chamber and greenhouse experiments on aspen (*Populus tremuloides*), pine (*Pinus ponderosa*), sagebrush (*Artemisia tridentata*) and three genotypes of *Brassica rapa* (turnip CGN10995, the oilseed R500, and r46 and r301, recombinant inbred lines from the R500xImb211 BraIRRi population). Well-watered and mild to extreme droughted conditions are represented. All symbols in Figure 12.1 represent single measurements taken using actinic light ranging 50 μmol photons

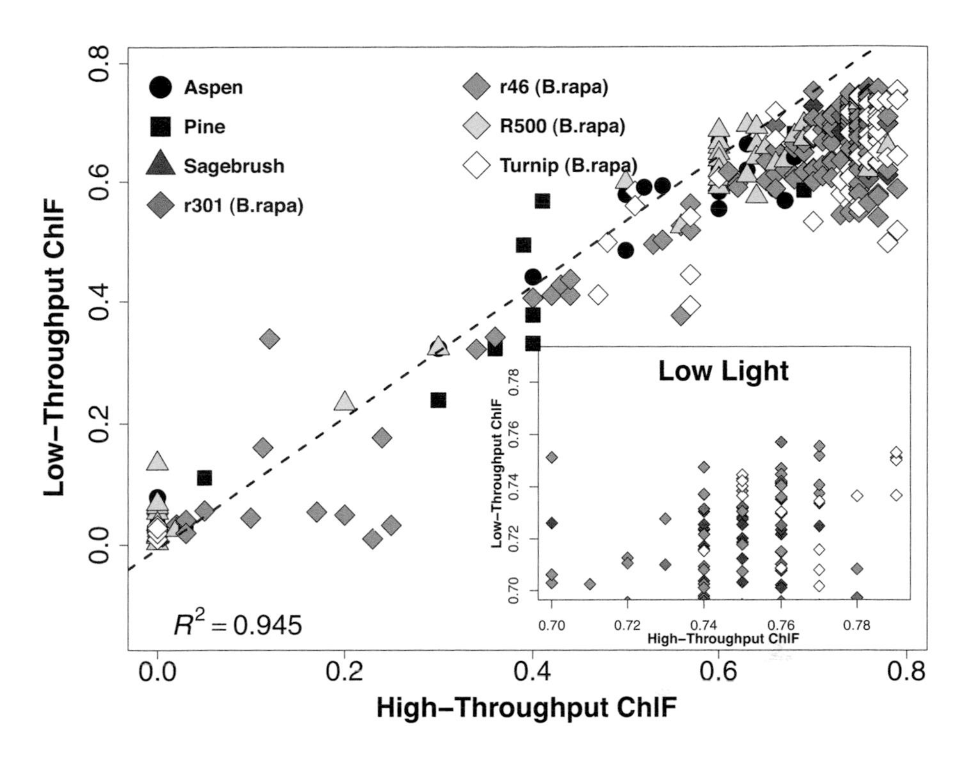

FIGURE 12.1 Chlorophyll *a* fluorescence (ChlF) as the efficiency of PSII in light conditions (F_v' / F_m') recorded with high-throughput (PSI FluorPen) and low-throughput (LI-COR6400-40) instruments.

(m^{-2} s^{-1}) ≤ PPFD ≤ 500 μmol photons (m^{-2} s^{-1}) and a saturating pulse ≥ 3000 μmol photons (m^{-2} s^{-1}) with 0.800 s duration and $\lambda = 470$ nm. The inset panel represents a closeup view of *Brassica rapa* evaluations at the lowest light conditions, PPFD ≤ 200 μmol photons (m^{-2} s^{-1}). We found that the high and low-throughput instruments linearly correlated with a coefficient greater than 0.3 (*R*) ($p < 0.001$). Interestingly, the slope of this relationship was overall not significantly different from 1 but at the lowest light intensities experienced in the growth chambers (Photosynthetic Photon Flux Density, PPFD < 200 μmol photons m^{-2} s^{-1}), when there was no correlation, and the LI-COR fluorometer seemed to lose accuracy, most probably for the longer time of collection (Table 12.2). A linear model considering all light conditions (dotted blue line, main panel) results in $x = -0.006781 + 1.2275y$ with $R^2 = 0.945$ and $p < 0.005$, while a model that does not consider the data at low light conditions (inset panel) shows $R^2 = 0.938$ and $p < 0.005$ for $x = 0.009515 + 0.993908y$. As a matter of fact, with the LI-COR 6400-40 the leaf needs to equilibrate with the environment in the chamber head before a reliable measurement can be taken. Overall, the gas analyzer chamber can be considered more intrusive than the very light-weight leaf clip of the FluorPen. This might have led to less accurate measurements at the lowest photosynthetic photon flux density (PPFD) when the switching between light sources is less drastic, leading to possible overestimations for electron rates and Q_A reduction.

We showed that across very diverse species, large genotypic diversity, and stress, ChlF measurements are intercomparable. This is extremely promising for future phenotyping work because ChlF may be one of the best examples of an appropriate compromise between high-throughput and mechanistically relevant phenotyping.

12.3 APPLICATIONS

12.3.1 Genotypic Variation

Characterizing genetic variation to identify genotypes with adaptive stress responses is one of the major aims of current plant science research (Debieu et al. 2018; Ray et al. 2019; Sircar and Parekh 2015). It is known that in the same species, genotypes may harbor different alleles or show allelic sensitivity at causal loci with altered responses to stress (Edwards et al. 2011; Matsui and Ehrenreich 2016; Saltz et al. 2018). As a consequence, the ability to discriminate between individual genotypes in pre-stressed conditions can explain genetic effects that may not be related to the trait of interest, ultimately resolving the genetic structure of populations for breeding and phenotyping purposes (Brown and Weir 1983; Cooper et al. 2019; Hubby and Lewontin 1966; Kumar et al. 2012).

We attempted to identify genetic variation within *Brassica rapa* types using ChlF. Specifically, we recorded F_v' / F_m' (efficiency of PSII in light conditions) using the FluorPen on twelve genotypes of *B. rapa* (a Chinese cabbage (CC); a vegetable turnip (VT); the oilseed R500 (R500) and Yellow Sarson (YS); the fast-cycling Wisconsin FastPlant self-compatible (FP) and its derived Imb211; r46, r301, r171, r213, r215, r325, recombinant inbred lines from the R500xImb211 *BraIRRi* population).

Despite the use of non-limiting growing conditions and full watering, ChlF captured some small, yet significant, genotypic differences in this particular panel (Figure 12.2). In the figure, histograms represent the frequency of occurrence for each F_v' / F_m' value (bins = 5 for each genotype; bins width = 0.008) for the twelve genotypes of *Brassica rapa*. Measurements were taken using actinic light ranging 100 μmol photons (m^{-2} s^{-1}) $\leq$ PPFD $\leq$ 300 μmol photons (m^{-2} s^{-1}) and a saturating pulse $\geq$ 3000 μmol photons (m^{-2} s^{-1}) with 0.800 s duration and $\lambda = 470$ nm.

Although all genotypes seemed to have the same highest frequency of occurrence, the overall distribution of the values was variable within the panel. With a few notable exceptions that are discussed below, the median values were not significantly different across genotypes. We evaluated the random effects of genotype finding small genotypic variation in well-watered conditions, except for r171 ($p < 0.005$) among the RIL lines which tended to show lower F_v' / F_m'. Among the other genotypes, FP ($p < 0.005$) showed lower median values compared to the group (i.e. all genotypes) (Figure 12.2). r46 among the RILs and VT among the other genotypes covered the biggest range of occurrence with an F_v' / F_m' between 0.650 and 0.795. Genotypes with high variability in PSII efficiency were characterized by larger variation in leaf size and thickness (data not shown) illustrating the connection between structure and function that should be a part of any phenomics program.

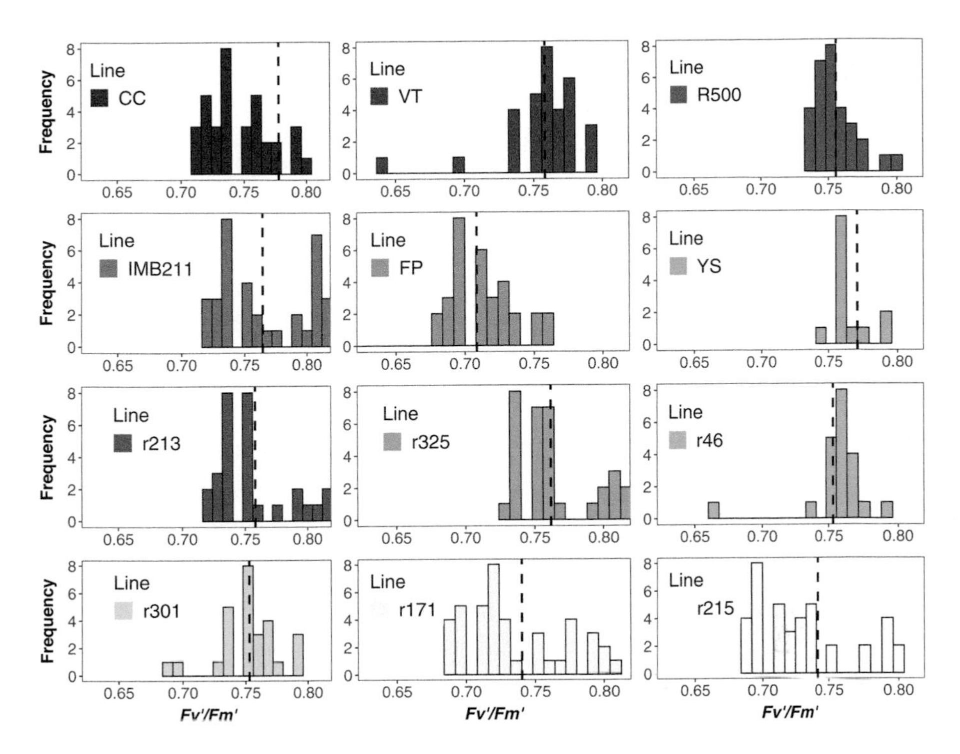

FIGURE 12.2 Genotypic variation in ChIF as the efficiency of PSII in light conditions (F_v' / F_m') recorded with a high-throughput instrument (PSI FluorPen).

Plants acclimate to environmental stress through a combination of fast physiological adjustments and long-term plasticity. Although these stress-driven changes increase plants' water uptake and turgor maintenance, they often result in final yield reduction (Passioura 2006; Torrion and Stougaard 2017). Consequently, there is a pressing need in phenotyping to quantify the relative magnitude of the genotype and genotype × environment interactions. Moreover, uninformed selection by breeders may favor undesired increased value in correlated traits (Edwards et al. 2016; Roff and Fairbain 2007; Sgro and Hoffmann 2004). An essential requirement to enable genetic gain in PSII efficiency within *B. rapa* is the identification of genetic variation. Stress response variation between *Brassica* species has been reported, with Annisa et al. (2013) reporting genetic variation within a global set of *B. rapa* accessions. If genetic variation within the primary gene pool responsible for PSII efficiency is confirmed, this may expedite the process of genetically improving stress response within *B. rapa*.

Screening genotypes for water-use and yield can be resource-intensive (Araus and Cairns 2014; Passioura 2012). Our results demonstrate that ChIF rapidly quantifies phenotypic and genotypic variation (Figure 12.2). ChIF analyses have the potential to enhance high-throughput phenomics and enable earlier selection decisions, with the overall implementation of breeding and agricultural procedures.

12.3.2 Stress Response

Changes in ChlF signal have been previously connected to the dynamics of PSII efficiency and cellular membrane failure (Baker and Roseqvist 2004; Woo et al. 2008). In a previous experiment, we confirmed ChlF to be a rapid screening to assess the degree of drought stress from mild to lethal, and we suggested the loss of variable fluorescence as an operational definition of plant mortality (Guadagno et al. 2017). While the 2017 work used *B. rapa* genotypes and bench-dried pine and spruce (*Pinus contorta* and *Picea engelmanii*, respectively) needles, here we expanded our results using plants naturally dried in pots from the highly drought-resistant species of pine (*P. ponderosa*) and sagebrush (*A. tridentata*) (Figure 12.3).

Using the high-throughput MultispeQ it was possible to follow the progression of drought from Day 1 to Day 58 as changes in ChlF as the efficiency of PSII in light conditions (F_v' / F_m'). Specifically, mild drought corresponds to days between 4 and 15, moderate to days between 16 and 28, severe to days between 29 and 44, and extreme to days between 45 and 58. Lines represent single measurements on sagebrush (*Artemisia tridentata*) leaflets, in gray, and needle fascicles for pine (*Pinus ponderosa*), in black. All measurements were taken using actinic light ranging from 500 μmol photons (m^{-2} s^{-1}) ≤ PPFD ≤ 1000 μmol photons (m^{-2} s^{-1}) and a saturating pulse ≥ 3000 μmol photons (m^{-2} s^{-1}) with 0.800 s duration and λ = 470 nm. For both species, F_v' / F_m' started its decline after Day 28 of drought (Figure 12.3) corresponding to ~8% volumetric water content (VWC), at a soil water potential of −3.5 ± 0.5. ChlF values close to 0.1 (±0.05) which characterize heavy disruption of

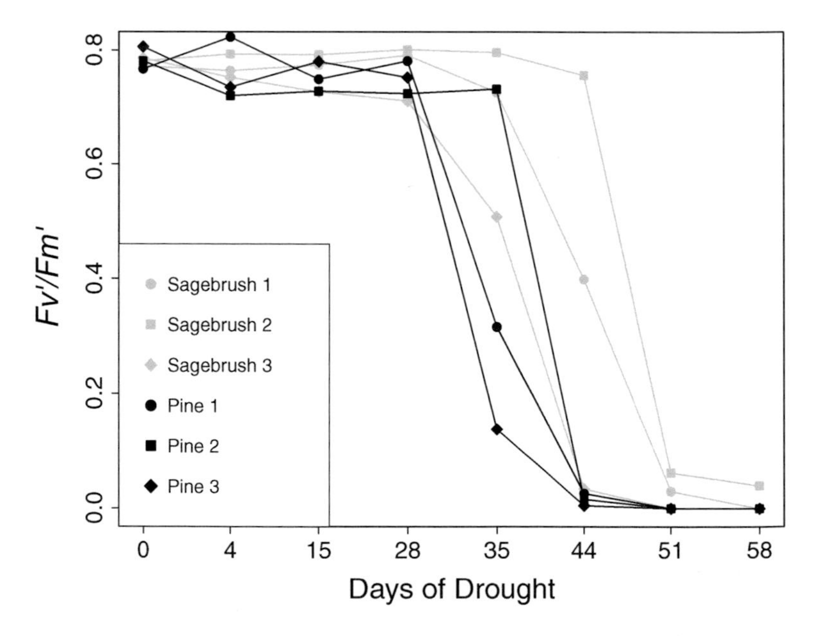

FIGURE 12.3 Impact of drought progression on ChlF as the efficiency of PSII in light conditions (F_v' / F_m') recorded with MultispeQ. Days of drought and individuals correspond to the ones pictured in Figure 12.4.

PSII functionality, were reached after Day 44 for both species, with sagebrush showing an even further decrease after Day 51 of drought.

We validated our findings using the FluorCAM next (Figure 12.4). Here, values of maximum ChlF in light condition (F'_m) were recorded *per image* on detached needles fascicles or leaflets from the same experimental plants used for the Multispeq measurements (Figure 12.3). Once again, we were able to dissect drought progression from mild to moderate to severe, and to the final extreme stage. Maximum chlorophyll *a* fluorescence in light conditions (F'_m) was recorded (PSI, FluorCam) using an actinic light ~800 mmol photons m^{-2} s^{-1} and a saturating pulse of 0.800 s duration, 6000 mmol photons m^{-2} s^{-1} intensity and $\lambda = 470$ nm (Figure 12.4). Drought progression (mild, 4–15 days of drought, moderate, 16–28 days of drought, severe, 29–44 days of drought and extreme, 45–58 days of drought) can be followed as loss of the fluorescence signal (from top to bottom panels).

We also compared pine and sagebrush to the less drought resistant aspen tree (*P. tremuloides*) and to the annual crop *B.rapa*, specifically to two different genotypes, R500 (oilseed) and VT (turnip) (Figure 12.5). We focused our attention on the variation in CO_2 gas exchange, as the net photosynthetic rate (CO_2 μmol m^{-2} s^{-1}), and ChlF as the efficiency of PSII in light conditions (F'_v / F'_m) in both well-watered conditions and during 14 days of drought.

Boxplots represent the spread of leaf-level CO_2 gas exchange (top and bottom left panels) and F'_v / F'_m recorded values (top and bottom right panels) for plants growing in greenhouse conditions. For each species and treatment $n > 60$, with gray panels representing well-watered conditions and white ones showing cumulative progressive (mild to moderate) drought conditions. All measurements were taken at midday (~10.00 h to 13.00 h) using a portable gas exchange system LI-COR6400XT and LI-COR6400-40 fluorometer with the following leaf cuvette settings: flow rate, 300 μmol s^{-1}; CO_2, 400 μmol^{-1} mol air; vapor pressure deficit (VPD), 1.2 ± 0.7 kPa, photosynthetically active radiation (PAR) 600 μmol photon m^{-2} s^{-1}, and block temperature set at 21°C, with cuvette fan on fast mode. Environmental PAR in the experimental greenhouse varied between 300 and 1000 μmol photon m^{-2} s^{-1}. As expected, at full watering ($26 \pm 2\%$ VWC), leaf photosynthetic rate was considerably different among species (Figure 12.5 (top left)), with the oilseed being significantly ($p < 0.01$) more efficient than sagebrush and both the trees. The extent of variation was minimal for sagebrush and pine with respect to *B. rapa* which has been previously shown to cover a large span of gas exchange at full watering in numerous settings (Edwards et al. 2012, 2016; Yarkhunova et al. 2016). In the same conditions, ChlF did not show any genotypic or species variation (Figure 12.5 (top right)). While we have shown that ChlF can highlight genotypic variation in growth chambers with PAR < 300 μmol photon m^{-2} s^{-1} (Figure 12.2), here the genotypic and the species difference were largely exceeded by the presence of a more variable and overall higher environmental PAR ($300 < PAR < 1000$ μmol photon m^{-2} s^{-1}) occurring in the experimental greenhouse. Nevertheless, ChlF largely increased the span of variation in the droughted ($10\% < VWC < 24\%$) cohorts of plants (Figure 12.5 (bottom right)). Remarkably, ChlF emphasized the high physiological variance occurring at the onset of drought among plants in a glasshouse environment (Li et al. 2018) while, although decreased with respect to well-watered conditions, the net photosynthetic

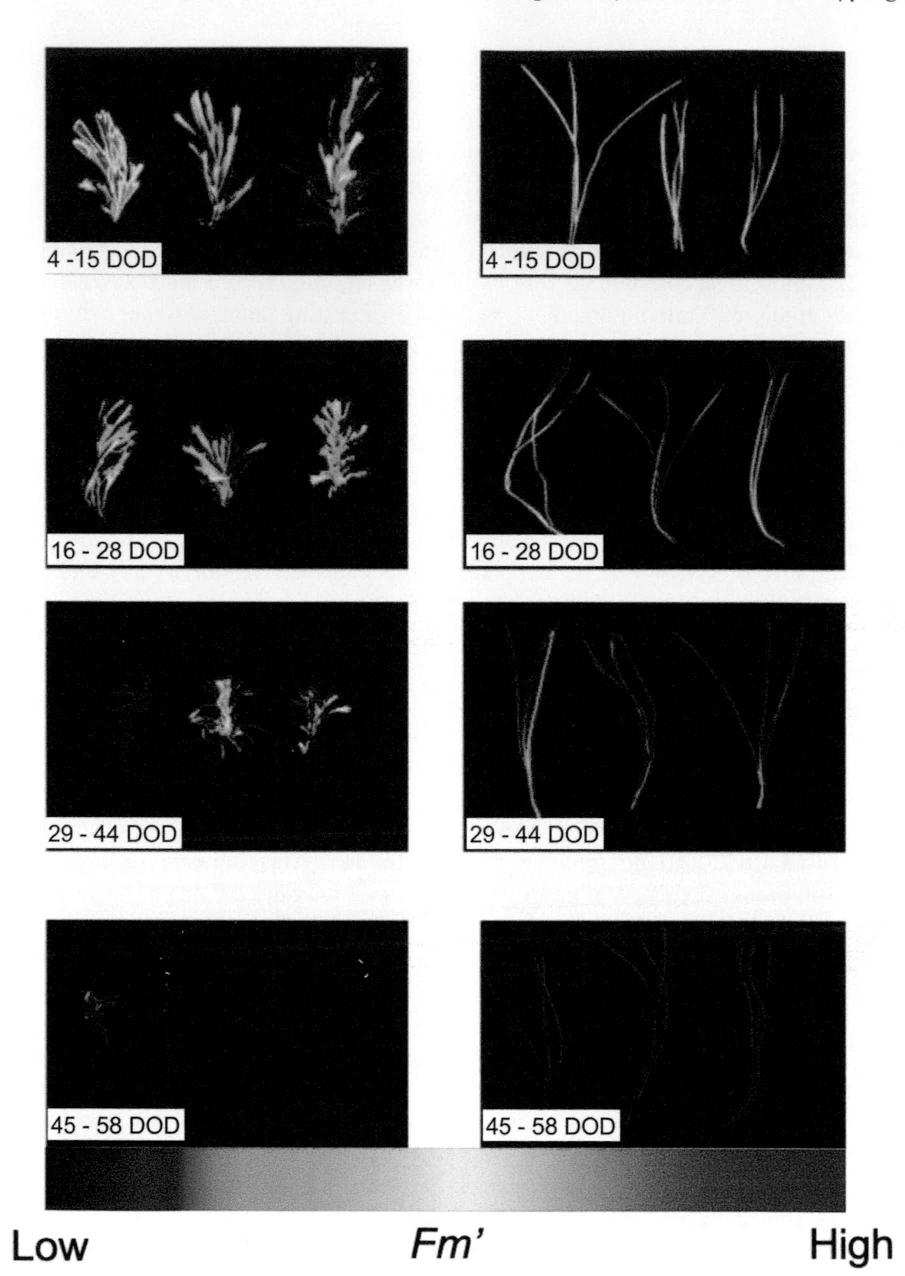

FIGURE 12.4 Drought progression (DOD, days of drought) monitored *via* imaging PAM fluorescence. At each time point, the left panels show 3 leaflets of sagebrush (Artemisia tridentata), whereas the right panels depict 3 needle fascicles from pine (Pinus ponderosa).

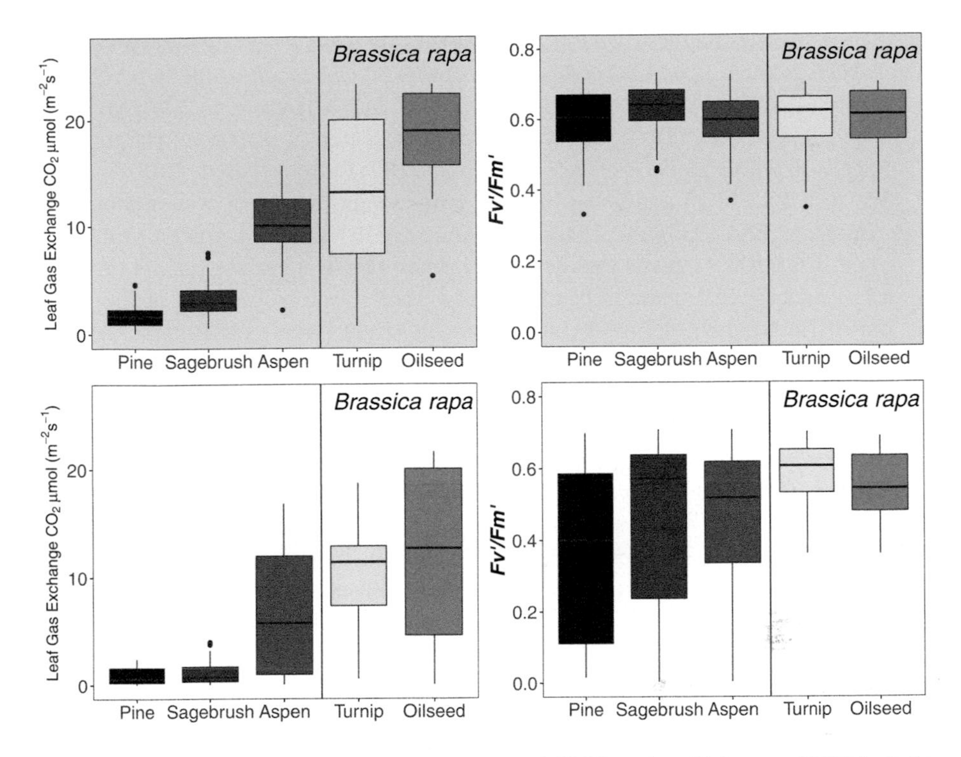

FIGURE 12.5 Variation in CO_2 gas exchange and ChlF as the efficiency of PSII in light conditions (as F_v' / F_m') measured using a portable gas exchange system LI-COR6400XT and LI-COR6400-40 fluorometer.

rate did not significantly change the span of variation among plants of the same species occurring in well-watered conditions (Figure 12.5 bottom left).

Our results corroborate previous work (Baker and Rosenqvist 2004; Lichtenthaler et al. 1986; Papageorgiou and Govindjee 2011; Woo et al. 2008) in showing ChlF signal (and its derived variables) to reliably mirror plant stress such as drought, for different duration/intensity and across a variety of species. Moreover, we demonstrate the highest responsiveness of ChlF in light conditions with F_m' and F_v' / F_m' to be the most robust variables to follow stress response dynamics from early sensing to mortality (Guadagno et al. 2018). In fact, while F_v/F_m values decrease significantly only when plants might have reached an unrecoverable state (Chen et al. 2015), F_v' / F_m' dynamics follow stress dynamics and the relative changes in photosynthetic efficiency from early sensing to death in both trees and herbaceous plants (Guadagno et al. 2017, 2018).

12.3.3 Unconventional Implementations

With the latest developments in technology and the ever-growing use of HTPPs in phenomics, exploration of possible phenotype dimensions is thriving (Pieruschka and Poorter 2012; Pratap et al. 2019; Zhao et al. 2019). Modern imaging systems have been largely shown to accurately measure plant traits from basic morphological ones, such as plant height, to complex physiological ones such as the hyperspectral signature (Das Choudhury et al. 2019; Liu et al. 2017; Mishra et al. 2017; Morais et al. 2019). These traits are nowadays routinely measured at different spatial scales, from cells to whole plants (Gehan and Kellogg 2017; Rouphael et al. 2018).

We tested ChlF in capturing drought progression in turnip organs monitored *via* imaging PAM (Figure 12.6). A mature well-watered *Brassica rapa* (turnip, VT, CGN10995) plant was de-topped when the drought started (Figure 12.6 (top left panel)). In Figure 12.6, the top row shows *in vivo* pictures of the turnip at day 0 (left panel), 15 (mid panel), and 38 (right panel) of drought, whereas the bottom row represents the corresponding fluorescence images for the same days of drought. Maximum chlorophyll *a* fluorescence in light conditions (F'_m) was recorded (PSI, FluorCam) using an actinic light ~800 mmol photons m^{-2} s^{-1} and a saturating pulse of 0.800 s duration, 6000 mmol photons m^{-2} s^{-1} intensity and $\lambda = 470$ nm. Unexpectedly, we were able to follow the changes in F'_m over time in the de-topped plants for 40 days. At the time of de-topping, mature turnips were well-watered, recording the highest values for F'_m, with the highest signal occurring in the innermost part of the organ (Figure 12.6 (bottom left panel)). After 15 days of complete water withholding, F'_m was less than 50% of the initial value; however, a high

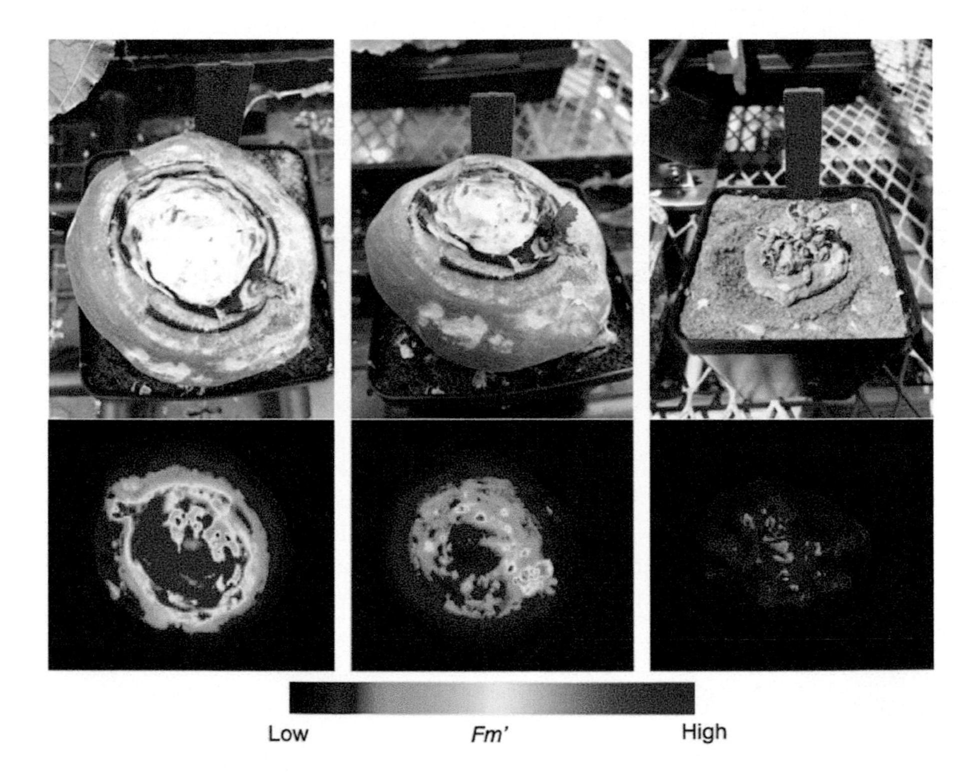

FIGURE 12.6 Drought progression and meristematic activity in turnip organ monitored *via* imaging PAM fluorescence using a FluorCam.

ChlF signal was recorded where re-sprouting initiated (Figure 12.6 (bottom mid panel)). Finally, on day 40 of drought, the re-sprouted turnip lost over 90% of the initial ChlF, and it only had a few live leaves occurring in the central part of the organ (Figure 12.6 (bottom right panel)). These results confirm our previous findings correlating the signal of ChlF to the presence of active meristematic tissue (Guadagno et al. 2017). Vital meristems are rich in protoplast where chlorophyll *a* has been demonstrated to be very abundant (Kobza et al. 1989; Pogson et al. 2015). In the detopped turnip, we speculate that the meristematic tissue is highly efficient to ensure fast re-sprouting, but it will incrementally lose its activity at the end of the drought and at death (Sala et al. 2010). It is not unlikely that sparse photosynthetic activity is occurring underneath the visible layer of other pigments, such as anthocyanins, which give turnips the characteristic purple color (Figure 12.6 (top-left panel)).

These surprising results, open an exciting prospect to the use of ChlF at scales never explored before, suggesting that this trait can be used to operationally define plant death and/or re-sprouting based on the possible detection of meristematic activity.

12.4 CONCLUSIONS

During the last decade, the plant science community has focused on understanding plant water and carbon relations with a particular interest in their dynamic response in stress conditions (Breshears et al. 2018; Lobell and Gourdji 2012; McDowell et al. 2019; Rosenzweig et al. 2014; Trumbore et al. 2015). However, despite intense research and countless phenotyping innovations (Pratap et al. 2019; Zhao et al. 2019), we still lack a generalizable plant vigor indicator, mechanistic enough to directly inform current biophysical models (Martinez-Vilalta et al. 2019).

ChlF, already widely used in modern phenomics, is an easy-to-measure, mechanistic, and scalable proxy for plant stress response (Adams and Demming-Adams 2004; Baker 2008; Murchie and Lawson 2013). In this chapter, we quantitatively compared low and high-throughput PAM methods and their tradeoffs, verifying the robustness of this technique. Our data highlighted crucial differences between the predictive value for ChlF-related traits (e.g., F_v' / F_m') and classic traits focused on H_2O relations (e.g., gas exchange, leaf water potential) at different stress levels. While fluorescence traits (Figure 12.7, shades of red) have a linear decrease, other traits (Figure 12.7, shades of blue-green) are not linearly correlated to stress increment. The black dotted double-ended arrow in Figure 12.7 indicates the shades of colors in all traits, representing the genotypic variation. Chlorophyll *a* fluorescence traits have a linear and thus consistent decline with drought while preserving more genotypic variation. In contrast, water relations traits do not change as systematically with drought and often have less genotypic impacts. Whereas changes in water relations are critical indicators of stress, they do not change as systematically with stress (e.g., drought) and are often unrelated to meristematic activity, resulting in inadequate assessments of mortality-threatening stress. ChlF, instead, is an integrative trait that takes into account spatiotemporal variations in the stressor itself and ChlF-traits follow the dynamics of stress from their onset to death. Remarkably, we also showed ChlF-traits have more room to quantify genotypic variation across

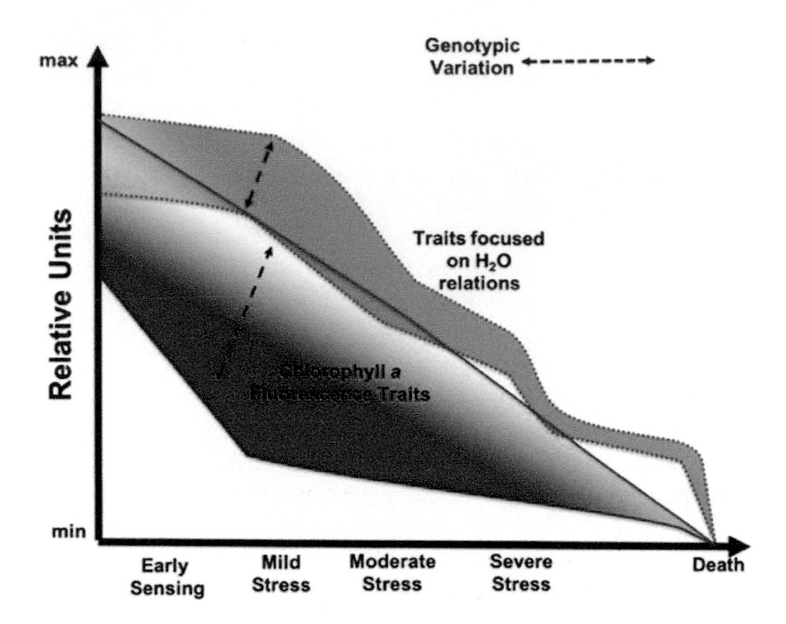

FIGURE 12.7 Conceptual representation of the predictive value for chlorophyll *a* fluorescence-related traits (e.g., F_v' / F_m') and classic traits focused on H_2O relations (e.g., gas exchange, leaf water potential) at different stress levels.

the spectrum of stress response (Figure 12.7), suggesting its key role in the future genome to phenome programs.

Focusing on the analyses of PAM measurements will likely improve the mechanistic underpinnings of parameters and test biophysical model predictions, ultimately moving forward the exploration of multidimensional phenotypes in a changing environment.

ACKNOWLEDGMENTS

We are grateful to Dr. Jonathan R. Pleban, Dr. Diane R. Wang, and Daniel P. Beverly for technical assistance, data collection, and constructive discussion of the data. We thank all the students and technicians for their precious assistance in data collection. NSF funding IOS-1025965, IOS-1547796, and IOS-1444571 supported this research.

REFERENCES

Adams, W. W. III, and B. Demmig-Adams. 2004. Chlorophyll fluorescence as a tool to monitor plant response to the environment. In *Advances in Photosynthesis and Respiration*, eds. G. C. Papageorgiou and Govindjee, 19:583–604. Springer, Berlin.

Anderegg, W. R. L., L. D. L. Anderegg, C. Sherman, and D. S. Karp. 2012. Effects of widespread drought-induced aspen mortality on understory plants. *Conservation Biology* 26(6):1082–1090.

Annisa, S. Chen, N. C. Turner, and W. A. Cowling. 2013. Genetic variation for heat tolerance during the reproductive phase in brassica rapa. *Journal of Agronomy and Crop Science* 199(6):424–435.

Araus, J. L., and J. E. Cairns. 2014. Field high-throughput phenotyping: The new crop breeding frontier. *Trends in Plant Science* 19(1):52–61.

Ashraf, M., and S., Mehmood. 1990. Response of four *Brassica* species to drought stress. *Environmental and Experimental Botany* 30(1):93–100.

Baker, N. R. 2008. Chlorophyll fluorescence: A probe of photosynthesis in vivo. *Annual Review of Plant Biology* 59:89–113.

Baker, N. R., and E. Rosenqvist. 2004. Applications of chlorophyll fluorescence can improve crop production strategies: An examination of future possibilities. *Journal of Experimental Botany* 55(403):1607–1621.

Baldocchi, D. D., E. Falge, L. Gu, R. Olson, D. Hollinger et al. 2001. FLUXNET: A new tool to study the temporal and spatial variability of ecosystem-scale carbon dioxide, water vapor, and energy flux densities. *Bulletin of the American Meteorological Society* 82(11):2415–2434.

Barr, A. G., T. A. Black, E. H. Hogg, T. J. Griffis, K. Morgenstern, N. Kljun, A. Theede, and Z. Nesic. 2007. Climatic controls on the carbon and water balance of a boreal aspen forest, 1994–2003. *Global Change Biology* 13(3):561–576.

Beverly, D. P., C. R. Guadagno, M. Bretfeld, H. N. Speckman, S. E. Albeke, and B. E. Ewers. 2019. Hydraulic and photosynthetic responses of big sagebrush to the 2017 total solar eclipse. *Scientific Reports* 9(1):8839.

Björkman, O., and B. Demmig. 1987. Photon yield of O_2 evolution and chlorophyll fluorescence at 77k among vascular plants of diverse origins. *Planta* 170(4):489–504.

Blais, F. 2004. Review of 20 years of range sensor development. *Journal of Electronic Imaging* 13(1):231–243.

Bolger, A. M., H. Poorter, K. Dumschott et al. 2019. Computational aspects underlying genome to phenome analysis in plants. *Plant Journal* 97(1):182–198.

Breshears, D. D., C. J. W. Carroll, M. D. Redmond et al. 2018. A dirty dozen ways to die: Metrics and modifiers of mortality driven by drought and warming for a tree species. *Frontiers in Forests and Global Change* 1:4.

Brooks, M. D., and K. K. Niyogi. 2011. Use of a pulse-amplitude modulated (PAM) chlorophyll fluorometer to study the efficiency of photosynthesis in Arabidopsis plants. *Methods in Molecular Biology* 775:299–310.

Brown, A. H. D., and B. S. Weir. 1983. Measuring genetic variability in plants population. *Developments in Plant Genetics and Breeding* 1:219–239.

Buras, A., and A. Menzel. 2018. Projecting tree species composition changes of European forests for 2061–2090 under RCP 4.5 and RCP 8.5 scenarios. *Frontiers in Plant Science* 9:1986.

Butler, W. L. 1978. Energy distribution in the photochemical apparatus of photosynthesis. *Annual Review of Plant Physiology* 29(1):345–378.

Campbell, P. K. E., K. F. Huemmrich, E. M. Middleton, L. A. Ward, T. Julitta, C. Daughtry, A. Burkart, A. Russ, and W. Kustas. 2019. Diurnal and seasonal variation in chlorophyll fluorescence associated with photosynthesis at leaf and canopy scale. *Remote Sensing* 11(5):488–515.

Challinor, A. J., J. Watson, D. B. Lobell, S. M. Howden, D. R. Smith, and N. Chhetri. 2014. A meta-analysis of crop yield under climate change and adaptation. *Nature Climate Change* 4(4):287–291.

Chen, D., S. Wang, B. Xiong, B. Cao, and X. Deng. 2015. Carbon/nitrogen imbalance associated with drought-induced leaf senescence in Sorghum bicolor. *PLOS ONE* 10(8):e0137026.

Choat, B., T. J. Brodribb, C. R. Brodersen, R. A. Duursma, R. López, and B. E. Medlyn. 2018. Triggers of tree mortality under drought. *Nature* 558(7711):531–539.

Confalonieri, R., S. Bregaglio, and M. Acutis. 2016. Quantifying uncertainty in crop model predictions due to the uncertainty in the observations used for calibration. *Ecological Modelling* 328:72–77.

Cook, B. I., J. E. Smerdon, R. Seager, and S. Coats. 2014. Global warming and 21st century drying. *Climate Dynamics* 43(9–10):2607–2627.

Cook, B. I., T. R. Ault, and J. E. Smerdon. 2015. Supplementary materials for unprecedented 21st century drought risk in the American Southwest and Central Plains. *Science Advances* 1:1–7.

Cooper, H. F., K. C. Grady, J. A. Cowan, R. J. Best, G. J. Allan, and T. G. Whitham. 2019. Genotypic variation in phenological plasticity: Reciprocal common gardens reveal adaptive responses to warmer springs but not to fall frost. *Global Change Biology* 25(1):187–200.

Costa, C., U. Schurr, F. Loreto, P. Menesatti, and S. Carpentier. 2019. Plant phenotyping research trends, a science mapping approach. *Frontiers in Plant Science* 9:1933.

Das Choudhury, S., A. Samal, and T. Awada. 2019. Leveraging image analysis for high-throughput plant phenotyping. *Frontiers in Plant Science* 10:508.

Daubenmire, R. F. 1968. Soil moisture in relation to vegetation distribution in the mountains of northern Idaho. *Ecology* 49(3):431–438.

Debieu, M., B. Sine, S. Passot et al. 2018. Response to early drought stress and identification of QTLs controlling biomass production under drought in pearl millet. *PLOS ONE* 13(10):e0201635.

De Sousa Mendes, M. M., and T. S. de Oliveira. 2014. Hydraulic redistribution. *Global Advanced Research Journal of Agricultural Science* 3:394–399.

Edwards, C. E., and C. Weinig. 2010. The quantitative-genetic and QTL architecture of trait integration and modularity in *Brassica rapa* across simulated seasonal settings. *Heredity* 106(4):661.

Edwards, C. E., B. E. Ewers, D. G. Williams, Q. Xie, P. Lou, X. Xu, C. R. McClung, and C. Weinig. 2011. The genetic architecture of ecophysiological and circadian traits in Brassica rapa. *Genetics* 189(1):375–390.

Edwards, C. E., B. E. Ewers, C. R. McClung, P. Lou, and C. Weinig. 2012. Quantitative variation in water-use efficiency across water regimes and its relationship with circadian, vegetative, reproductive, and leaf gas-exchange traits. *Molecular Plant* 5(3):653–668.

Edwards, C. E., B. E. Ewers, and C. Weinig. 2016. Genotypic variation in biomass allocation in response to field drought has a greater effect on yield than gas exchange or phenology. *BMC Plant Biology* 16(1):185.

Fahlgren, N., M. A. Gehan, and I. Baxter. 2015. Lights, camera, action: High-throughput plant phenotyping is ready for a close-up. *Current Opinion in Plant Biology* 24:93–99.

Fiorani, F., and U. Schurr. 2013. Future scenarios for plant phenotyping. *Annual Review of Plant Biology* 64:267–291.

Furbank, R. T., and M. Tester. 2011. Phenomics – Technologies to relieve the phenotyping bottleneck. *Trends in Plant Science* 16(12):635–644.

Gehan, M. A., and E. A. Kellogg. 2017. High-throughput phenotyping. *American Journal of Botany* 104(4):505–508.

Genty, B., J.-M. Briantais, and N. R. Baker. 1989. The relationship between the quantum yield of photosynthetic electron transport and quenching of chlorophyll fluorescence. *Biochimica et Biophysica Acta* 990(1):87–92.

Genty, B., Y. Goulas, B. Dimon, J. M. Peltier, and I. Moya. 1992. Modulation of the efficiency of primary conversion in leaves, mechanisms involved at PSII. In *Research in Photosynthesis* 4:603–610. Kluwer Academic Publishers, Dordrecht, Netherlands.

Govindjee, G. 2004. Chlorophyll a fluorescence: A bit of basics and history. In *Chlorophyll A Fluorescence: A Signature of Photosynthesis* 1–42. Springer, Dordrecht, Netherlands.

Greenham, K., C. R. Guadagno, M. Gehan, T. Mockler, C. Weinig, B. E. Ewers, and C. R. McClung. 2017. Temporal network analysis identifies early physiological and transcriptomic indicators of mild drought in *Brassica rapa*. *eLife* 6:e29655.

Großkinsky, D. K., J. Svensgaard, S. Christensen, and T. Roitsch. 2015. Plant phenomics and the need for physiological phenotyping across scales to narrow the genotype-to-phenotype knowledge gap. *Journal of Experimental Botany* 66(18):5429–5440.

Guadagno, C. R., A. Virzo De Santo, and N. D'Ambrosio. 2010. A revised energy partitioning approach to assess the yields of non-photochemical quenching components. *Biochimica et Biophysica Acta* 1797(5):525–530.

Guadagno, C. R., B. E. Ewers, H. N. Speckman, T. L. Aston, B. J. Huhn, S. B. DeVore, J. T. Ladwig, R. N. Strawn, and C. Weinig. 2017. Dead or Alive? Using membrane failure and chlorophyll a fluorescence to predict plant mortality from drought. *Plant Physiology* 175(1):223–234.

Guadagno, C. R., B. E. Ewers, and C. Weinig. 2018. Circadian rhythms and redox state in plants: Till stress do us part. *Frontiers in Plant Science* 9:247.

Hammer, G., C. Messina, E. van Oosterom et al. 2016. Molecular breeding for complex adaptive traits: How integrating crop ecophysiology and modelling can enhance efficiency. In *Crop Systems Biology* 147–162. Springer, Cham.

Harbinson, J., B. Genty, and N. R. Baker. 1990. The relationship between CO2 assimilation and electron transport in leaves. *Photosynthesis Research* 25(3):213–224.

Horton, P. 1983. Relations between electron transport and carbon assimilation; simultaneous measurement of chlorophyll fluorescence, trans-thylakoid pH gradient and O_2 evolution in isolated chloroplasts. *Proceedings of the Royal Society of London, Series B* 217(1209):405–416.

Horton, P., and A. Hague. 1988. Studies on the induction of chlorophyll fluorescence in isolated barley protoplasts IV. Resolution of non-photochemical quenching. *Biochimica et Biophysica Acta* 932:107–115.

Houle, D., D. R. Govindaraju, and S. Omholt. 2010. Phenomics: The next challenge. *Nature Reviews. Genetics* 11(12):855–866.

Hubby, J. L., and R. C. Lewontin. 1966. A molecular approach to the study of genic heterozygosity in natural populations. I. The number of alleles at different loci in *Drosophila pseudoobscura*. *Genetics* 54(2):577–594.

Iniguez-Luy, F., L. Lukens, M. Farnham, R. Amasino, and T. Osborn. 2009. Development of public immortal mapping populations, molecular markers and linkage maps for rapid cycling *Brassica rapa* and *B. oleracea*. *Theoretical and Applied Genetics* 120(1):31–43.

IPCC. 2014. *Climate Change 2014: Impacts, Adaptation, and Vulnerability. Part A: Global and Sectoral Aspects*. Cambridge University Press, New York, NY.

Jiao, W.-B., and K. Schneeberger. 2017. The impact of third generation genomic technologies on plant genome assembly. *Current Opinion in Plant Biology* 36:64–70.

Kalaji, H. M., G. Schansker, M. Brestic, et al. 2016. Frequently asked questions about chlorophyll fluorescence, the sequel. *Photosynthesis Research* 1:13–66.

Kautsky, H., and A. Hirsch. 1931. Neue Versuche zur Kohlensäureassimilation. *Naturwissenschaften* 19(48):964.

Kautsky, H., and W. Zedlitz. 1941. Fluoreszenzkurven von Chloroplasten-Grana. *Naturwissenschaften* 29(7):101–102.

Kautsky, H., W. Apel, and H. Amann. 1960. Chlorophyllfluoreszenz und Kohlens¨aureassimilation. Die Fluoreszenkurve und die Photochemie der Pflanze. *Biochem ZEIT* 322:277–292.

Knapp, A. K., J. M. Briggs, S. L. Collins et al. 2008. Shrub encroachment in North American grasslands: Shifts in growth form dominance rapidly alters control of ecosystem carbon inputs. *Global Change Biology* 14(3):615–623.

Kobza, J., B. D. Moore, and J. R. Seemann. 1989. Isolation of photosynthetically active protoplasts and intact chloroplasts from *Phaseolus vulgaris*. *Plant Science* 65(2):177–182.
Kolb, P. F., and R. Robberecht. 1996. High temperature and drought stress effects on survival of Pinus ponderosa seedlings. *Tree Physiology* 16(8):665–672.
Kolb, T., S. Dore, and M. Montes-Helu. 2013. Extreme late-summer drought causes neutral annual carbon balance in southwestern ponderosa pine forests and grasslands. *Environmental Research Letters* 8(1):015015.
Kuhlgert, S., G. Austic, R. Zegarac et al. 2016. MultispeQ beta: A tool for large-scale plant phenotyping connected to the open PhotosynQ network. *Royal Society Open Science* 3(10):160592.
Kumar, B., A. Abdel-Ghani, J. Reyes-Matamoros, F. Hochholdinger, and T. Lübberstedt. 2012. Genotypic variation for root architecture traits in seedlings of maize (*Zea mays* L.) inbred lines. *Plant Breeding* 131(4):465–478.
Kumar, J., A. Pratap, and S. Kumar. 2015. Plant phenomics: An overview. In *Phenomics in Crop Plants: Trends, Options and Limitations* 1–10. Springer, New Delhi.
Kurz, W. A., C. C. Dymond, G. Stinson, G. J. Rampley, E. T. Neilson, A. L. Carroll, T. Ebata, and L. Safranyik. 2008. Mountain Pine beetle and forest carbon feedback to climate change. *Nature* 452(7190):987–990.
Lazár, D. 2015. Parameters of photosynthetic energy partitioning. *Journal of Plant Physiology* 175:131–147.
Li, G., L. Tang, X. Zhang, J. Dong, and M. Xiao. 2018. Factors affecting greenhouse microclimate and its regulating techniques: A review. In *IOP Conference Series: Earth and Environmental Science* 167:012019.
Liang, X.-Z., Y. Wu, R. G. Chambers, D. L. Schmoldt, W. Gao, C. Liu et al. 2017. Determining climate effects on US total agricultural productivity. *PNAS* 114(12):E2285–E2292.
Lichtenhaler, H. K., C. Buschmann, U. Rinderle, and G. Shmuck. 1986. Application of chlorophyll fluorescence in ecophysiology. *Radiation and Environmental Biophysics* 25(4):297–308.
Liu, S., L. M. Acosta-Gamboa, X. Huang, and A. Lorence. 2017. Novel low-cost 3D surface model reconstruction system for plant phenotyping. *Journal of Imaging* 3(3):39.
Lobell, D. B., and S. M. Gourdji. 2012. The influence of climate change on global crop productivity. *Plant Physiology* 160(4):1686–1697.
Lorenz, K., and R. Lal. 2005. The depth distribution of soil organic carbon in relation to land use and management and the potential of carbon sequestration in subsoil horizons. *Advances in Agronomy* 88:35–66.
Marko, D., N. Briglia, S. Summerer, A. Petrozza, F. Cellini, and R. Iannacone. 2018. High-throughput phenotyping in plant stress response: Methods and potential applications to polyamine field. In *Methods in Molecular Biology* 1694:373–388. Humana Press, New York, NY.
Martinez-Vilalta, J., W. R. L. Anderegg, G. Sapes, and A. Sala. 2019. Greater focus on water pools may improve our ability to understand and anticipate drought-induced mortality in plants. *New Phytologist* 223(1):22–32.
Matesanz, S., E. Gianoli, and F. Valladares. 2010. Global change and the evolution of phenotypic plasticity in plants. *Annals of the New York Academy of Sciences* 1206:35–55.
Matsui, T., and I. M. Ehrenreich. 2016. Gene-environment interactions in stress response contribute additively to a genotype-environment interaction. *PLOS Genetics* 12(7):e1006158.
Maxwell, K., and G. N. Johnson. 2000. Chlorophyll fluorescence—A practical guide. *Journal of Experimental Botany* 51(345):659–668.
McDowell, N. G., C. Grossiord, H. D. Adams et al. 2019. Mechanisms of a coniferous woodland persistence under drought and heat. *Environmental Research Letters* 14(4):045014.

Mencuccini, M. 2003. The ecological significance of long-distance water transport: Short-term regulation, long-term acclimation and the hydraulic costs of stature across plant life forms. *Plant, Cell and Environment* 26(1):163–182.

Michaelian, M., E. H. Hogg, R. J. Hall, and E. Arsenault. 2011. Massive mortality of aspen following severe drought along the southern edge of the Canadian Boreal forest. *Global Change Biology* 17(6):2084–2094.

Mishra, P., M. Shahrimie, M. Asaari, A. Herrero-Langreo, S. Lohumi, B. Diezma, and P. Scheunders. 2017. Close range hyperspectral imaging of plants: A review. *Biosystems Engineering* 164:49–67.

Morais, C. L., H. J. Butler, M. R. McAinsh, and F. L. Martin. 2019. Plant hyperspectral imaging. In *eLS*. John Wiley & Sons, Ltd (Ed.). doi: 10.1002/9780470015902.a0028367.

Murchie, E. H., and T. Lawson. 2013. Chlorophyll fluorescence analysis: A guide to good practice and understanding some new applications. *Journal of Experimental Botany* 64(13):3983–3998.

Oxborough, K., and N. R. Baker. 1997. Resolving chlorophyll a fluorescence images of photosynthetic efficiency into photochemical and non-photochemical components – Calculation of qP and F_v'/F_m'; without measuring F_o'. *Photosynthesis Research* 54(2):135–142.

Papageorgiou, G. C., and G. Govindjee. 2011. Photosystem II fluorescence: Slow changes–scaling from the past. *Journal of Photochemistry and Photobiology* 104(1–2):258–270.

Parmesan, C., and M. E. Hanley. 2015. Plants and climate change: Complexities and surprises. *Annals of Botany* 116(6):849–864.

Passioura, J. B. 2006. Increasing crop productivity when water is scarce-from breeding to field management. *Agricultural Water Management* 80(1–3):176–196.

Passioura, J. B. 2012. Phenotyping for drought tolerance in grain crops: When is it useful to breeders? *Functional Plant Biology* 39(11):851–859.

Perala, D. A. 1990. *Populus tremuloides* Michx. Quaking aspen. In *Silvics of North America, 2. Hardwoods* 555–569. Agriculture Handbook No. 654. USDA Forest Service, Washington, DC.

Pieruschka, R., and H. Poorter. 2012. Phenotyping plants: Genes, phenes and machines. *Functional Plant Biology* 39(11):813–820.

Pleban, J. R., C. R. Guadagno, D. S. Mackay, C. Weinig, and B. E. Ewers. 2020. Rapid chlorophyll a fluorescence light response curves mechanistically inform photosynthesis modeling. *Plant Physiology* 183(2):602–619.

Pogson, M., A. Hastings, and P. Smith. 2012. Sensitivity of crop model predictions to entire meteorological and soil input datasets highlights vulnerability to drought. *Environmental Modeling and Software* 29(1):37–43.

Pogson, B. J., D. Ganguly, and V. Albrecht-Borth. 2015. Insights into chloroplast biogenesis and development. *Biochimica et Biophysica Acta – Series Bioenergetics* 1847(9):1017–1024.

Power, A. G. 2010. Ecosystem services and agriculture: tradeoffs and synergies. *Philosophical Transactions of the Royal Society B: Biological Sciences* 365(1554):2959–2971. doi:10.1098/rstb.2010.0143.

Pratap, A., S. Gupta, R. M. Nair et al. 2019. Using plant phenomics to exploit the gains of genomics. *Agronomy* 9(3):126–151.

Ray, D. K., P. C. West, M. Clark, J. S. Gerber, A. V. Prishchepov, and S. Chatterjee. 2019. Climate change has likely already affected global food production. *PLOS ONE* 14(5):e0217148.

Roff, D. A., and D. J. Fairbairn. 2007. The evolution of trade-offs: Where are we? *Journal of Evolutionary Biology* 20(2):433–447.

Rosenqvist, E., and O. Van Kooten. 2003. Chlorophyll fluorescence: A general description and nomenclature. In *Practical Applications of Chlorophyll Fluorescence in Plant Biology* 31–37. Springer, Boston, MA.

Rosenzweig, C., J. Elliott, D. Deryng, A. C. Ruane, C. Müller, A. Arneth et al. 2014. Global multi-model crop-climate impact assessment. *PNAS* 111:3268–3273.

Rouphael, Y., L. Spíchal, K. Panzarová, R. Casa, and G. Colla. 2018. High-throughput plant phenotyping for developing novel biostimulants: From lab to field or from field to lab? *Frontiers in Plant Science* 9:1197.

Rungrat, T., M. Awlia, T. Brown et al. 2016. Using phenomic analysis of photosynthetic function for abiotic stress response gene discovery. *Arabidopsis Book* 14:e0185.

Sala, A., F. Piper, and G. Hoch. 2010. Physiological mechanisms of drought-induced tree mortality are far from being resolved. *New Phytologist* 186(2):274–281.

Saltz, J. B., A. M. Bell, J. Flint, R. Gomulkiewicz, K. A. Hughes, and J. Keagy. 2018. Why does the magnitude of genotype-by-environment interaction vary? *Ecology and Evolution* 8(12):6342–6353.

Schreiber, U. 2004. Pulse-Amplitude Modulation (PAM) fluorometry and saturation pulse method: An overview. In *Chlorophyll A Fluorescence: A Signature of Photosynthesis* 279–319. Springer, Dordrecht, Netherlands.

Schreiber, U., U. Schliwa, and W. Bilger. 1986. Continuous recording of photochemical and non-photochemical chlorophyll fluorescence quenching with a new type of modulation fluorometer. *Photosynthesis Research* 10(1–2):51–62.

Sgro, C. M., and A. A. Hoffmann. 2004. Genetic correlations, tradeoffs and environmental variation. *Heredity* 93(3):241–248.

Sircar, S., and N. Parekh. 2015. Functional characterization of drought-responsive modules and genes in Oryza sativa: A network-based approach. *Frontiers in Genetics* 6:256.

Stinziano, J. R., D. K. McDermitt, D. J. Lynch, A. J. Saathoff, P. B. Morgan, and D. T. Hanson. 2019. The rapid A/Ci response (RACiR): A guide to best practices. *New Phytologist* 221(2):625–627.

Tietz, S., C. C. Hall, J. A. Cruz, and D. M. Kramer. 2017. NPQ(T): A chlorophyll fluorescence parameter for rapid estimation and imaging of non-photochemical quenching of excitons in photosystem-II-associated antenna complexes. *Plant, Cell and Environment* 40(8):1243–1255.

Torrion, J. A., and R. N. Stougaard. 2017. Impacts and limits of irrigation water management on wheat yield and quality. *Crop Science* 57(6):3239–3251.

Trumbore, S., P. Brando, and H. Hartmann. 2015. Forest health and global change. *Science* 349(6250):814–818.

United Nations, Department of Economic and Social Affairs, Population Division. 2017. *World Population Prospects: The 2017 Revision, Key Findings and Advance Tables.* ESA/P/WP/248.

van Bezouw, R. F. H. M., J. J. B. Keurentje, J. Harbinson, and M. G. M. Aarts. 2019. Converging phenomics and genomics to study natural variation in plant photosynthetic efficiency. *The Plant Journal* 97(1):112–133.

Verduzco, V. S., E. R. Vivoni, E. A. Yépez, J. C. Rodríguez, C. J. Watts, T. Tarin, J. Garatuza-Payán, A. Robles-Morua, and V. Y. Ivanov. 2018. Climate change impacts on net ecosystem productivity in a subtropical shrubland of Northwestern México. *Journal of Geophysical Resources: Biogeoscience* 123(2):688–711.

Wallach, D., and P. J. Thorburn. 2017. Estimating uncertainty in crop model predictions: Current situation and future prospects. *European Journal of Agronomy* 88:A1–A7.

Wang, D., C. R. Guadagno, X. Mao, D. Mackay, J. Pleban, R. L. Baker, C. Weinig, J.-L. Jannink, and B. E. Ewers. 2019. A framework for genomics-informed process-based modeling in plants. *Journal of Experimental Botany* 70(9):2561–2574.

Wilske, B., H. Kwon, L. Wei et al. 2010. Evapotranspiration (ET) and regulating mechanisms in two semiarid *Artemisia*-dominated shrub steppes at opposite sides of the globe. *Journal of Arid Environments* 74(11):1461–1470.

Woo, N. S., M. R. Badger, and B. J. Pogson. 2008. A rapid, non-invasive procedure for quantitative assessment of drought survival using chlorophyll fluorescence. *Plant Methods* 4:27–41.

Yarkhunova, Y., C. E. Edwards, B. E. Ewers, R. L. Baker, T. L. Aston, C. R. McClung, P. Lou, and C. Weinig. 2016. Selection during crop diversification involves correlated evolution of the circadian clock and ecophysiological traits in *Brassica rapa*. *New Phytologist* 210(1):133–144.

Zhao, C., Y. Zhang, J. Du, X. Guo, W. Wen, S. Gu, J. Wang, and J. Fan. 2019. Crop phenomics: Current status and perspectives. *Frontiers in Plant Science* 10:714.

Ziska, L. H., and J. A. Bunce. 2007. Predicting the impact of changing CO2 on crop yields: Some thoughts on food. *New Phytologist* 175(4):607–618.

13 Predicting Yield by Modeling Interactions between Canopy Coverage Image Data, Genotypic and Environmental Information for Soybeans

Diego Jarquin, Reka Howard, Alencar Xavier, and Sruti Das Choudhury

CONTENTS

13.1 INTRODUCTION

To meet the food needs of the increasing world population (Whitford et al. 2013) and to satisfy people's dietary requirements, it is important to increase food production. Plant breeding is a science that focuses on genetic improvement of plants to produce varieties that benefit people. Plant breeders have a task to develop high-yielding cultivars with specific traits suited to particular target environments, using a variety of tools to reach one of their major goals: increase in genetic gains. Throughout history, people produced and improved their crop varieties by selecting for superior progeny, based on phenotypic selection. As molecular techniques evolved, it gave breeders the opportunity to aid their selection, based on genotypic information. The fundamental idea of plant breeding is that the desired phenotype needs to be improved, and the phenotype can be explained with the equation $P = G + E + G \times E + e$, where P represents the phenotypic performance of an individual/line/variety, G is the contribution of the genotype, E is the effect due to the environment, $G \times E$ is the interaction between the genotype and the environmental effect, and e is the measurement error that cannot be explained by G, E, or $G \times E$. Hence, it is important to develop models that accurately account for the variation in the factors that contribute to phenotypic variation.

Soybean is a major oilseed crop in the United States, where more than 90% of oilseed production comes from soybean (Zulauf et al. 2017). Since soybeans have low production costs, there is potential to increase yield, and increased soybean yield will reduce the per-bushel production cost. Soybeans are cultivated mostly as part of a crop rotation with maize. The cultivatable land area for soybean cannot be significantly increased, so we must sustainably improve the yield potential of soybean to enhance food production. Thus, there is a need to develop methods that enable us to increase soybean yield. Nowadays, modern platforms can be used for monitoring large planted regions intensively in time and space, delivering accurate information about specific (physiological) conditions of the plants, allowing a better characterization of the response of the plants to specific stress stimuli. In this sense, we can now characterize genotypes, using high-dimensional marker information and high-dimensional phenotypic information. The markers are the genetic locations on which we have information, and which are used in prediction models.

The procedure of developing a new cultivar consists of generating genetic variability by selecting parental lines to cross, testing the progeny from the crosses in field trials, and developing improved commercial varieties by testing the best crosses in multiple environments. New advances in high-throughput genotyping and sequencing technologies have enabled us to obtain massive amounts of genomic molecular marker information (e.g., single-nucleotide polymorphisms, SNPs). Genomic prediction (GP) is a methodology that uses all available genotypic marker information to aid the process of selecting superior genotypes in plant breeding.

GP techniques have become an important part of plant breeding programs, due to their advantages compared with traditional phenotypic and pedigree-based selection (de los Campos et al. 2010). GP is a technique that aids selection for yield and quality-related traits, and it has been shown (Heffner et al. 2010) that it has the potential to lead to a three-fold increase in genetic gain compared to marker-assisted selection (using a few selected major genes).

In GP, marker and phenotypic information on individuals in the training set is used to model the relationship between phenotype and genotype. Then, the model is used to predict the phenotype of individuals in the testing set, for which only marker information is available. GP was first introduced by Meuwissen et al. (2001), and, since then, a large number of models have been developed for phenotype prediction, incorporating marker information (Endelman, 2011; Gianola et al. 2006; Habier et al. 2011). Early applications of GP for selection of soybean varieties were introduced by Jarquín et al. (2014a).

The prediction accuracy of GP methods can be improved by including the genotype-by-environment ($G \times E$) component, and allowing breeders to borrow information from related plant materials and from correlated environments. Jarquín et al. (2014b) utilized the reaction norm model for genomic prediction, where the genetic and environmental values are replaced, respectively, by the regression on the markers and on the interaction between the markers and the environmental covariates. Dealing with high-throughput phenotypic information, several authors (Aguate et al. 2017; Crain et al. 2018; Montesinos-Lopez et al. 2017a) have shown improvements in predictive ability with the inclusion of these sources of information in models for wheat and maize. Montesinos-Lopez et al. (2017b) showed that accounting for the band (hyperspectral image data)-by-environment interaction also improves yield predictability in wheat, compared with those models that did not include this component.

In this chapter, we extend the reaction norm model for genomic prediction, using canopy coverage image data. Xavier et al. (2017) incorporated phenomic canopy coverage image data into a selection scheme by studying the large-effect quantitative trait loci (QTLs) associated with canopy coverage. Since canopy coverage is a trait with high estimated heritability and a positive correlation with grain yield (Xavier et al. 2017), it has the potential to improve prediction models when information on canopy coverage is included in these models.

We used data from the soybean nested association mapping (SoyNAM) population (soybase.org/SoyNAM/) to develop our prediction models. The SoyNAM data consisted of phenotypic yield data on 5600 F_5-derived recombinant inbred lines (RILs), over 4000 SNP markers, and ground-based and/or aerial imagery data, depending on the year.

The chapter is organized as follows. First, we provide a brief description of the SoyNAM phenotypic and marker data that were used in this study. Then, we describe how the canopy coverage image data were collected, and why they have the potential to increase prediction accuracy when included in the prediction models, compared with traditional GP models. In the next section, the statistical models and cross-validation (CV) schemes used for genomic-enabled prediction are described. We compared nine prediction models, with three different CV schemes for yield and date to maturity (DTM), for the SoyNAM data set. In addition, we compared the effects on prediction models when considering canopy data captured in only the early stages of the growing season (days 14–33) instead of the entire growing season. Finally, we discuss the results and some future research avenues.

13.2 MATERIAL AND METHODS

13.2.1 SoyNAM Phenotypic and Genotypic Data

The predictions were conducted using a SoyNAM, containing 5600 RILs created by crossing a common high-yielding parent (IA3023) to each of 40 parents. For a detailed description of the structure of the SoyNAM population, the reader can refer to https://www.soybase.org/SoyNAM/ and Xavier et al. (2016). For the analysis, 39 families and 5143 RILs were retained after alignment and the application of quality control on the phenotypic and genomic data.

Phenotypic data were collected on grain yield (in kg ha^{-1}), days to maturity (in days after planting), plant height (in cm), lodging (score 1–5: almost all plants erect [low lodging] to all plants down [high lodging]), seed size (mass of 100 seeds in g), and seed composition (as % of the grain) of protein, oil, and fiber content. For the study described in this chapter, the focus was on grain yield and days to maturity.

Regarding the genomic data, 5303 SNP markers, derived from the SoyNAM 6K BreadChip array (Song et al. 2017), were available for analysis. Markers with minor allele frequency of less than 0.15 were removed, and missing markers were imputed using the Random Forest algorithm (Stekhoven and Buhlmann 2012). After quality control, the final set of markers consisted of 4240 SNPs.

For more information about the SoyNAM parents, RILs, the experimental design, the collected phenotypes, and the corresponding genotypes, the reader can refer to Xavier et al. (2016).

13.2.2 Canopy Coverage and Imagery Data Collection

Grain yield in soybean is influenced by genetic and environmental factors and their interactions. Among the environmental factors, drought, salinity, and temperature have the largest impact on the agronomic traits (e.g., yield) (Wang et al. 2003). Many of these factors are hard to measure, so that, even though they have a large influence on the trait, it is not cost-effective to incorporate them into prediction models. Canopy light interception (LI) is one of these important factors that influences yield and growth in soybean, but it is difficult to measure with conventional methods (Purcell 2000). Canopy coverage, which is the proportion of ground area that is covered by the soybean plants in a plot, is a LI-related trait that can be measured more easily than LI, and which can therefore be used as a replacement for canopy LI measurements. High-throughput phenotyping platforms allow us to capture information about different traits in a non-destructive way. It also permits the collection of time-series measurements at a relatively low cost. This is essential in order to follow the growth change of plants (Fahlgren et al. 2015), and helps to quantify the genotype-by-phenotype-by-environmental interactions.

The canopy data used for the analyses were obtained in the form of either ground-based imagery (in 2013 and 2014) or aerial imagery data (in 2014 and 2015), as described by Xavier et al. (2017). For 2014, canopy data were collected using both methods (ground-based and aerial) for the same genotypes. Xavier et al. (2017) noted that the correlation between the ground-based data and the aerial data was high ($r^2=0.87$) for that year. Since only aerial imagery data were collected for 2015 and because the correlation between ground-based and aerial data was high, in our analysis, we used data from the aerial platform for 2014 and 2015, together with the ground-based data collected in 2013. Thus, our models were calibrated using ground data for 2013 and aerial data for 2014 and 2015. The canopy data were collected at different time points at regular intervals ranging from two to eight weeks after planting, to provide information on the different phases of growth, development, and crop physiology. The collected data were classified to determine the canopy coverage, which was defined as the proportion of image pixels that were canopy pixels to the total number of pixels for any given field plot. The canopy coverage data of ground-based imageries were determined using the software SigmaScan Pro®, which uses the method described by Karcher and Richardson (2005). The canopy coverage data of aerial imageries were obtained by the software ENVI 5.0™, using a binomial model.

13.2.3 Relationship Between Canopy Coverage Data and Grain Yield

In order to successfully use canopy coverage data to increase the predictive ability of the models, canopy coverage should show some type of relationship with those traits to be predicted. In this study, for each year (2013, 2014, and 2015)-by-acquisition method (ground and aerial) combination, we computed the correlation between the interpolated canopy values over the entire season (days 14–71) and the corresponding grain yield values (Figure 13.1). The four datasets showed different patterns; however, in all cases, the correlation coefficient with grain yield decreased five weeks after planting. The 2013-ground and 2014-ground datasets achieved the highest correlation (0.27 and 0.34) on days 33 and 32, respectively. For the aerial method, the highest correlations (0.33 and 0.44) were reached on days 45 and 41, respectively. To assess the usefulness of canopy coverage data in the early stages of crop growth, the information from days 14–33 (for the three datasets) was included in the prediction models, and the results were compared with those obtained using canopy information from the entire season (days 14–71).

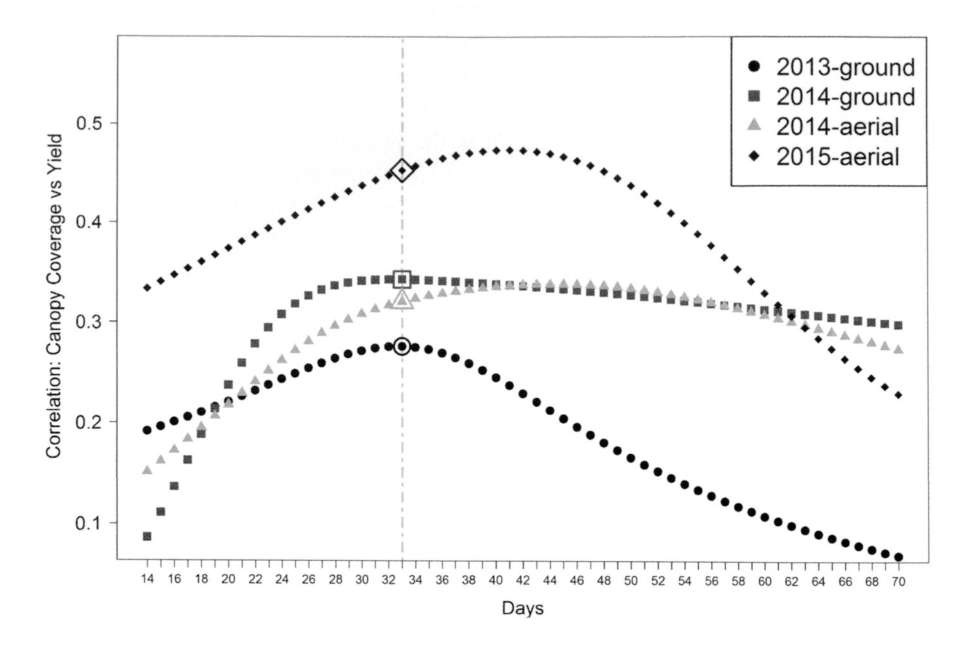

FIGURE 13.1 Correlation between the interpolated daily canopy coverage values and yield for datasets 2013-ground, 2014-ground, 2014-aerial and 2015-aerial canopy coverage data.

13.2.4 Statistical Prediction Models

In this study, we evaluated nine prediction models: four main effects models, that included some combination of line, environment, marker genotype, and canopy coverage image information, four models with two-way interaction(s) among the components, and one model with a three-way interaction among the environmental information, marker genotypes, and the canopy coverage data. The interaction components were modeled using the reaction norm model (Jarquín et al. 2014b). All models assume that the components are random effects. For each of the models, we considered two responses: grain yield (yield) and the number of days to maturity (DTM). Since the canopy coverage imagery data were either ground-based or collected from the air, we incorporated the two types of canopy coverage image data in the models. In the statistical models described below, the term, CC, denotes the canopy coverage, where the ground-based imagery data are included from 2013 and the aerial canopy coverage image data are included from 2014 and 2015.

13.2.5 Main Effects Models

13.2.5.1 Model 1: Environment + Line Model

The response of the jth genotype in the ith environment $\{y_{ij}\}$ can be written as:

$$y_{ij} = \mu + E_i + L_j + e_{ij} \quad (13.1)$$

where μ is the overall mean, E_i $(i=1,\ldots,I)$, is the random effect of the ith environment assuming $E_i \overset{iid}{\sim} N\left(0,\sigma_E^2\right)$, with $N(.,.)$ denoting a normal density, where *iid* stands for independent and identically distributed observations, and σ_E^2 represents the variance component of environments; L_j represents the random effect of the jth line $(j=1,\ldots,J)$, assuming $L_j \overset{iid}{\sim} N\left(0,\sigma_L^2\right)$, where σ_L^2 is the variance of the line; and e_{ij} is the random error term, where $e_{ij} \overset{iid}{\sim} N\left(0,\sigma_e^2\right)$, with σ_e^2 as the residual variance. All the models described below include the terms specified for Model 1 with additional components.

13.2.5.2 Model 2: Environment + Line + CC Model

The only difference between this model and Model 1 is that the *CC* (canopy coverage) component is also included here. The model can be described as follows:

$$y_{ij} = \mu + E_i + L_j + CC_{ij} + e_{ij} \tag{13.2}$$

where μ, E_i, L_j, and e_{ij} are defined as before, and CC_{ij} is the canopy coverage imagery data effect, where CC_{ij} is defined as $CC_{ij} = \sum_{k=1}^{K} P_{ijk}c_k$, where P_{ijk} is the kth canopy measurement for genotype j in environment i, and c_k is the effect of the kth canopy measurements, such that $c_k \overset{iid}{\sim} N\left(0,\sigma_c^2\right)$, with σ_c^2 acting as the corresponding variance component. In general, $CC \overset{iid}{\sim} N\left(0,\frac{PP'}{K}\sigma_c^2\right)$, and, for later derivations, we define $\omega_{cc}=\frac{\boldsymbol{PP'}}{K}$, where $\boldsymbol{P}$ is the standardized (by columns) matrix of canopy measures across environments. After the extrapolation of missing daily measures, the number of total canopy measurements for a genotype in an environment was 58 (=71–13).

13.2.5.3 Model 3: Environment + Line + Marker Model

In this model the response of the jth genotype in the ith environment $\{y_{ij}\}$ can be modeled as a linear function of the environmental effect, line effect, marker effect, and a random residual:

$$y_{ij} = \mu + E_i + L_j + g_j + e_{ij} \tag{13.3}$$

where μ, E_i, L_j, and e_{ij} are defined as before, and g_j is the linear combination of p markers and the corresponding marker effects, such that $g_j = \sum_{m=1}^{p} x_{jm}b_m$. The marker effects are assumed to be normally distributed, such that $b_m \overset{iid}{\sim} N\left(0,\sigma_b^2\right)$ for $m=1,\ldots,p$, where σ_b^2 is the marker effect variance, and x_{jm} is the genotypic value of the mth marker for the jth genotype. If we write the genomic values in a vector fashion of the form $\mathbf{g} = (g_1, \ldots, g_J)$, then its respective covariance matrix can be expressed as $Cov(\mathbf{g}) = \mathbf{G}\sigma_g^2$, where $\mathbf{G} = \frac{\boldsymbol{XX'}}{p}$ is the genomic relationship matrix (van Raden 2008), $\boldsymbol{X}$ is the centered and standardized (by columns) genotype matrix, and $\sigma_g^2 = p \times \sigma_b^2$. Combining the aforementioned results, we have $\mathbf{g} = \{g_j\} \sim N(0, \mathbf{G}\sigma_g^2)$.

13.2.5.4 Model 4: Environment + Line + Marker + CC Model

This model is the combination of Model 2 and Model 3, as both the marker and the CC canopy components are included, in addition to the environment and line components:

$$y_{ij} = \mu + E_i + L_j + g_j + CC_{ij} + e_{ij} \tag{13.4}$$

where each of the terms has been defined before (Models 2 and 3).

13.2.6 Two-Way Interaction Models

13.2.6.1 Model 5: Environment + Line + CC + (Env × CC) Interaction Model

This model is an extension of Model 2, where the interaction term between canopy and environments is added:

$$y_{ij} = \mu + E_i + L_j + CC_{ij} + CCE_{ij} + e_{ij} \tag{13.5}$$

where each of the terms, except CCE_{ij}, is as defined before (for Model 2). CCE_{ij} is the environment × canopy measurement interaction term with $\boldsymbol{CCE} = \{CCE_{ij}\} \sim N(0, (\boldsymbol{Z_E Z_E'}) \circ (\boldsymbol{w_{cc}}) \sigma_{CCE}^2)$, where $\boldsymbol{Z_E}$ is the incidence matrix for the environments, which connects phenotypes to environments, σ_{CCE}^2 is the variance component for the interaction term, ω_{cc} is as defined previously, and '∘' stands for the Hadamard or Schur (element-by-element) product between two matrices.

13.2.6.2 Model 6: Environment + Line + Marker + (Marker × Environment) Interaction Model

This model accounts for the environmental, line, and genomic main effects, as in Model 3, but it also incorporates the genotype × environment interaction *via* co-variance structures, as for the environment × canopy interaction in Model 5. In this case, the model can be written as

$$y_{ij} = \mu + E_i + L_j + g_j + gE_{ij} + e_{ij} \tag{13.6}$$

with $\boldsymbol{gE} = \{gE_{ij}\} \sim N(0, (\mathbf{Z_L} \boldsymbol{G} \mathbf{Z_L'}) \circ (\boldsymbol{Z_E Z_E'}) \sigma_{gE}^2)$, where $\mathbf{Z_L}$ is the incidence matrix that connects phenotypes with genotypes, σ_{gE}^2 is the variance component of the gE_{ij} interaction component, and $\mathbf{G}$ is the additive relationship matrix defined previously.

13.2.6.3 Model 7: Environment + Line + Marker + CC + (Marker × CC) Interaction Model

This model is an extension to Model 5 because it accounts not only for the main effects of the environment, line, marker, and the canopy coverage, but it also includes the interaction between the marker and the canopy coverage:

$$y_{ij} = \mu + E_i + L_j + g_j + CC_{ij} + gCC_{ij} + e_{ij} \quad (13.7)$$

where each of the terms, except gCC_{ij}, has been defined before (for Model 6). gCC_{ij} is the marker × canopy measurement interaction term, with $\mathbf{gCC} = \{gCC_{ij}\} \sim N\left(0, \left(\mathbf{Z}_L \mathbf{G} \mathbf{Z}_L'\right) \circ \left(\omega_{cc}\right) \sigma^2_{\mathbf{g}CC}\right)$, where $\mathbf{Z}_E$ is the incidence matrix for the genotypes, which connects the genotypes and the phenotypes, $\sigma^2_{\mathbf{g}CC}$ is the variance component for the genotype × canopy coverage interaction term, and ω_{cc} is as previously defined.

13.2.6.4 Model 8: Environment + Line + Marker + CC + (Marker × Environment) Interaction + (Environment × CC) Interaction Model

This model is a combination of Models 6 and 7, as it accounts for the main effects of environment, line, marker, and canopy coverage, and the interactions between marker and environment and between environment and the canopy coverage:

$$y_{ij} = \mu + E_i + L_j + g_j + CC_{ij} + gE_{ij} + CCE_{ij} + e_{ij} \quad (13.8)$$

where each of the terms is as defined previously.

13.2.7 Three-Way Interaction Model

13.2.7.1 Model 9: Environment + Line + Marker + CC + (Marker × Environment) Interaction + (Environment × CC) Interaction + (Marker × Environment × CC) Interaction Model

This three-way interaction model is an extension of Model 8, with the addition of the marker × environment × canopy coverage interaction:

$$y_{ij} = \mu + E_i + L_j + g_j + CC_{ij} + gE_{ij} + CCE_{ij} + gCCE_{ij} + e_{ij} \quad (13.9)$$

where the main effects and the two-way interaction terms are as defined previously, and $\mathbf{g}CCE = \{gCCE_{ij}\} \sim N\left(0, \left(\mathbf{Z}_E \mathbf{Z}_E'\right) \circ \left(\mathbf{Z}_L \mathbf{G} \mathbf{Z}_L'\right) \circ \left(\omega_{cc}\right) \sigma^2_{Egc}\right)$, where $\mathbf{Z}_E$ and $\mathbf{Z}_L$, $\mathbf{G}$, and ω_{cc} are as defined previously, and $\sigma^2_{\mathbf{g}CCE}$ is the variance component for the three-way interaction term.

13.2.8 Description of Cross-Validation Schemes Implemented for Assessing Predictive Ability

The performance of the models was compared, based on the within-environments Pearson correlation coefficient between the predicted and the observed phenotypic values. Three cross-validation schemes (CV2, CV1, and CV0) were implemented and compared when predicting yield and number of days to maturity (DTM). These cross-validation schemes mimic real plant breeding situations.

CV2 is the case where some lines are observed in some environments, but not in others, and we are interested in predicting the performance of these unobserved line × environment combinations. This scheme mimics the situation of predicting incomplete field trials. The graphical representation of CV2 is shown in Figure 13.2a. It is a simplified representation of CV2, where we have observed values for five lines in five environments, and the goal is to predict the phenotype of the lines that were

	E_1	E_2	E_3	E_4	E_5
$Line_1$	Y_{11}	NA	Y_{13}	Y_{14}	Y_{15}
$Line_2$	Y_{21}	Y_{22}	NA	Y_{24}	Y_{25}
$Line_3$	Y_{31}	Y_{32}	Y_{33}	Y_{34}	NA
$Line_4$	Y_{41}	Y_{42}	Y_{43}	NA	Y_{45}
$Line_5$	NA	Y_{52}	Y_{53}	Y_{54}	Y_{55}

(a)

	E_1	E_2	E_3	E_4	E_5
$Line_1$	Y_{11}	Y_{12}	Y_{13}	Y_{14}	Y_{15}
$Line_2$	Y_{21}	Y_{22}	Y_{23}	Y_{24}	Y_{25}
$Line_3$	NA	NA	NA	NA	NA
$Line_4$	Y_{41}	Y_{42}	Y_{43}	Y_{44}	Y_{45}
$Line_5$	Y_{51}	Y_{52}	Y_{53}	Y_{54}	Y_{55}

(b)

	E_1	E_2	E_3	E_4	E_5
$Line_1$	Y_{11}	Y_{12}	NA	Y_{14}	Y_{15}
$Line_2$	Y_{21}	Y_{22}	NA	Y_{24}	Y_{25}
$Line_3$	Y_{31}	Y_{32}	NA	Y_{34}	Y_{35}
$Line_4$	Y_{41}	Y_{42}	NA	Y_{44}	Y_{45}
$Line_5$	Y_{51}	Y_{52}	NA	Y_{54}	Y_{55}

(c)

FIGURE 13.2 Graphical representation of the cross-validation schemes.

not observed in a particular environment. Y_{ij} corresponds to the phenotypic value of the ith line in the jth environment, with $i = 1,\ldots,5$ and $j = 1,\ldots,5$, and NA represents the lines for which the phenotype needs to be predicted (testing set).

CV1 refers to the case of predicting untested lines across environments. The graphical representation of CV1 is shown in Figure 13.2b. Here, we are interested in predicting the performance of a newly developed line (Line 3) in all environments.

CV0 is the cross-validation scheme where the performance of lines is evaluated in some environments, and the goal is to predict the performance of the already-tested lines in untested environments. For determining the training and testing sets for the CV0 scheme, the leave-one-environment-out scheme was implemented, and, since there is no random partitioning involved in the CV0 method, it was implemented only once. The graphical representation of CV0 is shown in Figure 13.2c.

For both the CV1 and CV2 schemes, five-fold random partitions were used for determining the training and testing sets, i.e., 80% of the data were used as the training set, and the remaining 20% were used as the testing set, and the partitioning was repeated 50 times.

13.3 RESULTS AND DISCUSSION

13.3.1 Analysis of Variance Components

The percentages of the estimated variance components explained for the model terms of the nine models for yield are shown in Table 13.1, and for DTM in Table 13.2. For both traits, the greater amount of the total variability was accounted for by the environmental term (E), but it was significantly reduced when marker and canopy information were included in the model. Also, the residual variation was reduced when the interaction components were added in the models. For yield, the percentage

TABLE 13.1
Estimated Variance Components for the Nine Models for Yield

Model	Model No.	Estimated Variance Components								
		E	L	G	GE	CC	CCE	GCC	GECC	R
E+L	1	59.0	10.3							30.7
E+L+G	2	49.9	2.4	14.9						32.8
E+L+G+GE	3	46.5	4.0	10.4	10.4					28.7
E+L+CC	4	51.7	10.1			7.0				31.3
E+L+CC+CCE	5	63.4	8.9			0.1	0.1			27.5
E+L+G+CC	6	48.1	1.6	13.7		8.8				27.8
E+L+G+CC+GCC	7	42.8	1.9	15.7		8.1		3.4		28.1
E+L+G+CC+GE+CCE	8	55.1	2.4	8.3	7.8	5.0	0.4			20.9
E+L+G+CC+GE+CCE+GECC	9	44.3	2.8	11.7	7.8	5.0	0.6		2.8	25.0

E = environment, L = line, G = genotype, GE = genotype × environmental interaction, CC = canopy information, CCE = canopy × environment interaction, GCC = genotype × canopy interaction, GECC = genotype × environment × canopy interaction.

TABLE 13.2
Estimated Variance Components for the Nine Models for Days to Maturity

	Model	Estimated Variance Components								
Model	**No.**	**E**	**L**	**G**	**GE**	**CC**	**CCE**	**GCC**	**GECC**	**R**
E+L	1	45.1	32.0							22.9
E+L+G	2	19.8	5.2	61.2						13.8
E+L+G+GE	3	17.9	6.4	61.0	2.7					12.0
E+L+CC	4	31.5	37.5			3.0				28.0
E+L+CC+CCE	5	35.8	37.1			0.1	0.1			26.9
E+L+G+CC	6	15.2	5.4	63.4		1.1				14.9
E+L+G+CC+GCC	7	12.3	6.0	64.6		1.2		1.3		14.5
E+L+G+CC+GE+CCE	8	18.8	5.7	52.1	2.2	0.0	10.4			10.7
E+L+G+CC+GE+CCE+GECC	9	13.3	6.0	54.1	2.3	0.0	12.8		0.7	10.9

E=environment, L=line, G=genotype, GE=genotype×environmental interaction, CC=canopy information, CCE=canopy×environment interaction, GCC=genotype×canopy interaction, GECC=genotype×environment×canopy interaction.

of the variability explained by the residual term was reduced from 30.7% to 20.9%. DTM also showed a decrease in the residual variance, from 22.9% to 10.7%.

13.3.2 Assessment of Predictive Ability

In this study, nine models were evaluated in terms of prediction accuracy. The main objective was to assess whether incorporating CC data into prediction models led to an increase in prediction accuracy, and whether the inclusion of interaction terms would further improve these results. Jarquín et al. (2014b) showed the advantage of including the genotype×environmental interaction in the prediction models, and, here, we wanted to determine whether the inclusion of interaction terms among the environment, the genotype, and the CC measurements enhanced the predictive accuracy.

Out of the nine models, only three did not include any CC measures: Model 1: Environment+Line model; Model 2: Environment+Line+Marker model; and Model 3: Environment+Line+Marker+(Marker×Environment) Interaction model. We evaluated the models, using SNP marker information, CC measurements, and phenotypic information collected on yield and DTM on the SoyNAM recombinant inbred lines. The phenotypic information and the CC measurements were collected in 2013, 2014, and 2015. Each of the prediction accuracies was calculated by implementing the CV2, CV1, and CV0 cross-validation schemes.

Figures 13.3a and 13.4a show the results for the nine models, using CV2 for yield and DTM, respectively. Similarly, Figures 13.3b and 13.4b show the same for CV1, and Figures 13.3c and 13.4c for CV0, respectively.

The performance of the CV1 scheme depends mostly on the genetic similarities between the training and testing sets, whereas CV2 and CV0 also account for

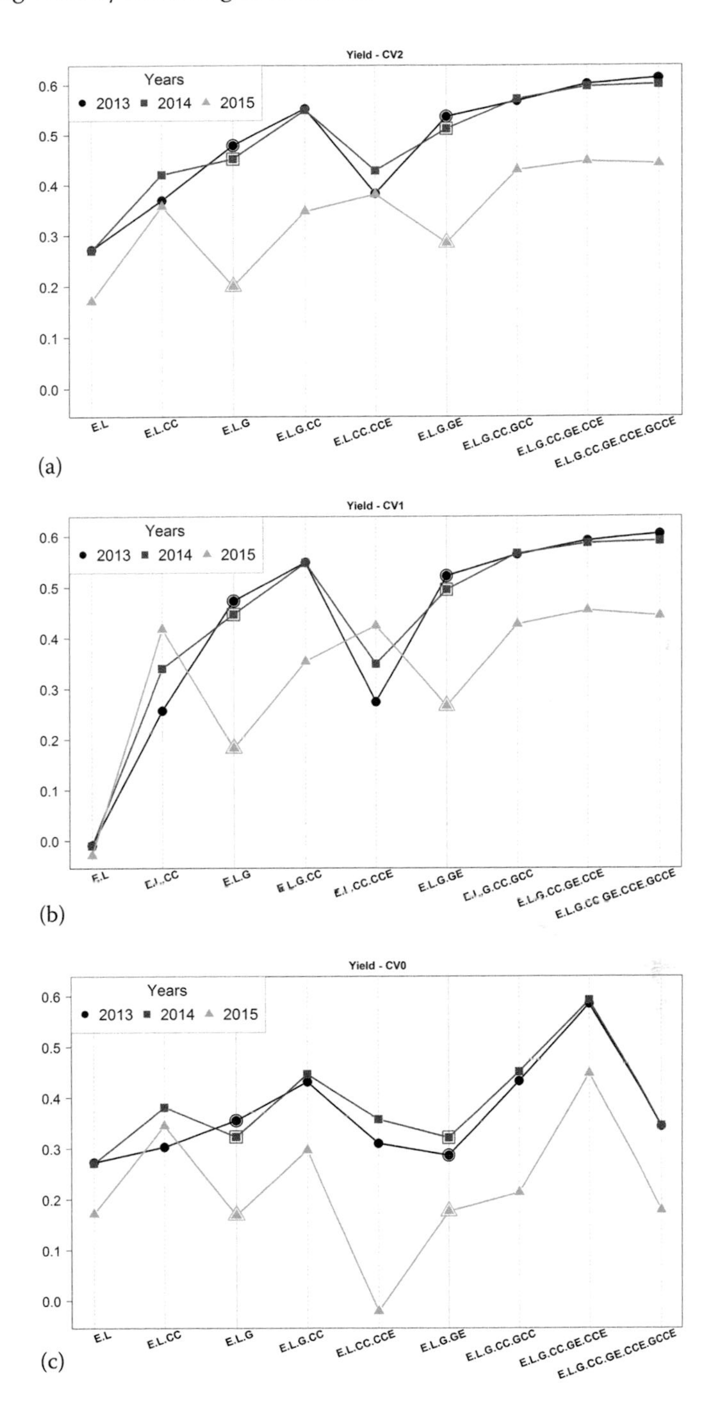

FIGURE 13.3 Average prediction accuracy obtained for the nine models for yield, using the different cross-validation schemes. The three different symbols represent the years for which the prediction was carried out.

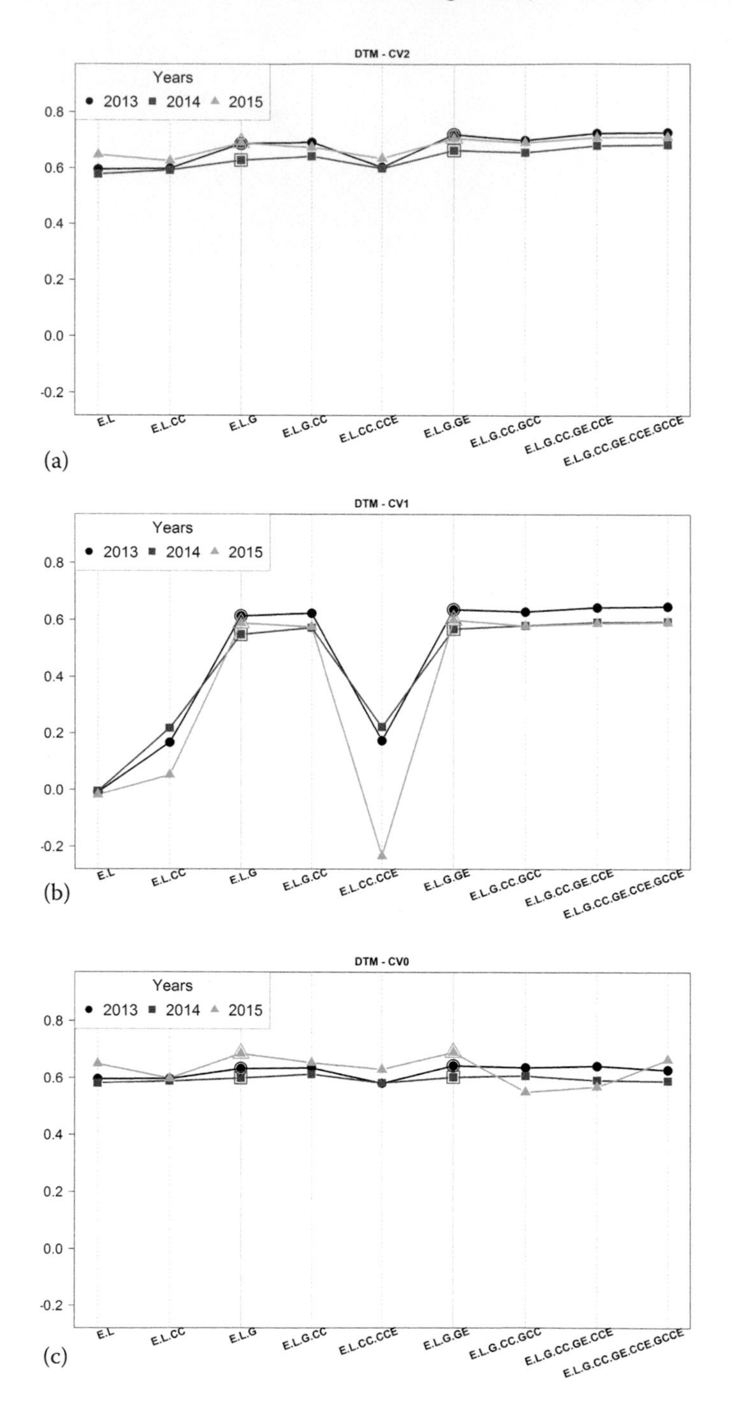

FIGURE 13.4 Average prediction accuracy obtained for the nine models for days to maturity (DTM) using the different cross-validation schemes. The three different symbols represent the years for which the prediction was carried out.

the environmental variations *via* replicates of the same genotypes observed in other environments.

When we considered yield as the trait to be predicted, for the CV2 and CV1 cross-validation schemes, Models 7–9 performed better than those models that did not combine marker and canopy data in interaction with the environments. In these cases, the correlation between the predicted and observed values was around 0.6 for 2013 and 2014, and 0.42 for 2015. These values gradually increased in response to the addition of terms in the models. The common feature among the models was that each of these models (Models 7–9) included the genotypic information, the CC information, and at least one interaction component involving CC and marker data. For CV0, Model 8 performed better, reaching the same level of predictive ability that was shown for the CV2 and CV1 schemes.

For each of the cross-validation techniques, most of the models had the lowest mean prediction accuracy when the prediction was carried out for 2015. In addition, similar patterns were shown when comparing results from 2013 and 2014. In general, each of the cross-validation schemes showed significant variations among models, in terms of predictive ability. However, in each case, predictive ability was improved when the interactions involving canopy data were included.

When we evaluated the models for DTM, and when CV2 and CV0 were implemented, we did not see a marked variation among the models in terms of prediction accuracy. The predictive accuracy of all models ranged from 0.55 to 0.65. For CV1, Models 1, 2, and 5 had significantly lower prediction accuracy than the rest of the models. These were the only models that did not include genotype as a main effect term or as an interaction with environments. For DTM, we could not identify a year where models exhibited the highest predictive accuracy across the cross-validation schemes. For each of the three cross-validation schemes, Model 6 – which is the Environment + Line + Marker + (Marker × Environment) interaction model – performed the best, but, for CV0 and CV2, the difference in prediction accuracy between Model 6 and some of the other models was not significant.

When predicting yield, we clearly see the advantage of including the CC measurements and the interactions among markers, environment, and CC measurement. The highest predictive abilities for CV2 and CV1 were delivered by Model 9 (the three-way interaction model), whereas, for CV0, Model 8 produced the greatest predictive accuracy values.

For predicting DTM, the advantage of including these terms was not as evident as for yield, but including the CC measurement as a main effect or as part of interactions did not hurt the model performance when the markers were also included in the model.

13.3.3 Effectiveness of Canopy Data from Early Stages

As mentioned earlier, another objective of this study was to compare the usefulness, as a predictive tool, of canopy data collected in the early stages (days 14–33) of the growing season with the entire season's data set (days 14–71). Figures 13.5 and 13.6 a, b, and c show the correlation between predicted and observed values for yield and DTM, respectively, for the nine models using the reduced (days 14–33) and entire canopy datasets (days 14–71) for the different cross-validation schemes. For yield, in schemes CV2 and CV1, the entire and reduced datasets showed similar results. The entire set always performed slightly better, in terms of predictive accuracy, than the reduced set. On the other hand, with CV0, in most of the cases, the reduced dataset provided better results than the entire dataset. For DTM, the entire and reduced datasets performed similarly for CV2 and CV1, whereas, for CV0, there were a few cases where the reduced dataset slightly outperformed the results of the entire set (Figures 13.5 and 13.6).

Since the results obtained from the reduced set of canopy values were similar to those derived from the entire set (days 14–71), we are confident of the applicability of this technique to use the reduced dataset to improve the predictive ability of genomic prediction models, based on information from the very early stages of the growing season. An important practical implication of these results is that we can reach the same degree of predictive accuracy using CC coverage data collected in the very early stages of the growing season instead of needing to collect canopy information throughout the entire growing season.

13.3.4 Usefulness of Including Canopy Data in Prediction Models in Real Scenarios, and Future Directions

Since the ultimate goal of performing predictions of unobserved genotypes is to save time and resources for developing new and improved cultivars, the implementation of phenomics might not be straightforward in terms of shortening breeding cycles, to avoid spending money or resources to phenotype these lines. However, as was shown, the integration of CC data and marker data improved model performance. This might allow a more accurate selection of superior breeding lines, as pointed by Crain et al. (2018). From a more practical perspective, the use of the CC information collected over the early stages of the growing season could aid the selection process when no phenotypes are available, in those cases where breeders prefer phenotypic selection. In this scenario, when an extreme hydro-climatic event (e.g., a hurricane, tornado, etc.) occurs, it might partially or completely destroy the breeding material before harvest, leaving no chance to perform phenotypic selection. Thus, the use of CC data might allow us to recover valuable information on these destroyed cultivars, improving the accuracy of selection. Some future work may attempt to investigate the use of canopy information for predicting the performance of future generations. In this case, our proposed solution is to predict canopy values of untested lines and use these predicted values as covariates for predicting yield and/or DTM.

ABBREVIATIONS

- GP: genomic prediction.
- CV0: cross-validation, predicting the performance of previously tested lines in untested locations.
- CV1: cross-validation, evaluating the performance of new developed lines, or lines that have not been evaluated in any of the observed environments.

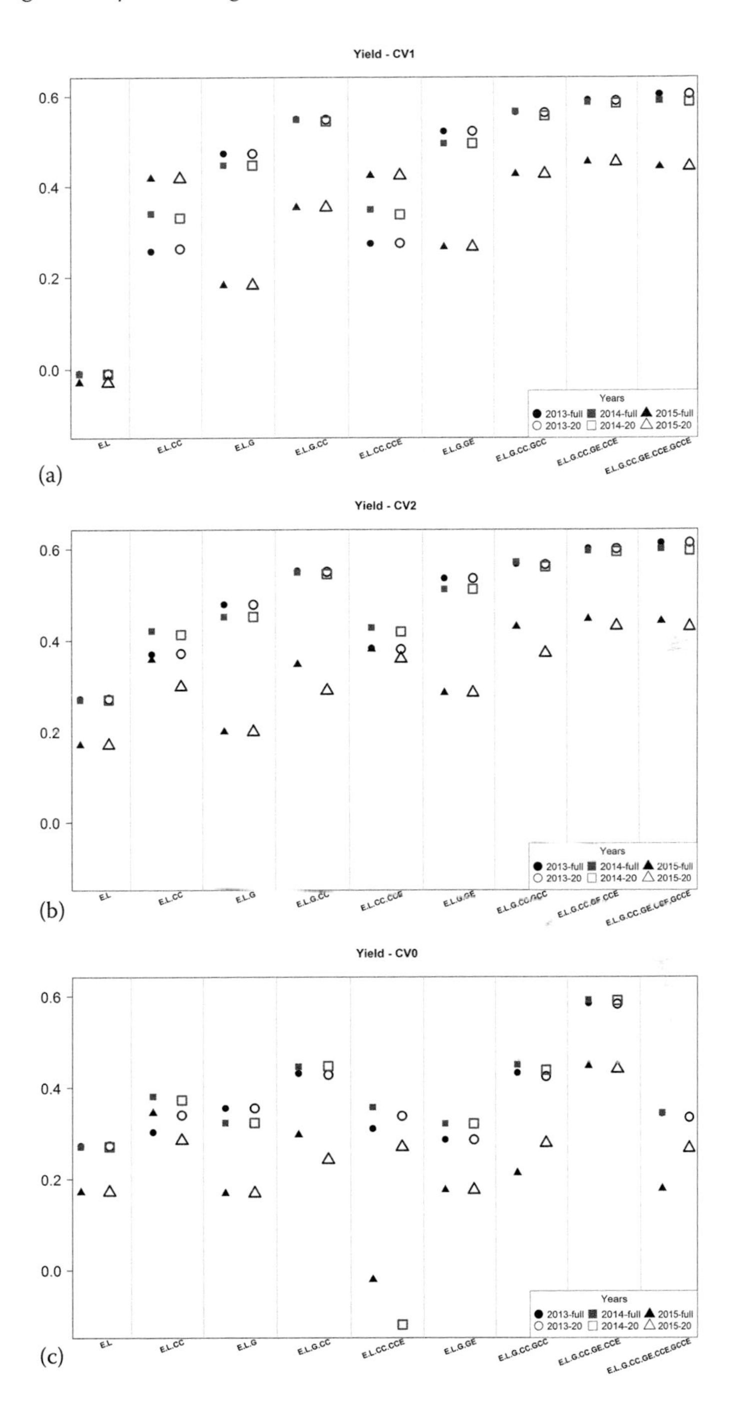

FIGURE 13.5 Average prediction accuracy obtained from the nine models for yield *using* the different cross-validation schemes for the entire (days 14–71) and reduced (days 14–33) canopy datasets. The three different symbols represent the years for which the prediction was carried out.

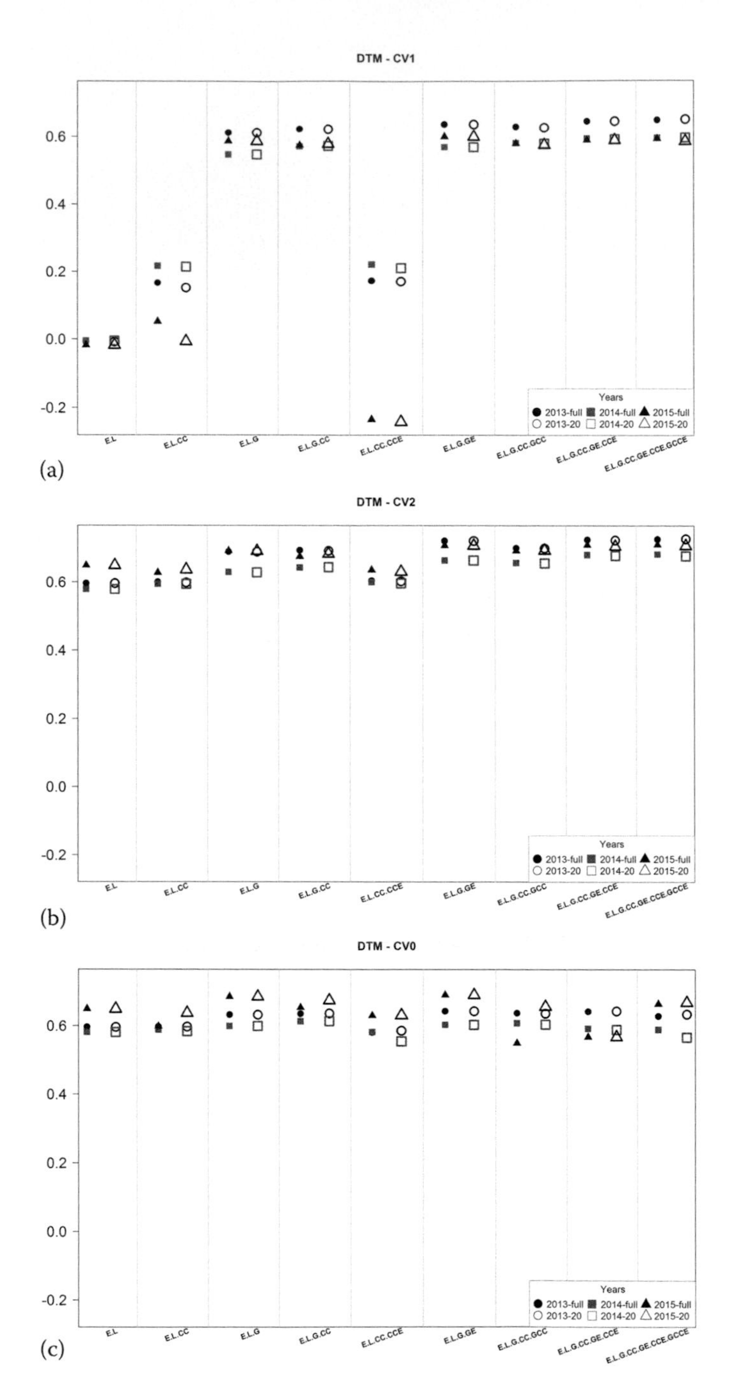

FIGURE 13.6 Average prediction accuracy obtained from the nine models for DTM *using* the different cross-validation schemes for the entire (days 14–71) and reduced (days 14–33) canopy sets. The three different symbols represent the years for which the prediction was carried out.

- CV2: cross-validation, evaluating the performance of lines that have been evaluated in some environments but not in others, i.e., incomplete field trials.
- SoyNAM: soybean nested association mapping.

REFERENCES

Aguate, F. M., S. Trachsel, L. G. Pérez, J. Burgueño, J. Crossa, M. Balzarini, D. Gouache, M. Bogard, and G. de los Campos. 2017. Use of hyperspectral image data outperforms vegetation indices in prediction of maize yield. *Crop Science* 57(5):2517–2524. doi: 10.2135/cropsci2017.01.0007.

Crain, J., S. Mondal, J. Rutkoski, R. P. Singh, and J. Poland. 2018. Combining high-throughput phenotyping and genomic information to increase prediction and selection accuracy in wheat breeding. *Plant Genome* 11(1):1. doi: 10.3835/plantgenome2017.05.0043.

de los Campos, G., D. Gianola, G. J. M. Rosa, K. A. Weigel, and J. Crossa. 2010. Semi-parametric genomic-enabled prediction of genetic values using reproducing kernel Hilbert spaces methods. *Genetics Research* 92(4):295–308.

Endelman, J. B. 2011. Ridge regression, and other kernels for genomic selection with R package rrBLUP. *Plant Genome* 4(3):250–255.

Fahlgren, N., M. A. Gehan, and I. Baxter. 2015. Lights, camera, action: High-throughput plant phenotyping is ready for a close-up. *Current Opinion in Plant Biology* 24:93–99.

Gianola, D., R. R. Fernando, and A. Stella. 2006. Genomic-assisted prediction of genetic value with semiparametric procedures. *Genetics* 173(3):1761–1776.

Habier, D., R. L. Fernando, K. Kizilkaya, and D. J. Garrick. 2011. Extension of the Bayesian alphabet for genomic selection. *BMC Bioinformatics* 12:186.

Heffner, E. L., A. J. Lorenz, J. L. Jannink, and M. E. Sorrells. 2010. Plant breeding with genomic selection: Potential gain per unit time and cost. *Crop Science* 50(5):1681–1690.

Jarquín, D., K. Kocak, L. Posadas, K. Hyma, J. Jedlicka, G. Graef, and A. Lorenz. 2014a. Genotyping by sequencing for genomic prediction in a soybean breeding population. *BMC Genomics* 15:740. doi: 10.1186/1471-2164-15-740.

Jarquín, D., J. Crossa, X. Lacaze et al. 2014b. A reaction norm model for genomic selection using high dimensional genomic and environmental data. *Theoretical and Applied Genetics* 127:595–607. doi: 10.1007/s00122-013-2243-1.

Karcher, D. E., and M. D. Richardson. 2005. Batch analysis of digital images to evaluate turfgrass characteristics. *Crop Science* 45(4):1536–1539. doi: 10.2135/cropsci2004.0562.

Meuwissen, T. H. E., B. J. Hayes, and M. E. Goddard. 2001. Prediction of total genetic value using genome-wide dense marker maps. *Genetics* 157(4):1819–1829.

Montesinos-Lopez, O. A., A. Montesinos-Lopez, J. Crossa, G. De los Campos, G. Alvarado, S. Mondal, J. Rutkoski, L. Gonzalez-Perez, and J. Burgueño. 2017a. Predicting grain yield using canopy hyperspectral reflectance in wheat breeding data. *Plant Methods* 13:4.

Montesinos-López, A., O. A. Montesinos-Lopez, J. Cuevas et al. 2017b. Genomic Bayesian functional regression models with interactions for predicting wheat grain yield using hyper spectral image data. *Plant Methods* 13:62.

Purcell, L. C. 2000. Soybean canopy coverage and light interception measurements using digital imagery. *Crop Science* 40(3):834–837. doi: 10.2135/cropsci2000.403834x.

Song, Q., L. Yan, C. Quigley et al. 2017. Genetic characterization of the soybean nested association mapping population. *Plant Genome* 10(2):2. doi: 10.3835/plantgenome2016.10.0109.

Stekhoven, D. J., and P. Buhlmann. 2012. MissForest—Non-parametric missing value imputation for mixed-type data. *Bioinformatics* 28(1):112–118.

VanRaden, P. M. 2008. Efficient methods to compute genomic predictions. *Journal of Dairy Science* 91(11):4414–4423. doi: 10.3168/jds.2007-0980.

Wang, W., B. Vinocur, and A. Altman. 2003. Plant responses to drought, salinity and extreme temperatures: Towards genetic engineering for stress tolerance. *Planta* 218(1):1–14.

Whitford, R., D. Fleury, J. C. Reif, M. Garcia, T. Okada, V. Korzun, and P. Langridge. 2013. Hybrid breeding in wheat: Technologies to improve hybrid wheat seed production. *Journal of Experimental Botany* 64(18):5411–5428. doi: 10.1093/jxb/ert333.

Xavier, A., W. M. Muir, and K. M. Rainey. 2016. Assessing predictive properties of genome-wide selection in soybeans. *G3: Genes, Genomes, Genetics* 6(8):2611–2616. doi: 10.1534/g3.116.032268.

Xavier, A., B. Hall, A. A. Hearst, K. A. Cherkauer, and K. M. Rainey. 2017. Genetic architecture of phenomic-enabled canopy coverage in Glycine max. *Genetics* 206(2):1081–1089. doi: 10.1534/genetics.116.198713. Epub 2017 Mar 31.

Zulauf, C., J. Coppess, N. Paulson, and G. Schnitkey. 2017. U.S. Oilseeds: Production and policy comparison. *Farmdoc Daily* 7:28.

14 Field Phenotyping for Salt Tolerance and Imaging Techniques for Crop Stress Biology

Shayani Das Laha, Amlan Jyoti Naskar, Tanmay Sarkar, Suman Guha, Hossain Ali Mondal, and Malay Das

CONTENTS

14.1 INTRODUCTION

Stresses are biotic and abiotic factors that may negatively affect a crop plant's growth and productivity (Mittler 2006). Therefore, hundreds of laboratories across the globe, spanning academia as well as industry, are engaged in research into understanding the response of crop plants to diverse stresses (Beacham et al. 2017; Mangin et al. 2017; Vij and Tyagi 2007). The major abiotic stresses include inadequate (drought) or excessive (hypoxia) water availability (Shinozaki and Yamaguchi-Shinozaki 2007), soil salinity (Negrão et al. 2017), high (heat stress) or low (cold stress) temperatures (Ohama et al. 2017), deficiency of major nutrients in the soil, including nitrogen, potassium, phosphorus, calcium, and magnesium (Forieri et al. 2017), exposure to either ultraviolet ray or ozone, and suboptimal soil pH (Lager et al. 2010). In addition to these abiotic factors, biotic stresses such as bacteria, fungi, pests, viruses, and insects may significantly limit crop yields. It has been predicted that soil salinity and drought may reduce the available agricultural land to 50% of the current area by 2050 (Wang et al. 2003), while the world population may increase to somewhere within the range 9.6–12.3 billion (Gerland et al. 2014) by 2100. Therefore, understanding the impact of different stresses on crops and finding ways to address global food security in the foreseeable future is critical. In this chapter, we review the current state of understanding of crop stress phenotyping and describe how recent progress in high-throughput plant phenotyping (HTPP) can aid in identifying stress-tolerant species or genotypes.

14.1.1 CURRENT STATUS OF STRESS PHENOTYPING ON *BRASSICA* CROP PLANTS

To understand the impact of stresses on plants, stress phenotyping experiments are typically performed on plants grown under controlled environment conditions (Shi et al. 2016) or in greenhouses (Siddiqui et al. 2009), where plants can be exposed to different concentrations/levels/durations of specific stresses. The impact of stress application is then assessed at the morphological (Waller et al. 2008), biochemical (Sajedi et al. 2012), metabolite (Akula and Ravishankar 2011), transcript (Jamil et al. 2011), or protein levels (Akhzari and Pessarakli 2016). For instance, the sensitivity of 11 germplasm accessions of *Brassica napus* to heat stress was assessed based on (a) reproductive traits, like duration of flowering, ratio of pod *vs.* flower numbers, number of pollen grains produced, and number of seeds per silique, or (b) vegetative traits like dry matter biomass. The study showed that the majority of these phenotypes were negatively affected as a consequence of heat treatment (Koscielny et al. 2018). To study the tolerance of *B. napus* to waterlogging, three different growth conditions, i.e., seedlings grown in the laboratory, seedlings in the greenhouse, or mature plants grown under field conditions, were used to compare the responses of 25 accessions. The study revealed that the accessions which were predicted to be tolerant by laboratory or greenhouse assays were not shown to be the most tolerant ones in the field study (Zou et al. 2014). Compared to controlled environment studies, only a limited number of stress phenotyping experiments have been conducted under field conditions. One such study assessed the sensitivity of 796 accessions of *Brassica juncea* to heat stress, and identified several tolerant accessions, namely DRMR-1574, DRMR-1624, DRMR-1600, DRMR-1799, and 'Urvashi' (Ram et al. 2014).

14.1.2 Salt Stress Assay on Brassica Genotypes Grown in the Field

To date, the majority of phenotyping studies related to crop stress response have been conducted under controlled laboratory environment conditions. However, the environmental conditions in the field are much more divergent than the laboratory conditions, and therefore, the knowledge generated from the laboratory-based studies is not always applicable to the field settings, so field-level screening can be particularly valuable. For example, to evaluate the response of 44 wheat accessions to drought, field studies were performed for four consecutive years, measuring each of eleven quantitative traits, and each of the accessions showed significant variation for each of the traits studied across the years (Kumar et al. 2018). Therefore, a pilot field study was conducted by our research group to identify the challenges associated with the prediction of salt-tolerant *Brassica* accessions, based on the phenotypes obtained from plants grown under field conditions. A total of 104 *Brassica* genotypes were employed in this study (Table 14.1) and their responses to salt stress were assessed using a range of morphological and biochemical markers.

14.2 MATERIALS AND METHODS

14.2.1 Field Trial of Salt Stress Response

A total of 104 genotypes of *Brassica* were selected for the field study, out of which 100 were *Brassica juncea* and four were *Brassica rapa* (Table 14.1). The experiments were conducted in saline as well as in non-saline fields. The saline field was located in Gosaba (22.1652° N, 88.8079° E), West Bengal, India, whereas the control (non-saline) field was located in Baruipur (22.3597° N, 88.4318° E), West Bengal, India. These experiments were initiated in December 2017 and continued until the first week of April 2018. The experimental design was a completely randomized replicated block design (RBD), with three biological replicates per genotype in each of the saline and non-saline fields. Seeds were sown directly in the fields during the first and second week of December.

14.2.2 Measurements of Eight Morphological and Physiological Phenotypes

The morphological parameters selected for our analysis were: fresh and dry shoot biomass, chlorophyll concentration, number of leaves, number of branches, number of siliques per plant, number of seeds per silique, and silique length (Table 14.2). To measure the dry biomass of shoots, all of the above-ground shoot tissue was collected and oven-dried overnight at 70°C (Sharma et al. 2013). The chlorophyll concentrations of the leaves were measured using a Chlorophyll Meter SPAD-502Plus (Minolta, Tokyo, Japan). Values obtained by SPAD metrics are equivalent to the concentration of total chlorophyll in the leaf (Rodriguez and Miller 2000).

14.2.3 Estimation of Proline Concentration in Leaves

Approximately 3 weeks after germination, vegetative tissues were collected for estimation of proline concentration in the leaves. Estimation of proline concentration

TABLE 14.1
A List of *Brassica* Genotypes Used in this Study

Sample Identity	*Brassica* species	Genotypes	Sample Identity	*Brassica* Species	Genotypes
1	*B. rapa* cv. Toria	B-54 (Agrahi)	3	*B. rapa* cv. Yellow Sarson	B9 (Binoy)
4	*B. rapa* cv. Yellow Sarson	YSB-19-7-C (subinoy)	5	*B. rapa* cv. Yellow Sarson	NC-1 (jhumka)
6	*B. juncea*	B-85 (Seeta)	7	*B. juncea*	RW-4C-6-3 (sanjukta Asech)
8	*B. juncea*	RW-85-59 (Sarna)	9	*B. juncea*	RW-351 (Bhagirathi)
11	*B. juncea*	NPJ-194	13	*B. juncea*	Rohini (SC)
14	*B. juncea*	KMR-15-4	15	*B. juncea*	PR-2012-9
16	*B. juncea*	Divya-88	17	*B. juncea*	RL-JEB-52
18	*B. juncea*	Kranti-NC	19	*B. juncea*	DRMRIJ-15-85
20	*B. juncea*	RH1202	21	*B. juncea*	NPJ-196
22	*B. juncea*	RMM-09-10	23	*B. juncea*	JMM-927-RC
24	*B. juncea*	RRN-871	25	*B. juncea*	KM-126
26	*B. juncea*	SKM-1313	27	*B. juncea*	RB-77
28	*B. juncea*	DRMR-15-5	29	*B. juncea*	KMR-53-3
30	*B. juncea*	RL-JEB-84	31	*B. juncea*	Ganga
32	*B. juncea*	RGN-73-JC	33	*B. juncea*	RH-1209
34	*B. juncea*	PR-2012-12	35	*B. juncea*	RGN-385
36	*B. juncea*	NPJ-195	37	*B. juncea*	Maya-C
38	*B. juncea*	SKJM-05	39	*B. juncea*	SVJ-64
40	*B. juncea*	Sitara-sreenagar	41	*B. juncea*	RH-0923
42	*B. juncea*	DRMR-15-16	43	*B. juncea*	NPJ-198
44	*B. juncea*	JMM-927-RC	45	*B. juncea*	DRMR-15-47
46	*B. juncea*	RGN-389	47	*B. juncea*	RAURD-214
48	*B. juncea*	DRMR-15-14	49	*B. juncea*	DRMR-4001
50	*B. juncea*	RGN-384	51	*B. juncea*	NPJ-197
52	*B. juncea*	RB-81	53	*B. juncea*	NPJ-200
54	*B. juncea*	DRMR-15-9	55	*B. juncea*	KMR-L-15-6
56	*B. juncea*	PRD-2013-9	57	*B. juncea*	DRMRIJ-15-66
58	*B. juncea*	RH-1368	59	*B. juncea*	RH-1325
60	*B. juncea*	RGN-386	61	*B. juncea*	RNWR-09-3
62	*B. juncea*	PRD-2013-2	63	*B. juncea*	RH-781
64	*B. juncea*	Swaran Jyoti (RH 9801)	65	*B. juncea*	RGN 48
66	*B. juncea*	RGN 13	67	*B. juncea*	CS 52
68	*B. juncea*	CS 54	69	*B. juncea*	CS 56
70	*B. juncea*	CS 58	71	*B. juncea*	Varuna

(*Continued*)

TABLE 14.1 (CONTINUED)
A List of *Brassica* Genotypes Used in this Study

Sample Identity	*Brassica* species	Genotypes	Sample Identity	*Brassica* Species	Genotypes
72	*B. juncea*	Kranti	73	*B. juncea*	Krishna
74	*B. juncea*	Pusa bold	75	*B. juncea*	CS 2000-61
76	*B. juncea*	CS 2000-106	77	*B. juncea*	CS 2000-189
78	*B. juncea*	CS 2000-195	79	*B. juncea*	CS 2004- 105
80	*B. juncea*	CS 2004- 191	81	*B. juncea*	CS 2004- 114
82	*B. juncea*	CS 2005-124	83	*B. juncea*	CS 2005-125
84	*B. juncea*	CS 2005-127	85	*B. juncea*	CS 2005- 196
86	*B. juncea*	CS 2005-134	87	*B. juncea*	CS 2005- 138
88	*B. juncea*	CS 2005- 197	89	*B. juncea*	CS 2007- 154
90	*B. juncea*	CS 2009-105	91	*B. juncea*	CS 2009-118
92	*B. juncea*	CS 2009-119	93	*B. juncea*	CS 2009- 142
94	*B. juncea*	CS 2009- 145	95	*B. juncea*	CS 2009- 229
96	*B. juncea*	CS 2009- 256	97	*B. juncea*	CS 2009-261
98	*B. juncea*	CS 2009-332	99	*B. juncea*	CS 2009- 347
100	*B. juncea*	CS 2009-401	101	*B. juncea*	CS 2009- 440
102	*B. juncea*	CS 2013-10	103	*B. juncea*	CS 2013-19
104	*B. juncea*	CS 2013-27	105	*B. juncea*	CS 2013-41
106	*B. juncea*	CS 2013-50	107	*B. juncea*	IC401570

in the leaves was performed following the method described by Bates et al. (1973). Approximately 100 mg of frozen leaf tissues were homogenized in 1.5 ml of 3% sulfosalicylic acid. Centrifugation was carried out at 13,000 × g for 5 minutes. Approximately 300 µl of the supernatant was mixed with 2 ml of glacial acetic acid and 2 ml acid ninhydrin (1.25 g ninhydrin warmed in 30 ml glacial acetic acid and 20 ml 6 M phosphoric acid until dissolved) in test tubes at 100°C in a boiling water bath for 1 h. The reaction was stopped by plunging the tubes into ice. The reaction mixture was extracted with 1 ml toluene by vigorous mixing for 10–30 s. The chromophore-containing toluene was pipetted away from the aqueous phase, warmed at room temperature and the absorbance was recorded at 520 nm in a Bio Spectrometer (Eppendorf, USA), using toluene as the blank, and the concentration of proline was determined, using a standard curve. Three biological replicates were used per sample.

14.2.4 Estimation of Na^+ and K^+ Concentrations in Leaves

Approximately 3 weeks after germination, vegetative tissues were collected for determination of Na^+ and K^+ concentrations in the leaves. To measure Na^+ and K^+ concentrations, 100 mg of leaf tissues were homogenized in 0.1% HNO_3 to prepare an extract. One ml of the extract was then diluted with deionized water to a working

TABLE 14.2

Top Ten Salt-Tolerant *Brassica* Genotypes Predicted on the Basis of Eight Morphological and Three Biochemical Phenotypes

Phenotype	Type of Phenotype	Effect of Salinity	Salt Tolerance Index (STI) Ranking									
			1	2	3	4	5	6	7	8	8	10
Fresh biomass	Morphological	Decrease	RW-85-59 (Sarna)	RW-4C-6-3 (sanjukta Asech)	**RW-351 (Bhagirathi)**	**B-85 (Seeta)**	Maya-C	**CS 58**	DRMRIJ-15-66	RH-0923	DRMR-15-16	KMR-53-3
Dry Biomass	Morphological	Decrease	NPJ-200	KMR-53-3	B-54 (Agrahi)	JNM-927-RC	PRD-2013-9	CS 2000-106	CS 2004- 114	**B-85 (Seeta)**	KMR-15-4	RW-4C-6-3 (sanjukta Asech)
No. of leaves	Morphological	Decrease	**RW-351 (Bhagirathi)**	RW-85-59 (Sarna)	CS 2005-134	**B-85 (Seeta)**	NPJ-194	KMR-53-3	CS 2005- 138	**CS 58**	RW-4C-6-3 (sanjukta Asech)	RH-0923
No. of branches	Morphological	Decrease	NPJ-194	**RW-351 (Bhagirathi)**	RL-JEB-84	**B-85 (Seeta)**	PR-2012-9	Rohini (SC)	RL-JEB-52	RW-4C-6-3 (sanjukta Asech)	DRMR-15-5	Kranti-NC
No. of silique	Morphological	Decrease	Sitara-sreenagar	DRMR-15-5	CS 2009-118	**B-85 (Seeta)**	JNM-927-RC	DRMRIJ-15-85	Kranti-NC	DRMR-15-16	PRD-2013-2	RB-81
Silique length	Morphological	Decrease	Divya-88	RH-1325	DRMR-15-16	YSB-19-7-C (subinoy)	CS 2005-134	PRD-2013-2	DRMRIJ-15-85	CS 2009- 347	**CS 58**	JNM-927-RC
No. of seeds/siliques	Morphological	Decrease	**RW-351 (Bhagirathi)**	RH1202	CS 2005-134	**B-85 (Seeta)**	Maya-C	RMM-09-10	CS 2007- 154	Kranti-NC	KM-126	JMM-927-RC
Chlorophyll	Morphological	Decrease	RGN-385	DRMR-15-16	**CS 58**	SKM-1313	RH-1325	RH-0923	Sitara-sreenagar	RGN-384	RMM-09-10	KMR-15-4
Proline	Biochemical	Increase	NC-1 (jhumka)	**CS 58**	CS 2009-229	RH1202	**RW-351 (Bhagirathi)**	B-54(Agrahi)	KM-126	JMM-927-RC	RW-85-59(Sarna)	RH-1368
Sodium content	Biochemical	increase	JNM-927-RC	DRMR-15-9	CS 2000-106	PRD-2013-9	RH-0923	NPJ-200	KMR-53-3	RH-1325	RB-81	CS 2009-261
Potassium content	Biochemical	Decrease	B54 (Agrahi)	Divya-88	RH1202	DRMRIJ-15-85	PR-2012-9	RL-JEB-52	Swaran Jyoti (RH 9801)	CS 2007-154	**RW-351 (Bhagirathi)**	**CS 58**

The three predicted tolerant genotypes of *B. juncea* are B-85 (Seeta; bold and underlined), RW-351 (Bhagirathi; bold and highlighted in gray), and CS 58 (bold), which appeared in the top ten list in six out of 11 phenotypes used in this study.

volume of 5 ml. A few drops of the diluted extract were placed into the sensor of a LAQUA Twin Potassium Ion Meter (Spectrum Technologies, Inc.) to determine the potassium concentration. A similar method was adopted to measure sodium concentration in the leaves, using a LAQUA Twin Sodium Meter (Spectrum Technologies, Inc.; Gangaiah et al. 2015).

14.2.5 Statistical Analysis

For statistical analysis, 15 plants (biological replicates) were sampled randomly for each of the 104 genotypes of *Brassica* for each of the saline and control (non-saline) treatments. Use of multivariate statistical analysis on data with both continuous variables (e.g., fresh biomass and chlorophyll SPAD value) and discontinuous variables (e.g., number of branches, and number of silique), using a joint model-based data analysis, is challenging. Although multivariate GLM- type models (Berridge and Crouchley 2011) seem appropriate; it is made complex by the fact that univariate modeling of individual phenotypes is itself difficult. Classical models like one-way ANOVA and two-way ANOVA were sufficient for previous studies conducted in controlled laboratory or greenhouse settings. However, for the present study, which was conducted in the field, these models showed a poor fit, even with different transformations, e.g., logarithm or square root transformations (Box and Cox 1964). Usual assumptions, like homogeneity of variance and normality, were violated and model-based statistical analysis remained infeasible (Nelson and Dudewicz 2002). So, instead of modeling, summary statistics were calculated to understand the differential salt tolerance ability of different genotypes for each of the 11 phenotypic characters studied. The salt tolerance index (STI) was calculated, based on the difference in the phenotypes between the saline and non-saline treatments. For an attribute, j, and genotype, i, the salt tolerance index is defined as the relative difference between the median variable value for the two treatments. It is given by

$$STI_{i,j} = \frac{y_{i,j}^{s} - y_{i,j}^{n}}{y_{i,j}^{n}} \times 100 \quad 1 \leq i \leq 104 \text{ and } 1 \leq j \leq 11$$

where, $y_{i,j}^{s}$ represents the median value of the jth phenotype among the (15) replicates for the ith germplasm under saline conditions, and $y_{i,j}^{n}$ represents the median value of the jth phenotype among the replicates for the ith genotype under non-saline conditions. In the definition of the salt tolerance index, the median of the values of all replicates is used, instead of the mean, since preliminary graphical analysis of the data strongly indicated large departures from the normal distribution.

However, it should be noted that the interpretation of the salt tolerance index depends on the phenotype. For example, the index for the shoot dry biomass is, in general, negative, and a smaller negative value indicates a more tolerant germplasm. In contrast, the index for proline concentration in leaves is generally positive, and plants of more tolerant genotypes have larger, positive $STI_{i,j}$ values. Based on these indexes, we ranked all 104 genotypes, from most tolerant to most susceptible.

14.3 RESULTS

14.3.1 Evaluation of Salinity Tolerance of *Brassica* Genotypes

The effect of salinity on the 104 accessions of *Brassica* under the field conditions was studied (Figure 14.1). The ranking for tolerance of the 104 genotypes was based on the phenotypic traits evaluated at the morphological, physiological, and biochemical levels (Table 14.2). The genotypes were ranked, based on their salt tolerance index for each phenotype (Table 14.2). However, as mentioned in Section 14.2.5, the interpretation of the salt tolerance index differs, depending upon how the phenotype is measured. In general, tolerant values for traits like shoot fresh and dry biomass, number of leaves, branches, siliques per plant, seeds per silique, silique length, and chlorophyll and potassium concentration tend to decrease in response to salinity. This reflects the

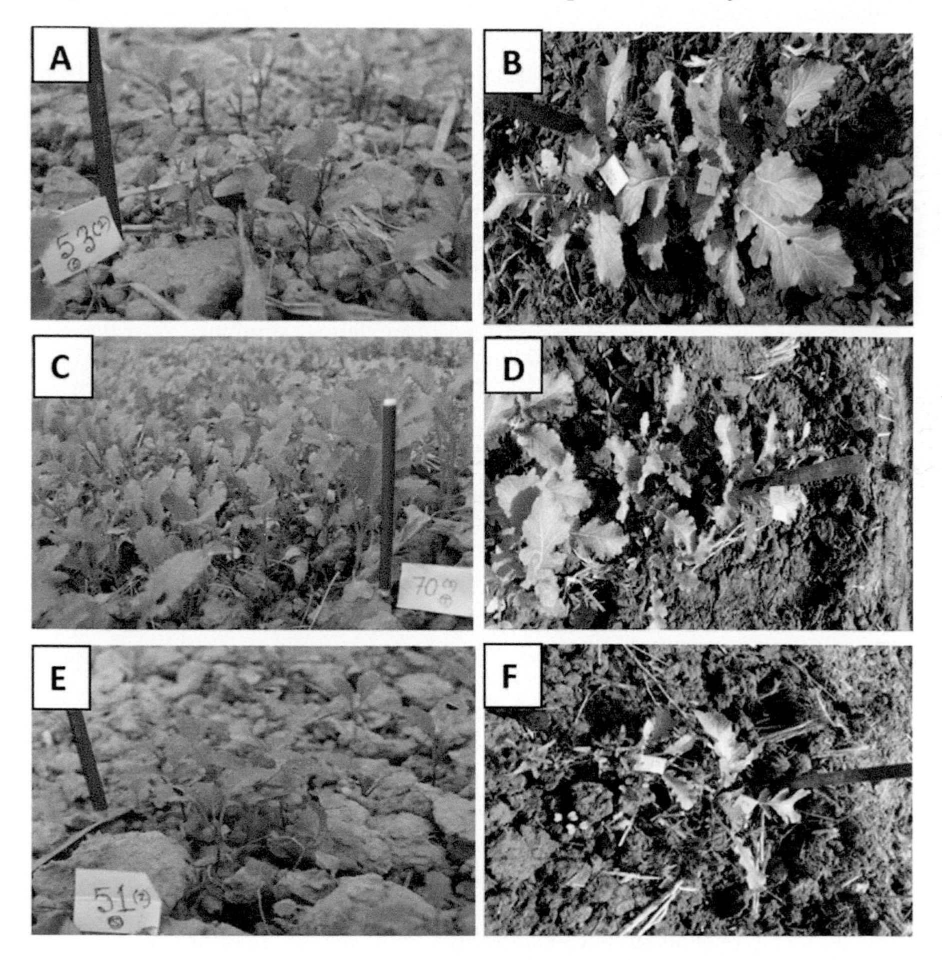

FIGURE 14.1 Screening of *Brassica* genotypes in the field. Predicted tolerant *Brassica juncea* NPJ-200 grown in (A) saline and (B) non-saline soil; predicted tolerant *B. juncea* CS 58 grown in (C) saline and (D) non- saline soil; predicted susceptible *B. juncea* NPJ-197 grown in (E) saline and (F) non-saline soil.

effect of salinity on plants. The rate of decrease was used as a criterion to determine the tolerance of germplasm. For example, *B. juncea* RW-85-59 (Sarna) exhibited the lowest rate of decrease in fresh biomass in response to salinity and was ranked as the most tolerant germplasm in this category (Table 14.2). However, for a phenotype like proline concentration, tolerance would be associated with an increased value in response to salinity; hence, increased proline concentration is expected in the tolerant plants, e.g. *B. juncea NC-1*(jhumka). The sodium concentration inside cells will increase in response to saline treatment, but the rate of increase will be less in the tolerant genotypes (e.g., *B. juncea* JNM-927-RC), compared with the susceptible ones.

To predict the most tolerant genotypes on the basis of the phenotypes, the top ten genotypes from each phenotypic category were obtained and compared (Table 14.2). As mentioned, different phenotypes predicted different genotypes to be the tolerant ones (Figure 14.2). However, *B. juncea* B-85 (Seeta), *B. juncea* RW-351 (Bhagirathi),

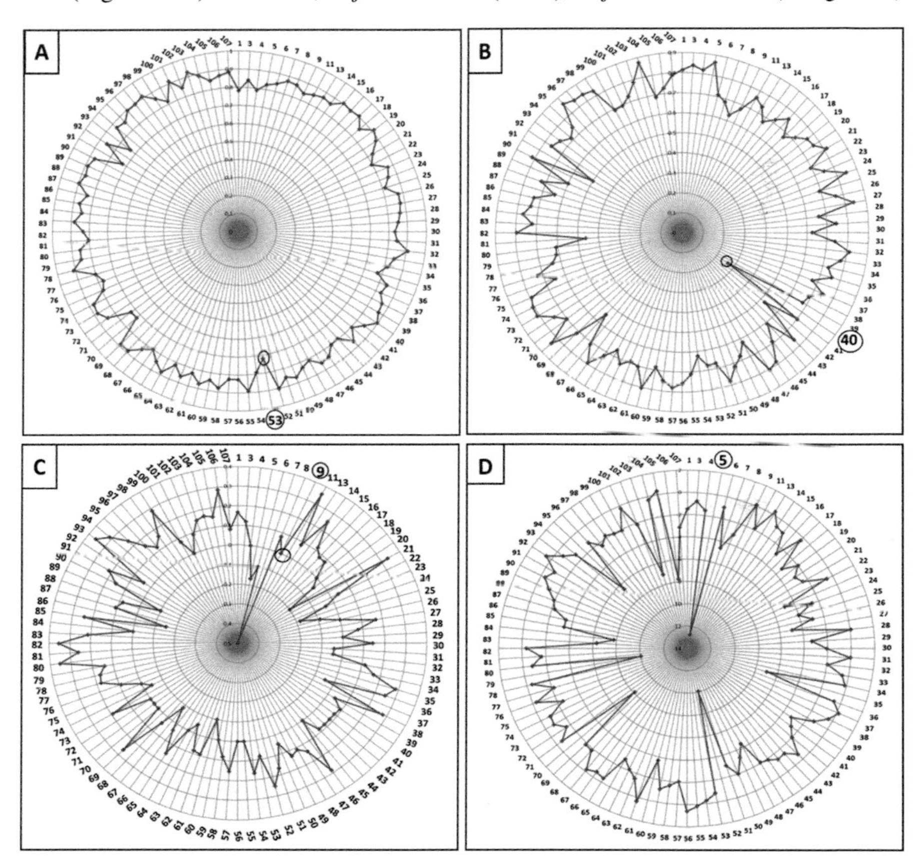

FIGURE 14.2 A radar-plot presentation of the salt tolerance indices for different phenotypes obtained from 104 genotypes of *Brassica rapa* and *Brassica juncea*, which were exposed to salt stress. (A) Dry biomass, (B) number of siliques, (C) number of seeds (D) proline concentration in leaves. The numbers represent the different genotypes studied (see Table 14.1). Each value represents a salt tolerance index for each genotype for a particular phenotype. The circled ones are the most tolerant genotypes, predicted on the basis of the respective phenotype.

and *B. juncea* CS 58 appeared in the top ten list for six out of 11 phenotypes used in this study, and hence were predicted to be salt-tolerant genotypes (Table 14.2).

14.3.2 Challenges in Field Phenotyping

It is evident from the results presented here and elsewhere (Farshadfar et al. 2012; Kumawat et al. 2017; Mansilla-Córdova et al. 2018) that identification of stable phenotypic markers linked to a specific stress-tolerance trait is quite challenging in the field setting, for the following reasons:

a) Since environmental conditions such as temperature, rainfall, and humidity may vary from season to season (Cramer et al. 2011), the observed phenotype may also vary, thereby posing a serious challenge to the identification of stress-linked phenotypes that could be reproducible across seasons (Wang et al. 2018).
b) The evaluation of yield and related phenotypes is time-consuming, and, therefore, the majority of stress phenotyping assays are typically dependent on vegetative growth, which is invasive in nature. In a typical field setting, collection of such large bodies of data is often prone to errors, since impact assessment of stresses on individual plants, such as the appearance of leaf yellowing, growth stunting, etc., varies among different researchers.
c) Multiple studies have highlighted that reliance on one or two phenotypic characters often failed to identify truly tolerant genotypes of crop plants, such as tomato plants tolerant to tomato chlorosis virus (ToCV) (Mansilla-Córdova et al. 2018) and wheat plants tolerant of drought (Farshadfar et al. 2012).

Many of these above-mentioned challenges of field phenotyping can be resolved by high-throughput plant phenotyping (HTPP). In HTPP platforms, plants can be monitored throughout their life cycle non-destructively with greater potential for identifying tolerant phenotypes at a much earlier time point than any of the conventionally used field phenotyping methods.

14.4 APPLICATIONS OF IMAGE-BASED HTPP IN PLANT STRESS BIOLOGY

Image-based HTPP provides enhanced sensitivity to predict tolerance or susceptibility of a genotype or species toward a certain stress. The use of imaging techniques in HTPP permits the phenotyping traits to be measured with greater accuracy and speed than is possible by conventional phenotyping methods. Since HTPP is non-destructive, plants can be monitored throughout their life cycle. Developmental stages before and after the application of stress and the consequent changes in plant growth and development parameters, such as shoot biomass reduction, photosynthesis, pigmentation, and transpiration rate can be measured. The application of this technique may significantly reduce time and improve the accuracy and reproducibility of data for phenotypic characterization of crop plants (Singh et al. 2016). Therefore, a number of research facilities and networks have been developed to cater to the growing needs of the plant phenotyping research community. These include the European Plant Phenomics Network (http://www.plant-phenotyping-network.eu/), and phenotyping centers in Germany (http://www.dppn.de/dppn/EN/Home/home_node.html), Australia (https://www.plantphenomics.org.au), Italy (http://www.phen-italy.it/index.php), France (https://www.phenome-fppn.fr/phenome_eng/) and the United Kingdom (http://www.ukppn.org.uk/). All these centers are part of the International Plant Phenotyping Network (IPPN; http://www.plant-phenotyping.org/), which promotes the growth of plant phenotyping research across the world. In addition, several international companies such as Qubit Phenomics (https://qubitphenomics.com/), Lemnatech (https://www.lemnatec.com/), and Photon System Instruments (http://psi.cz/products/) are developing cutting-edge phenotyping sensors that can be used in plant science and agricultural research. However, the majority of these phenotyping methods have been standardized for only a few model plants, such as *Arabidopsis* (Awlia et al. 2016), rice (Yang et al. 2013), maize (Yendrek et al. 2017), and sorghum (Wang et al. 2018). Therefore, extensive research efforts are needed to evaluate the efficiency of these platforms for other plants with different growth characteristics and architectures. In the following sections, we review the recent progress in image-based plant phenotyping technologies, with special emphasis on their potential with respect to better understanding of plant stress biology.

14.4.1 RGB Imaging

Red-green-blue (RGB) imaging is one of the simplest imaging techniques, which enables the image to be captured in two dimensions at visible wavelengths of light (400–750 nm). The unprocessed data from an image is visualized in red (~600 nm), green (~550 nm), and blue (~450 nm) bands of visible light. Images can be analyzed to measure morphological characteristics, e.g., shoot biomass, leaf structure, seedling length, root structure, and yield (Li et al. 2014). RGB imaging has been successfully used to study the sensitivity of nine *Arabidopsis* accessions toward drought stress by measuring the rosette-leaf area and automated weighing of pots to assess the rate of transpiration (Granier et al. 2006). RGB imaging was used to assess the root architecture of *Brassica napus* plants and its correlation with shoot biomass under phosphate (Pi)-deficient growth condition (Table 14.3; Shi et al. 2016). The effect of salt stress on wheat has also been studied using digital RGB imaging in a greenhouse, using an automated phenotyping platform. This study provided a deeper insight into the physiological processes governing salinity stress responses in wheat (Rajendran et al. 2009). In addition to abiotic stress research, RGB imaging is also useful in studying the effects of biotic stress on plants. In a wheat field, inoculated with the wheat streak mosaic virus, a study was conducted using RGB imaging to better understand the plant's responses to infection by the virus (Table 14.4; Casanova et al. 2014).

The RGB imaging technology is cheaper, compared with other imaging sensors used in phenotyping platforms, and therefore, plays an important role in encouraging agricultural scientists to focus on field crops. In spite of its many promising advantages, however, application of RGB technology in the field is

TABLE 14.3
A Summary of Different High-throughput Plant Phenotyping (HTPP) Methods Used in Abiotic Stress Phenotyping

Crop Plant	Stress	Imaging Technique	Plant Growth Conditions	Reference
Rapeseed (*Brassica napus*)	Phosphate deficiency	Visible	Controlled environment growth chambers	Shi et al. 2016
Tomato (*Solanum lycopersicum)*	Drought	Hyperspectral	Greenhouse and field grown	Römer et al. 2012
Maize (*Zea mays*)	Drought	Thermal	Field grown	Araus et al. 2012
Soybean (*Glycine max*)	Drought	Visible	Field grown	Hoyos-Villegas et al. 2014
Spinach (*Spinacia oleracea*)	Drought	Visible and thermal	Field grown	Raza et al. 2014
Rapeseed (*B. napus*)	Ozone	Fluorescence	Controlled environment growth chambers	Gielen et al. 2006

limited by several technical challenges, including separating the leaf from the background, automatically addressing the occlusion problem, and overcoming the difficulty of shadows in the computation of phenotypes for canopy plants (Li et al. 2014).

14.4.2 Thermal Imaging

Thermal imaging detects the infrared radiation reflected from an object. Since infrared radiation varies with temperature, thermography is a useful imaging technique to identify objects with variable temperatures (Li et al. 2014). Thermal imaging commonly reflects radiation in the wavelength range of 3–5 μm or 7–14 μm. In plant phenotyping, thermal imaging has been used to measure stomatal conductance as a measure of water content and the rate of transpiration in both greenhouse- and field-grown plants (Jones and Leinonen 2009). For example, screening of maize accessions for drought sensitivity has been successfully performed under field conditions (Table 14.3; Araus et al. 2012). A large number of accessions have been studied under salinity stress by capturing and automatically analyzing infrared images of seedlings (James and Sirault 2012). Also, the effects of biotic factors, e.g., *Alternaria* sp. on *B. juncea* plants have been studied (Table 14.4; Baranowski et al. 2015), using thermal imaging.

A major limitation of thermal imaging is that it needs calibration of the imaging sensor under individual environmental conditions. Another major challenge is how to accurately normalize the soil temperature from the plant temperature in scattered canopies.

14.4.3 Hyperspectral Imaging

Hyperspectral imaging uses a combination of digital imaging and spectroscopy. For each pixel in an image, the hyperspectral camera acquires the light intensity (radiance) for a large number (up to several hundred) of contiguous and narrow spectral bands. Every pixel in the image thus contains a continuous spectrum (in radiance or reflectance) and can be used to characterize the objects with great precision and in great detail (Li et al. 2014). The high spectral resolution of hyperspectral imaging has proved useful for detecting biotic stress damage caused by insects (Table 14.4; Huang et al. 2012; Yang et al. 2013) and for assessing drought stress at an early growth stage of barley (Behmann et al. 2014).

Predictions based on hyperspectral imaging provide better interpretations of early symptoms of stress responses, and thus help in identifying the appropriate stage as to when to collect samples for further molecular examination. Canopy reflectance was analyzed to identify symptoms of stress caused by cyst nematodes and *Rhizoctonia* (crown root rot disease) in sugar beet (Hillnhütter et al. 2011).

TABLE 14.4
A Summary of Different HTPP Methods Used in Biotic Stress Phenotyping

Crop Plant	Stress	Causal Organism	Imaging Technique	Plant Growth Conditions	Reference
Rapeseed (*Brassica napus*)	Fungus	Black spot disease (*Altenaria brassicicola*)	Hyperspectral and thermal	Controlled environment growth chambers	Baranowski et al. 2015
Tomato (*Solanum lycopersicum*)	Fungus	Powdery mildew (*Oidiopsis taurica*)	Thermal and stereo microscopy	Controlled environment growth chambers	Raza et al. 2014
Barley (*Hordeum vulgare*)	Fungus	Powdery mildew (*Blumeria graminis* f. sp. *hordei*)	Hyperspectral	Greenhouse	Thomas et al. 2018
Wheat (*Triticum aestivum*)	Virus	Wheat streak mosaic virus	Visible	Field	Casanova et al. 2014
Rice (*Oryza sativa*)	Caterpillar	Rice leaf-folder moth (*Cnaphalocrocis medinalis*)	Hyperspectral	Field	Huang et al. 2012
Chilli pepper (*Capsicum annuum*)	Virus	Yellow vein virus	Visible	Controlled environment growth chambers	González Pérez et al. 2013

A recent study conducted on maize (*Zea mays*) demonstrated that drought stress may be detected as early as three days after exposure to it (Asaari et al. 2018).

One potential limitation of this technology is the requirement for huge computational support for data storage and analysis. A large experiment using hyperspectral imaging would require terabytes of data.

14.4.4 Fluorescence Imaging

In fluorescence imaging, ultraviolet (UV) light, in the wavelength range of 340–360 nm, is reflected by different plant components as discrete wavelengths. Most of the fluorescence imaging studies are restricted either to single leaves or to the seedlings of model crops only, and primarily targeted to chlorophyll fluorescence, since this is directly associated with photosynthesis (Cen et al. 2017). In a recent study to address the effect of drought stress on *Arabidopsis*, kinetic chlorophyll fluorescence, in combination with multicolor fluorescence imaging, was used. The study demonstrated that combining kinetic chlorophyll fluorescence and multicolor fluorescence imaging, along with machine learning techniques, can be effective at identifying the morphological, physiological, and pathological phenotypes related to photosynthesis and can thereby efficiently assess the effects of drought stress on plants (Yao et al. 2018). Measuring the kinetics of chlorophyll fluorescence provides dynamic information on the impact of stresses on the photosynthesis of stress-affected plants. It provides emission signals in the red (690 nm) and far-red (740 nm) regions, which can be employed to analyze the photosynthetic activity of vegetative tissues in the field. On the other hand, multicolor fluorescence imaging can process a broad spectrum of wavelength, and hence achieve superior qualitative information about the plants. It can receive signals in the red (690 nm) and far-red (740 nm) regions, emitted from chlorophyll a, and in the blue (440 nm) and green (520 nm) regions, emitted from plant phenolic compounds.

14.5 CONCLUSIONS

Results obtained from previous studies and the case study presented in this chapter identify limitations of the conventional method of plant phenotyping, with respect to distinguishing between tolerant and susceptible genotypes, that are reproducible across seasons and agronomic practices. HTPP platforms, with RGB, fluorescence, thermal, or hyperspectral imaging modalities, offer a non-invasive, robust, and more sensitive alternative to addressing the existing shortcomings of traditional phenotyping platforms. In particular, integration of these technologies with unmanned aerial vehicles (UAVs) may significantly improve the scope of HTPP for collecting agricultural data over large geographical areas and longer time periods. UAVs equipped with different imaging sensors can capture high-resolution images of the plants in a field (Shi et al. 2016). The effects of cold stress on *Sorghum bicolor* plants have been studied, using imagery collected using UAVs (Wang et al. 2018).

Another emerging area in stress biology is the understanding of the development of root system architecture. This entails the investigation of soil profiles and root phenotypes like root length, root dry matter, root surface area, and root structure

(Chen et al. 2017). Since the root system plays a critical role in the initial response of a plant to its immediate environment, computing the root phenotypes would be a major step forward in understanding the plant's response to stress in greater spatial and temporal detail.

ACKNOWLEDGMENTS

The research was funded by the University Grant Commission (Grant no. MRP–MAJOR–BIOT–2013–18380), India and Faculty Research and Professional Development (FRPDF) grant of the Presidency University. The *Brassica* genotypes used in this study were provided by Uttar Banga Krishi Vishwavidyalaya, Cooch Behar, ICAR-Central Soil Salinity Research Institute, Karnal and ICAR-Directorate of Rapeseed Mustard Research, Bharatpur. The authors thank Bhabasindhu Mondal and Jayo Society, Gosaba, South 24 Pargana, West Bengal for providing help with the field studies.

REFERENCES

Akhzari, D., and M. Pessarakli. 2016. Effect of drought stress on total protein, essential oil content, and physiological traits of *Levisticum officinale* Koch. *Journal of Plant Nutrition* 39:1365–1371.

Akula, R., and G. A. Ravishankar. 2011. Influence of abiotic stress signals on secondary metabolites in plants. *Plant Signaling and Behavior* 6(11):1720–1731.

Araus, J. L., M. D. Serret, and G. O. Edmeades. 2012. Phenotyping maize for adaptation to drought. *Frontiers in Physiology* 3:305.

Asaari, M. S. M., P. Mishra, S. Merten, S. Dhondt, D. Inzé, N. Wuyts, and P. Scheunders. 2018. Close-range hyperspectral image analysis for the early detection of stress responses in individual plants in a high-throughput phenotyping platform. *ISPRS Journal of Photogrammetry and Remote Sensing* 138:121–138.

Awlia, M., A. Nigro, J. Fajkus, S. M. Schmoeckel, S. Negrão, D. Santelia, M. Trtílek, M. Tester, M. M. Julkowska, and K. Panzarová. 2016. High-throughput non-destructive phenotyping of traits that contribute to salinity tolerance in *Arabidopsis thaliana*. *Frontiers in Plant Science* 7:1414.

Baranowski, P., M. Jedryczka, W. Mazurek, D. B. Skowronska, A. Siedliska, and J. Kaczmarek. 2015. Hyperspectral and thermal imaging of oilseed rape (*Brassica napus*) response to fungal species of the genus *Alternaria*. *PLOS ONE* 10(3):e0122913.

Bates, L. S., R. P. Waldren, and I. D. Teare. 1973. Rapid determination of free proline for water-stress studies. *Plant and Soil* 39(1):205–207.

Beacham, A. M., P. Hand, D. A. Pink, and J. M. Monaghan. 2017. Analysis of *Brassica oleracea* early stage abiotic stress responses reveals tolerance in multiple crop types and for multiple sources of stress. *Journal of the Science of Food and Agriculture* 97(15):5271–5277.

Behmann, J., J. Steinrücken, and L. Plümer. 2014. Detection of early plant stress responses in hyperspectral images. *ISPRS Journal of Photogrammetry and Remote Sensing* 93:8–111.

Berridge, D. M., and R. Crouchley. 2011. *Multivariate Generalized Linear Mixed Models Using R*. CRC Press. Boca Raton, FL.

Box, G. E. P., and D. R. Cox. 1964. An analysis of transformations. *Journal of the Royal Statistical Society. Series B (Methodological)* 26(2):211–252.

Casanova, J. J., S. A. O'Shaughnessy, S. R. Evett, and C. M. Rush. 2014. Development of a wireless computer vision instrument to detect biotic stress in wheat. *Sensors* 14(9):17753–17769.

Cen, H., H. Weng, J. Yao, M. He, J. Lv, S. Hua, H. Li, and Y. He. 2017. Chlorophyll fluorescence imaging uncovers photosynthetic fingerprint of citrus Huanglongbing. *Frontiers in Plant Science* 8:1–11.

Chen, X., Y. Li, R. He, Q. Ding, Q. Sun, R. He, and Y. Li. 2017. Phenotyping for the dynamics of field wheat root system architecture for root foraging traits in response to environment×management interactions. *Scientific Reports* 7:37649.

Cramer, G. R., K. Urano, S. Delrot, M. Pezzotti, and K. Shinozaki. 2011. Effects of abiotic stress on plants: A systems biology perspective. *BMC Plant Biology* 11:163.

Farshadfar, E., M. Saeidi, and S. J. Honarmand. 2012. Evaluation of drought tolerance screening techniques among some landraces of bread wheat genotypes. *European Journal of Experimental Biology* 2:1585–1592.

Forieri, I., C. Sticht, M. Reichelt, N. Gretz, M. J. Hawkesford, M. Malagoli, M. Wirtz, and R. Hell. 2017. System analysis of metabolism and the transcriptome in *Arabidopsis thaliana* roots reveals differential co-regulation upon iron, sulfur and potassium deficiency. *Plant, Cell and Environment* 40(1):95–107.

Gangaiah, C., A. A. Ahmad, H. V. Nguyen, and T. J. K. Radovich. 2015. A correlation of rapid Cardy meter sap test and ICP spectrometry of dry tissue for measuring potassium (K+) concentrations in Pak Choi (*Brassica rapa* Chinensis Group). *Communications in Soil Science and Plant Analysis* 47:2046–2052.

Gerland, P., A. E. Raftery, H. Ševcíková et al. 2014. World population stabilization unlikely this century. *Science* 346(6206):234–237.

Gielen, B., K. Vandermeiren, N. Horemans, D. D'Haese, R. Serneels, and R. Valcke. 2006. Chlorophyll a fluorescence imaging of ozone-stressed *Brassica napus* L. plants differing in glucosinolate concentrations. *Plant Biology* 8(5):698–705.

González Pérez, J. L., M. C. Espino-Gudiño, J. Gudiño-Bazaldúa, J. L. Rojas-Rentería, V. Rodríguez-Hernández, and V. M. Castaño. 2013. Color image segmentation using perceptual spaces through applets for determining and preventing diseases in chili peppers. *African Journal of Biotechnology* 12:679–688.

Granier, C., L. Aguirrezabal, K. Chenu et al. 2006. PHENOPSIS, an automated platform for reproducible phenotyping of plant responses to soil water deficit in *Arabidopsis thaliana* permitted the identification of an accession with low sensitivity to soil water deficit. *New Phytologist* 169(3):623–635.

Hillnhütter, C., A. K. Mahlein, R. Sikora, and E. C. Oerke. 2011. Remote sensing to detect plant stress induced by *Heterodera schachtii* and *Rhizoctonia solani* in sugarbeet fields. *Fuel and Energy Abstracts* 22:70–77.

Hoyos-Villegas, V., J. H. Houx, S. K. Singh, and F. B. Fritschi. 2014. Ground-based digital imaging as a tool to assess soybean growth and yield. *Crop Science* 54(4):1756.

Huang, J., H. Liao, Y. Zhu, J. Sun, Q. Sun, and X. Liu. 2012. Hyperspectral detection of rice damaged by rice leaf folder (*Cnaphalocrocis medinalis*). *Journal of Computers and Electronics in Agriculture* 82:100–107.

James, R. A., and X. R. R. Sirault. 2012. Infrared thermography in plant phenotyping for salinity tolerance. *Methods in Molecular Biology* 913:173–189.

Jamil, A., S. Riaz, M. Ashraf, and M. R. Foolad. 2011. Gene expression profiling of plants under salt stress. *Critical Reviews in Plant Sciences* 30(5):435–458.

Jones, H. G., and I. Leinonen. 2009. Thermal imaging for the study of plant water relations. *Journal of Agricultural Meteorology* 59(3):205–217.

Koscielny, C. B., J. Hazebroek, and R. W. Duncan. 2018. Phenotypic and metabolic variation among spring *Brassica napus* genotypes during heat stress. *Crop and Pasture Science* 69(3):284–295.

Kumar, S., J. Kumari, R. Bansal et al. 2018. Multi-environmental evaluation of wheat genotypes for drought tolerance. *Indian Journal of Genetics and Plant Breeding* 78(1):26–35.
Kumawat, K. R., D. K. Gothwal, and D. Singh. 2017. Salinity tolerance of lentil genotypes based on stress tolerance indices. *Journal of Pharmacognosy and Phytochemistry* 6:1368–1372.
Lager, I., O. Andréasson, T. L. Dunbar, E. Andreasson, M. A. Escobar, and A. G. Rasmusson. 2010. Changes in external pH rapidly alter plant gene expression and modulate auxin and elicitor responses. *Plant, Cell and Environment* 33(9):1513–1528.
Li, L., Q. Zhang, and D. Huang. 2014. A review of imaging techniques for plant phenotyping. *Sensors* 14(11):20078–20111.
Mangin, B., P. Casadebaig, E. Cadic et al. 2017. Genetic control of plasticity of oil yield for combined abiotic stresses using a joint approach of crop modelling and genome-wide association. *Plant, Cell and Environment* 40(10):2276–2291.
Mansilla-Córdova, P. J., D. Bampi, N. V. Rondinel-Mendoza, P. C. T. Melo, A. L. Lourenção, and J. A. M. Rezende. 2018. Screening tomato genotypes for resistance and tolerance to Tomato chlorosis virus. *Plant Pathology* 67(5):1231–1237.
Mittler, R. 2006. Abiotic stress, the field environment and stress combination. *Trends in Plant Science* 11(1):15–19.
Negrão, S., S. M. Schmöckel, and M. Tester. 2017. Evaluating physiological responses of plants to salinity stress. *Annals of Botany* 119(1):1–11.
Nelson, P. R., and E. J. Dudewicz. 2002. Exact analysis of means with unequal variances. *Technometrics* 44(2):152–160.
Ohama, N., H. Sato, K. Shinozaki, and K. Yamaguchi-Shinozaki. 2017. Transcriptional regulatory network of plant heat stress response. *Trends in Plant Science* 22(1):53–65.
Rajendran, K., M. Tester, and S. J. Roy. 2009. Quantifying the three main components of salinity tolerance in cereals. *Plant, Cell and Environment* 32(3):237–249.
Ram, B., H. S. Meena, V. V. Singh, B. K. Singh, J. Nanjundan, A. Kumar, S. P. Singh, N. S. Bhogal, and D. Singh. 2014. High temperature stress tolerance in Indian mustard (*Brassica juncea*) germplasm as evaluated by membrane stability index and excised-leaf water loss techniques. *Journal of Oilseed Brassica* 1:149–157.
Raza, S. A., H. K. Smith, G. J. J. Clarkson, G. Taylor, A. J. Thompson, J. Clarkson, and N. M. Rajpoot. 2014. Automatic detection of regions in spinach canopies responding to soil moisture deficit using combined visible and thermal imagery. *PLOS ONE* 9(6):e97612.
Rodriguez, I. R., and G. L. Miller. 2000. Using a chlorophyll meter to determine the chlorophyll concentration, nitrogen concentration, and visual quality of St. Augustinegrass. *HortScience* 35(4):751–754.
Römer, C., M. Wahabzada, A. Ballvora et al. 2012. Early drought stress detection in cereals: Simplex volume maximisation for hyperspectral image analysis. *Functional Plant Biology* 39(11):878–890.
Sajedi, N. A., M. Ferasat, M. Mirzakhani, and M. M. Boojar. 2012. Impact of water deficit stress on biochemical characteristics of safflower cultivars. *Physiology and Molecular Biology of Plants* 18(4):323–329.
Sharma, P., V. Sardana, and S. S. Banga. 2013. Salt tolerance of Indian mustard (*Brassica juncea*) at germination and early seedling growth. *Environmental and Experimental Biology* 11:39–46.
Shi, Y., J. A. Thomasson, S. C. Murray et al. 2016. Unmanned aerial vehicles for high-throughput phenotyping and agronomic research. *PLOS ONE* 11(7):e0159781.
Shinozaki, K., and K. Yamaguchi-Shinozaki. 2007. Gene networks involved in drought stress response and tolerance. *Journal of Experimental Botany* 58(2):221–227.
Siddiqui, M. H., F. Mohammad, and M. N. Khan. 2009. Morphological and physio-biochemical characterization of *Brassica juncea* L. Czern. & Coss. genotypes under salt stress. *Journal of Plant Interactions* 4(1):67–80.

Singh, A., B. Ganapathysubramanian, A. K. Singh, and S. Sarkar. 2016. Machine learning for high-throughput stress phenotyping in plants. *Trends in Plant Science* 21(2):110–124.

Thomas, S., J. Behmann, A. Steier, T. Kraska, O. Muller, U. Rascher, and A. K. Mahlein. 2018. Quantitative assessment of disease severity and rating of barley cultivars based on hyperspectral imaging in a non-invasive, automated phenotyping platform. *Plant Methods* 14:45.

Vij, S., and A. K. Tyagi. 2007. Emerging trends in the functional genomics of the abiotic stress response in crop plants. *Plant Biotechnology Journal* 5(3):361–380.

Waller, D. M., J. Dole, and A. J. Bersch. 2008. Effects of stress and phenotypic variation on inbreeding depression in *Brassica rapa*. *Evolution* 62(4):917–931.

Wang, W., B. Vinocur, and A. Altman. 2003. Plant responses to drought, salinity and extreme temperatures: Towards genetic engineering for stress tolerance. *Planta* 218(1):1–14.

Wang, X., D. Singh, S. Marla, G. Morris, and J. Poland. 2018. Field-based high-throughput phenotyping of plant height in sorghum using different sensing technologies. *Plant Methods* 14:53.

Yang, W., L. Duan, G. Chen, L. Xiong, and Q. Liu. 2013. Plant phenomics and high-throughput phenotyping: Accelerating rice functional genomics using multidisciplinary technologies. *Current Opinion in Plant Biology* 16(2):180–187.

Yao, J., D. Sun, H. Cen, H. Xu, H. Weng, F. Yuan, and Y. He. 2018. Phenotyping of *Arabidopsis* drought stress response using kinetic chlorophyll fluorescence and multicolor fluorescence imaging. *Frontiers in Plant Science* 9:603.

Yendrek, C. R., T. Tomaz, C. M. Montes, Y. Cao, A. M. Morse, P. J. Brown, L. M. McIntyre, A. D. Leakey, and E. A. Ainsworth. 2017. High-throughput phenotyping of maize leaf physiological and biochemical traits using hyperspectral reflectance. *Plant Physiology* 173(1):614–626.

Zou, X., C. Hu, L. Zeng, Y. Cheng, M. Xu, and X. Zhang. 2014. A comparison of screening methods to identify waterlogging tolerance in the field in *Brassica napus* L. during plant ontogeny. *PLOS ONE* 9(3):e89731.

15 The Adoption of Automated Phenotyping by Plant Breeders

Lana Awada, Peter W. B. Phillips, and Stuart J. Smyth

CONTENTS

15.1 INTRODUCTION

The global food system is based primarily on the productivity of a handful of large-scale food crops, few of which are consumed directly, but rather serve as ingredients in some processed food items or are used as feed to raise animals for food. Greater output is forecast to be needed to feed a growing and more prosperous global population (Hunter et al. 2017). This will require greater plant productivity and shifts in agricultural research and development (Pardey et al. 2016). Plant breeding is at the heart of that mission. Traditional and modern breeding programs are being challenged to take up and use a variety of innovations in the breeding process. One place of particular focus is the selection process, whereby breeders select materials for further development, based on phenomic expression. While many breeders have adopted genomics and various advanced genetic breeding or selection processes, for the most part, plant breeders have not exploited many of the opportunities for mechanization or intensive analysis now available, as a result of enhanced sensing, imaging, and phenotype data collection.

A phenotype is the composite of an observable expression of a genome for traits in a given environment. Traits could be visible to the naked eye (conventional phenotype, CP), or visible by using technical procedures. Phenomics – the systematic genome-wide study of an organism's phenotype – is an emerging approach that aims to automate and standardize the phenotyping process by using a wide array of non-invasive and non-destructive imaging and remote sensing techniques, and high-throughput methods of data acquisition and analysis. This approach promises to provide accurate, precise, fast, large-scale, and accumulated data under controlled and varying environmental conditions for plant growth, composition, and performance (Kumar et al. 2015; Singh and Singh 2015). Interdisciplinary collaboration of expertise, including that from biologists, engineers, and computer scientists, is crucial for the implementation of phenomics/automated phenotyping (AP) into practice (Cobb et al. 2013; Kumar et al. 2015).

In plant breeding, AP could be used to screen germplasm collections for desirable traits (forward phenomics) and to dissect traits shown to be of value to reveal their mechanistic basis, including various physiological, biochemical, and biophysical processes, and genes controlling these traits (reverse phenomics) (Kumar et al. 2015; Singh and Singh 2015). The greatest benefit of AP would be achieved if this technology allowed breeders to select superior plants that would otherwise be rejected by using CP methods.

With the deluge of cheap high-throughput genotype data and the rapid developments in plant molecular-based breeding technologies, there is an increased interest in AP as an approach that will provide precise and correspondingly high-throughput phenotypic data to harness the potential of genomic investigations, involving mapping initiatives and training genomic selection (GS) models. This is especially important when breeding for quantitative traits (QT), such as yield and drought tolerance, as these traits usually show continuous phenotype variation, due to their polygenic inheritance and environmental influence, and, thus, need to be repeatedly measured during the life cycle of a plant under multi-environmental conditions (Bassi et al. 2016; Cobb et al. 2013; Desta and Ortiz 2014).*

Despite promising applications, the adoption of AP in plant breeding is still in its infancy (James et al. 2017; Kumar et al. 2015). It is still not clear to many plant breeders if or how much of the generated AP data could be usefully integrated into breeding programs. The enormous volume, diversity, and velocity of imaging and remote-sensing data generated by AP makes it a 'big data' problem. Much work is needed to address issues of ease of access, ease of use, and data management before AP technologies are likely to see widespread uptake in plant breeding. The development of ontology-based big data management is needed to facilitate the integration of metadata to establish genotype, phenotype, and environmental-data-point relationships that create meaningful insights and provide opportunities to enhance the breeding process (International Plant Phenotyping Network 2017; Kumar et al. 2015; Phenospex 2017).

* Traits can be grouped into one of two genetic categories: (1) qualitative traits or (2) quantitative traits (QTs). Qualitative traits are generally governed by one or a few major genes, called oligogenes, where each of these genes produces a large effect on the trait phenotype. Qualitative traits show Mendelian inheritance (high heritability), exhibit discrete/discontinuous phenotypic variation, and the phenotypic expression of oligogenes is generally little affected by the environment. Therefore, as their phenotypes are good indicators of their respective genotype loci, qualitative traits are easy to manipulate and evaluate in a breeding program (Singh and Singh 2015). QTs are governed by many genes, called polygenes. Each of these genes has a small effect on trait phenotype. The effect of each gene is too small to be individually identified and the effects of all polygenes affecting a trait are cumulative. QTs show continuous phenotypic variation (non-discrete) because of the polygenic inheritance and environmental influences, and thus, cannot be grouped into distinct phenotyping classes.

This chapter develops an adoption decision model to assess breeders' attitude towards the adoption of the AP approach; the assumption of breeder homogeneity is relaxed. Breeders' preferences towards technology adoption are assumed to be linked to the characteristics of the technology. Breeders, according to their preferences, choose the technology for which they obtain the highest return or profit. Thus, heterogeneous breeders are assumed to differ in the relative gain or profit they generate from breeding a cultivar under AP or CP. The model focuses on the evolving nature of technologies, assuming that breeders may expect a future technology improvement in AP, which could affect their decisions to adopt AP at present or in the future. Knowing that technology will improve, breeders may have an incentive to wait to adopt AP to mitigate the impacts of costs expended in the use of preliminary AP.

15.2 PLANT BREEDING AND THE NEED FOR AUTOMATED PHENOTYPING

The most common techniques used in plant breeding are presented in Table 15.1. These techniques may involve the introduction of new genetic variation and the identification and tracking of genes for key traits, to achieve greater genetic gain (ΔG) from selection and to accelerate the breeding cycle. Regardless of the technology used, the following general steps are associated with plant breeding: (1) defining the objective(s) of the breeding program: breeding objectives are defined based on factors such as improved farmer and/or processor productivity, new product attributes to satisfy consumer preferences, and improved environmental impacts; (2) developing population or germplasm collections that include the genetic variations of interest; and (3) identifying and selecting individuals with superior characteristics. If successful, selection produces a new population that is phenotypically and genetically different from and superior to the base population. The new population is then used to develop new varieties that eventually, after evaluation and certification, find their way to a farmer's field.

The most common selection methods used in plant breeding are presented in Table 15.2. Traditionally, selection of superior plants involves visual assessment for traits, a process otherwise known as conventional phenotype (CP) selection. In the field, breeders focus on a plant's appearance – they work in the field to visually study plant phenotypic expression under different environmental conditions and to select the type of lines they will use to identify traits that have the potential to improve agronomic features, offer higher yields, or produce specific qualities. However, CP methods are labor and time intensive, environmentally sensitive, and costly. The data collected are frequently subjectively encoded and can vary significantly. In addition, CP has a limited capacity to measure traits on very large genetic populations, particularly those with low heritability or which contain dynamic traits that have phenotypes that change with time and environment and, thus, require to be repeatedly measured during the life cycle of a plant in multi-environment trials.

Recently, advances in genomics technologies (i.e., next-generation sequencing technologies) have provided a better understanding of the genetic basis of a trait and improved the efficiency of selection through the use of marker-based selection

TABLE 15.1
Techniques Used in Plant Breeding

Traditional Techniques	Random Mutation Techniques	Genetically Modifying Techniques
• Emasculation	• Mutagenesis	• Oligonucleotide-directed mutagenesis (ODM)
• Hybridization	• Tissue culture	• Cisgenesis/intragenesis
• Wide crossing	• Haploidy	• Genome editing (e.g., zinc finger nuclease (ZFN)
• Chromosome doubling	• Isozyme markers	• TALENS and CRISPRs
• Chromosome counting	• *In situ* hybridization	• GM rootstock grafting
• Male sterility	• Molecular marker	• RNA-dependent DNA methylation (RdDM)
• Triploidy	• DNA sequencing	• Reverse breeding
• Linkage analysis	• Plant genomic analysis	• Agroinfiltration
• Statistical tools	• Bioinformatics	
	• Microarray analysis	
	• Primer design	
	• Plant transformation	

Sources: Lusser et al. (2011); Acquaah (2012); Benkeblia (2014); Mahesh (2016)

TABLE 15.2
Selection Methods in Plant Breeding

Conventional Selection Methods/ Conventional Phenotype Selection	Molecular Selection Methods/ Marker Selection Methods
• Pedigree selection	• Marker-based selection (MBS)
• Mass selection	• Marker-assisted selection (MAS)
• Pure line selection	• Marker-assisted backcrossing (MABC)
• Bulk population selection	• Marker-assisted recurrent selection (MARS)
• Single-seed descent selection	• Genomic selection (GS) or genome-wide selection (GWS)

Sources: Benkeblia (2014); Singh and Singh (2015)

methods (Table 15.2). The integration of molecular markers and conventional selection methods into the breeding programs can improve the precision of selection and can accelerate the breeding cycle, as markers can be screened before the plant is grown rather than measured in seeds or at the seedling stage. The ability of markers to predict the phenotype of a trait allows breeders to select for multiple generations each year in a greenhouse.*

* Note that the new molecular methods will supplement and extend, but will not replace, conventional breeding. The ultimate test of value of a genotype is its performance in the targeted environment. For instance, although molecular selection, based on molecular markers, can be used in early generations at, say, the F_2 generation, population size at the F_2 is often very large, and thus, it is not efficient to perform molecular selection at this stage. Breeders usually use conventional breeding based on phenotypic selection up until generation F_4, after which they use molecular selection in order to increase the frequency of desirable alleles (Bonnett et al. 2005; Richards et al. 2010).

The genomic selection (GS) or genome-wide selection model is seen as a promising molecular-based selection approach. GS generates marker effects, called genomic estimated breeding values (GEBVs), across the whole genome of a breeding population (BP), based on a statistical model developed in a training population (TP). The training population includes related individuals that have been both genotyped and phenotyped, whereas the breeding population includes the descendants of a TP that have only been genotyped but not phenotyped. GS can consider the effects of all markers spread across the genome, thus capturing more of the genetic variance in additive effects/breeding value. GS does not eliminate phenotyping selection but rather replaces many of the selections with whole-genome predictions. The greater the phenotypic similarity between the true breeding value in the TP and the expected GEBV in the BP, the greater the accuracy of the prediction model. Therefore, precision phenotyping is important for evaluating a training population because the resulting dataset serves as a foundation for the GS, on which to build the accuracy of the statistical models (Cobb et al. 2013; Desta and Ortiz 2014).

However, like conventional selection methods, the application of molecular methods has not measurably improved the selection of QTs. As previously indicated, QTs show continuous phenotypic variation due to the low level of inheritance of polygenes and the influence of environmental factors. These characteristics usually result in a deviation from the genetic variance of additive effects (breeding value) (), that are caused by the genetic variances of dominant effects (σ_D^2) and epistatic<<TNF-CH015_eqn_0001.eps>> effects (σ_P^2), and by the interaction between the genotype and environment ($G \times E$) variance (σ_{GE}^2), that influence their phenotypic expression.

Previous studies of quantitative trait loci (QTL) (e.g., Crossa et al. 2014; Nakaya and Isobe 2012; Singh and Singh 2015; Thomas 2010), suggest that, to overcome $G \times E$, it is important to accumulate and evaluate field phenotypes of QT by planting the mapping populations in replicated trials, conducted over different environments and years. To overcome dominant and epistatic effects, Singh and Singh (2015), Nakaya and Isobe (2012), and Lu et al. (2011) suggested the use of mapping causative loci, such as interval mapping, association mapping/linkage disequilibrium (LD), and genome-wide association studies (GWAS). Mapping causative loci involve phenotypic, genotypic, and pedigree data.

The discussion so far suggests that the application of genomic information may trigger a great leap forward in plant breeding, but only if linked to and integrated with corresponding phenomic information. In light of this challenge, several phenomic facilities (e.g., the Julich Plant Phenotyping Centre, Australian Plant Phenomics, and Canada Plant Phenotyping & Imaging Research Centre (P^{2}IRC)) and networks (e.g., the International Plant Phenotyping Network and the European Plant Phenotyping Network) have been established and are operating at the national and global level. Looking forward, improvements in digital phenotyping technology, computing capacity, and statistical methodology should pave the way to efficiently archive, retrieve, analyze, integrate, and interpret phenomic data. The compelling social science research question is, where will those innovations mostly likely be adopted and used?

15.3 BREEDERS' DECISION TO ADOPT AUTOMATED PLANT PHENOTYPING

15.3.1 Model Assumptions

In this model, we assume that a group of heterogeneous breeders, each of whom is trying to optimize their relative profit function, needs to decide whether to adopt AP or to continue using CP. At time t, a breeder with attribute A has the following profit function:

$$\pi_t^{AP} = R^{AP} - \left(C^{AP} + \delta A\right) - K_t \quad \text{if a cultivar is produced using AP} \tag{15.1}$$

$$\pi_t^{CP} = R^{CP} - C^{CP} \quad \text{if a cultivar is produced using CP} \tag{15.2}$$

where π_t^{AP} and π_t^{CP} are the per unit profits associated with breeding a cultivar using AP or CP, respectively. R^{AP} and R^{CP} are the economic revenue, and C^{AP} and C^{CP} are the costs of breeding a cultivar under AP and CP, respectively. C^{AP} and C^{CP} include the cost of the breeder's activities, input costs, cost of information/data, and costs of quality testing, variety registration, and release under each approach. Parameter A captures heterogeneous breeders' preferences and, thus, differences in their willingness to adopt AP. The sources of heterogeneity can reside in breeders' experience, education, management skills, and information sources, among other attributes. A is assumed to be uniformly distributed, with a unit density $f(A) = 1$ in an interval $A \in \left[0,1\right]$, such that the greater the differentiating A (i.e., A value is closer to 1), the lower is the breeder's preference for AP. The parameter δ is the cost enhancement factor that captures the degree of aversion to AP and is assumed to be a non-negative constant across all breeders. Thus, δA denotes the additional cost that a breeder with attribute A incurs when adopting AP. The term K_t is the sunk cost at time t, an expense that was typically incurred at the time of adoption, and which cannot be recovered once spent (e.g., training to operate the new technology, R&D, and specialized asset costs). Since AP is an emerging technology and many breeders are still uncertain about its integration into breeding programs, there is an option value to waiting before expending ('sinking') the costs of adoption. Thus, K_t can be seen as the loss that a breeder sustains by not waiting until the next period to adopt the technology and the delayed benefits from the subsequent higher rate of improvement.

In this study, the genetic gain (ΔG) is used to determine the revenue, R, from breeding a cultivar. Following Brennan (1989), the economic revenue from breeding a new cultivar of crop i is given by:

$$R_i = YES\left[\Delta G_{iy} W_{iy} + \left(1 + \frac{\Delta G_{iy}}{100}\right)\left(\Delta G_{iq} W_{iq}\right)\right] \tag{15.3}$$

where Y is the mean yield (tonne/ha) before the introduction of the new cultivar, E is the crop i total growing area (ha); and S is the share of area E sown to the new cultivar (the adoption rate of the new cultivar); ΔG_{iy} is the percentage genetic gain in traits affecting yield; W_{iy} is the unit value of ΔG_{iy}; ΔG_{iq} is the percentage genetic gain in traits affecting quality; and W_{iq} is the unit value of ΔG_{iq}.

Regardless of the technique or selection method employed, the genetic gain (ΔG) from selection serves as a universal concept for quantifying improvements in a cultivar. Hence, it is called the breeders' equation. ΔG is the predicted change in the mean value of a trait within a population that happens by selection, and results in the introduction of a new cultivar in characters affecting yield and/or quality. The genetic gain is given by:

$$\Delta G = h^2 \sigma_p i / L \tag{15.4}$$

where h^2 is the heritability parameter and represents the probability that a trait will be transmitted from parents to offspring. h^2 in Equation 15.4 is used in its narrow sense, representing the proportion of phenotypic variation due to additive genetic effects ($h^2 = \sigma_A^2 / \sigma_P^2$). The additive effect/breeding value is important in plant breeding as it represents what is transferred to offspring and can be changed by selection. As previously indicated, QTs are more difficult to breed for because of the large impact of dominance and epistatic effects, which reduce heritability. The term σ_p is the phenotypic variability in the original source population (parental population), which is positively associated with genetic diversity, and affected by the environment and the interactions between genotype and the environment ($G \times E$). The term i is the selection intensity, expressed in units of standard deviation from the mean, and represents the fraction of the current population retained and used as parents to produce the next generation (i.e., if the whole population is retained, i is zero). Finally, the term L is the length of the cycle interval, which is usually one generation (i.e., how quickly a generation can be completed and how many generations are possible per year) (Acquaah 2012; Moose and Mumm 2008).

Automated phenotyping enables breeders to improve ΔG by: (1) estimating h^2 through the use of large-scale selection of phenotype×genotype association data (GWAS) and a large training population in GS, that is phenotyped and genotyped to estimate the breeding value (GEBV). In addition, automated phenotyping can improve h^2 by determining the environmental effects on traits under multiple field environmental conditions (G×E); (2) increasing σ_p by introducing new genes, to result in a larger phenotypic variance, which would provide the breeder with a wide range of variability from which to select, resulting in a higher ΔG; (3) increasing selection intensity, i, by helping breeders to choose a lower proportion of individuals, having a mean superior to the population mean; and (4) shortening cycles (L), as individuals can be selected at an earlier growth stage.

15.3.2 Breeder's Decision Making

A breeder's adoption decision is determined by comparing the profit derived from producing a cultivar under CP and AP, so that the breeder with a differentiating attribute $\hat{A}_t = \dfrac{\left(R^{AP} - R^{CP}\right) - \left[\left(C^{AP} + K_t\right) - C^{CP}\right]}{\delta}$ (found by equating $\pi_t^{AP} = \pi_t^{CP}$) is indifferent between breeding a cultivar under AP or CP (Figure 15.1). Breeders with attributes $A \in \left[0, A_t\right)$ find it optimal to breed under AP, while breeders with attributes $A \in \left(\hat{A}_t, 1\right]$ breed using CP. Given that breeders are uniformly distributed in the interval [0,1], the indifferent breeder, $\hat{A}_t$, determines the share of AP and CP at time t, given by Equations 15.5 and 15.6, respectively.

$$x_t^{AP} = \hat{A}_t = \frac{(R^{AP} - R^{CP}) - \left[\left(C^{AP} + K_t\right) - C^{CP}\right]}{\delta} \tag{15.5}$$

$$x_t^{CP} = 1 - \hat{A}_t = \frac{\delta - (R^{AP} - R^{CP}) - \left[\left(C^{AP} + K_t\right) - C^{CP}\right]}{\delta} \tag{15.6}$$

Equation 15.5 shows that, for AP to have a positive adoption, the following conditions must hold: $R^{AP} > \left(\left(R^{CP} - C^{CP}\right) + \left(C^{AP} + K_t\right)\right)$ and/or $\left(C^{AP} + K_t\right) < \left(\left(R^{AP} - R^{CP}\right) - C^{CP}\right)$. Otherwise, the profit curve of the AP, π_t^{AP}, will lie below the profit curve of the CP, π_t^{CP}, for all A values, and all breeders will not adopt AP but will continue using CP (Figure 15.1).

Figure 15.1 depicts the profit curves, the adoption shares, and aggregate breeder welfare when the revenue, cost, and breeder preference parameters are such that both AP and CP enjoy a positive share. At time t, breeder welfare is given by the area below the π_t^{CP} curve plus the dotted area in Figure 15.1.

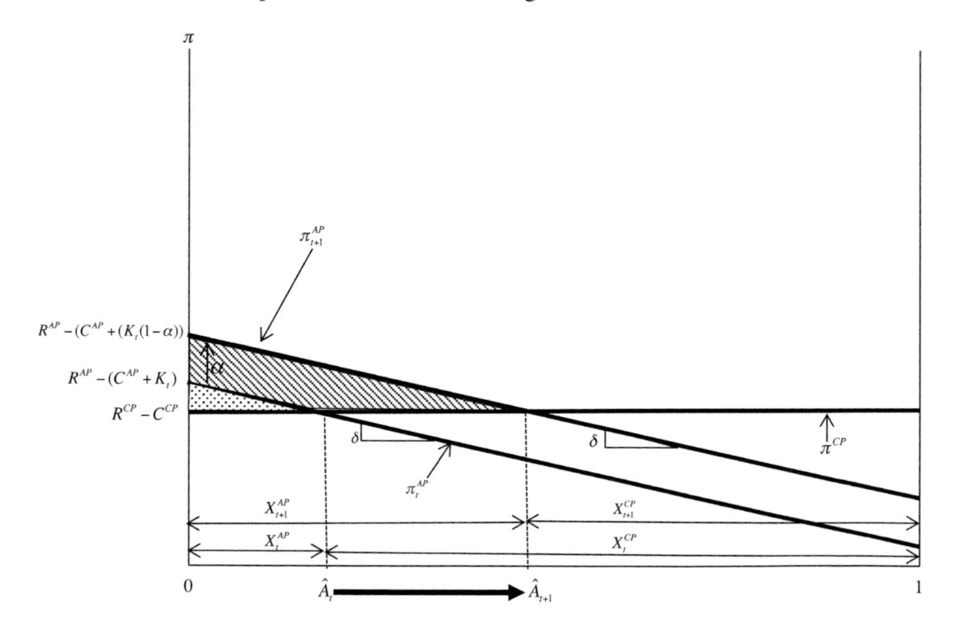

FIGURE 15.1 Breeder's decisions to adopt AP or continue using CP.

At time $t+1$, AP improves, and the sunk cost, K_t, decreases. Assuming that the technology improvement decreases K_t by a constant rate $\alpha \in (0,1)$ (α could address aspects such as: improvement in robotics and automation; big data management; training and education needed; collaborations; etc.) the profit function, *ceteris paribus*, of the breeder with attribute A from the adoption of AP at time $t+1$ is given by:

$$\pi_{t+1}^{AP} = \mathrm{R}^{AP} - \left(C^{AP} + \delta A\right) - K_t\left(1-\alpha\right) \quad \text{if a cultivar is produced using AP} \tag{15.7}$$

At time $t+1$, the indifferent breeder with attribute $\hat{A}_{t+1}$ is given by equating Equations 15.2 and 15.7: $\hat{A}_{t+1} = \dfrac{\left(R^{AP} - R^{CP}\right) - \left[\left(C^{AP} + K_t\left(1-\pm\right)\right) - C^{CP}\right]}{\delta}$.

In Figure 15.1, breeders with attributes $A \in \left[0, \hat{A}_{t+1}\right)$ find it optimal to adopt AP, whereas breeders with attributes $A \in \left(\hat{A}_{t+1}, 1\right]$ continue breeding using CP. The breeder with $\hat{A}_{t+1}$ determines the adoption/share of AP and CP at time $t+1$, given by Equations 15.8 and 15.9, respectively.

$$x_{t+1}^{AP} = \hat{A}_{t+1} = \frac{\left(R^{AP} - R^{CP}\right) - \left[\left(C^{AP} + K_t\left(1-\pm\right)\right) - C^{CP}\right]}{\delta} \tag{15.8}$$

$$x_{t+1}^{CP} = 1 - \hat{A}_{t+1} = \frac{\delta - \left(R^{AP} - R^{CP}\right) - \left[\left(C^{AP} + K_t\left(1-\pm\right)\right) - C^{CP}\right]}{\delta} \tag{15.9}$$

Equation 15.9 shows that the greater the rate of technology improvement (α), the lower the sunk cost, K_t, and, thus, the greater the proportion of breeders who use AP at time $t+1$. This is shown graphically in Figure 15.1 by shifting the profit curve of AP upward from π_t^{AP} to π_{t+1}^{AP}, indicating an increase in the return to breeders from using AP by $K_t(\alpha)$, and thus, an increase in the proportion of breeders who adopted AP by the interval $\left[\hat{A}_t, \hat{A}_{t+1}\right]$. Breeders who were using CP at time t decided to start using AP at time $t+1$ as a result of improvements in AP technology. At time $t+1$, breeder welfare is given by the area below the π_t^{CP} curve plus the dotted and dashed areas in Figure 15.1, indicating an increase in breeder welfare by the dashed area as a result of the technology improvement α.

In addition to a decrease in sunk cost, technology improvement, α, may increase the genetic gain (ΔG) from selection, which, in turn, improves the return, R^{AP}, from the adoption of AP. The result is an upward parallel shift in the profit curve of AP in Figure 15.1, and in an increase in the adoption of AP.

It is clear that the subsequent rate of improvement is an important determinant in advancing the adoption of AP by plant breeders. Progress in developing multidisciplinary technologies that empower AP is moving forward rapidly. Tremendous investments in phenomic projects – Awada et al. (2018) reported that more than 33 plant phenotyping facilities have been developed and seven networks established around the world – are underway to develop standards and ontologies for trait measurement and to facilitate the integration of the large volume of phenotypic data with other big data relevant to plant breeding.

An important parameter in the model is breeder aversion, δ, to AP. Lower breeder aversion (i.e., lower δ), *ceteris paribus,* leads to a decrease in the cost, δA, for all breeders (see Equation 15.1) and, thus, higher AP adoption, x_t^{AP}, (see Equation 15.5). In Figure 15.2, the decrease in δ causes the AP profit curve to rotate counter-clockwise from π_t^{AP} to $\pi_t^{AP'}$, resulting in an increase in the adoption of AP by the segment $\left[\hat{A}_t, \hat{A}_t'\right]$. Breeders' welfare increases by the dashed area in Figure 15.2. Note that, when δ is equal to one, the AP profit curve, π_t^{AP}, lies above the CP profit curve π_t^{CP} in Figure 15.2, and all breeders consider AP to be a better technology than CP.

Clearly, the next step to operationalize the model is to simulate a range of scenarios by collecting data on each of the model parameters. Data on the economic parameters involve the use of economic statistics concerning the collection, processing, compilation, dissemination, and analysis of economic information. However, collecting data on breeders' heterogeneity and their level of aversion to AP is a complex, largely empirical task. The nature and formation of a breeder's behavior involves a psychological process that breeders go through, in recognizing the need for AP, interpreting information, making an adoption decision, and implementing the technology in their programs. The process blends elements from four types of social sciences, namely psychology, sociology, anthropology, and economics. The identification and measurement of these elements imply the use of diverse research

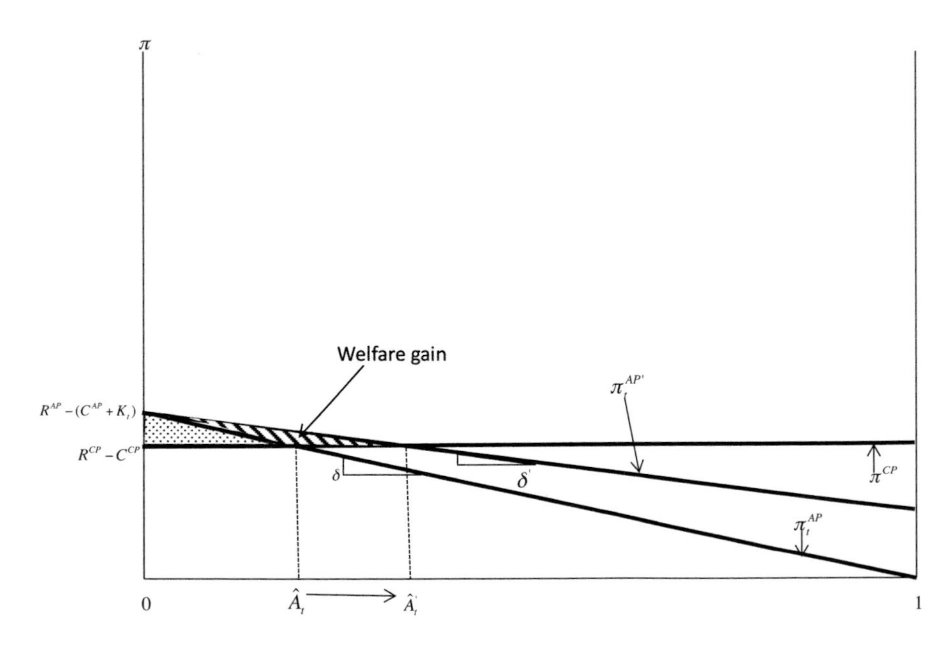

FIGURE 15.2 Impact of a decrease in breeder aversion on the adoption of AP.

methods, including survey research, interviews, statistical analysis, econometrics, social network analysis, case studies, behavioral experiments, and model building, among other approaches.

15.4 CONCLUSION

Some researchers believe that the adoption of the latest high-throughput genomics and phenomics technologies by plant breeders can deliver better new cultivars and accelerate the process of breeding. Whereas in the past two decades, we have witnessed large-scale adoption of genomic technologies, the adoption of automated phenotyping/phenomics by plant breeders is still in its infancy. The question remains as to whether widespread adoption of phenomics can happen. And if it does, under what conditions will it happen?

Adoption of new technologies is always difficult to anticipate. While the literature is rife with examples of innovators capturing first-mover advantage, including lock-in and network effects (Shapiro and Varian 1999), there are many real impediments to realizing that vision. Just because a new method improves output does not assure uptake and use. Sunk costs, individual characteristics, and preferences, and uncertainty about the evolution of the technology are all factors that could limit the adoption and use of such new technology. The emergence of AP emergence looks exciting but the enormous volume, diversity, and velocity of imaging and remote-sensing data generated by AP, and the difficulty of linking these data to genotypic and environmental data could delay the development of automated plant phenotyping. This chapter has explored one way to model this decision space. A theoretical model of heterogeneous breeders is built to analyze breeders' decision-making as they ponder whether to adopt automated phenotyping or to continue using conventional phenotyping. The model focuses on the evolving nature of technologies, which assumes that future R&D will improve the technology, so that breeders may have an incentive to wait to adopt AP to mitigate the impacts of sunk cost. The result of this model indicates that many interlocking factors are at work as breeders decide whether or not to adopt AP. We found that factors, including the expected return, adoption costs, the rate of technology improvement, and breeders' preferences and degree of aversion to AP, can affect the present and future adoption of AP. *A priori*, it is not possible to determine the adoption path for this technology – practical estimation of the model parameters and manipulation of the model is necessary to determine the likely path.

This chapter does not address the impact of the institutional policy framework on the adoption of AP in plant breeding programs. Future research may focus on the role that might be played by regulatory and governance models in facilitating or delaying the adoption and application of the phenomics technology in plant breeding.

REFERENCES

Acquaah, G. 2012. *Principles of Plant Genetics and Breeding*, 2nd ed. Wiley-Blackwell, Oxford.

Awada, L., P. W. B. Phillips, and S. J. Smyth. 2018. The adoption of automated phenotyping by plant breeders. *Euphytica* 214(8):148. doi: 10.1007/s10681-018-2226-z.

Bassi, F. M., A. Bentley, G. Charmet, R. Ortiz, and J. Crossa. 2016. Breeding schemes for the implementation of genomic selection in wheat (Triticum spp.). *Plant Science* 242:23–36. http://www.sciencedirect.com/science/article/pii/S0168945215300534#.

Benkeblia, N. 2014. *Omics Technologies and Crop Improvement*. CRC Press: Taylor & Francis Group. Boca Raton, FL.

Bonnett, D. G., G. J. Rebetzke, and W. Spielmeyer. 2005. Strategies for efficient implementation of molecular markers in wheat breeding. *Molecular Breeding* 15(1):75–85. https://link.springer.com/content/pdf/10.1007%2Fs11032-004-2734-5.pdf.

Brennan, J. P. 1989. An analytical model of a wheat breeding program. *Agricultural Systems* 31(4):349–366.

Cobb, J. N., G. Declerck, A. Greenbrg, R. Clark, and S. McCouch. 2013. Next-generation phenotyping: Requirements and strategies for enhancing our understanding of genotype-phenotype relationships and its relevance to crop improvement. *Theoretical & Applied Genetics* 126(4):867–887. https://link.springer.com/article/10.1007%2Fs00122-013-2066-0.

Crossa, J., P. Pérez, J. Hickey et al. 2014. Genomic prediction in CIMMYT maize and wheat breeding programs. *Heredity* 112(1):48–60. http://repository.cimmyt.org/xmlui/handle/10883/3441?locale-attribute=en.

Desta, Z. A., and R. Ortiz. 2014. Genomic selection: Genome-wide prediction in plant improvement. *Trends in Plant Science* 19(9):592–601. http://www.sciencedirect.com/science/article/pii/S1360138514001411#.

Hunter, M. C., R. G. Smith, M. E. Schipanski, L. W. Atwood, and D. A. Mortensen. 2017. Agriculture in 2050: Recalibrating targets for sustainable intensification. *BioScience* 67(4):386–391.

International Plant Phenotyping Network. 2017. A survey about the status of plant phenotyping: July 2016. http://www.plant-phenotyping.org/ippn-survey_2016.

James, W., J. Edwards, G. McDonald, and H. Kuchel. 2017. Australian Government, GRDC. The application of precision phenotyping technologies to a wheat breeding program. https://grdc.com.au/Research-and-Development/GRDC-Update-Papers/2017/02/The-application-of-precision-phenotyping-technologies-to-a-wheat-breeding-program.

Kumar, J., A. Pratap, and S. Kumar (eds). 2015. *Phenomics in Crop Plants: Trends, Options and Limitations*. Springer, New Delhi.

Lü, H. Y., X. F. Liu, S. P. Wei, and Y. M. Zhang. 2011. Epistatic association mapping in homozygous crop cultivars. *PLOS ONE* 6(3):3. http://journals.plos.org/plosone/article?id=10.1371/journal.pone.0017773.

Lusser, M., C. Parisi, D. Plan, and E. Rodríguez-Cerezo. 2011. *New Plant Breeding Techniques State-Of-The-Art and Prospects for Commercial Development*. Publications Office of the European Union, Luxembourg. http://ftp.jrc.es/EURdoc/JRC63971.pdf.

Mahesh, S. 2016. The state of art of new transgenic techniques in plant breeding: A review. *Journal of Advances in Biology & Biotechnology* 9(4):2394–1081.

Moose, S. P., and R. Mumm. 2008. Molecular plant breeding as the foundation for 21st century crop improvement. *Plant Physiology* 147(3):969–977.

Nakaya, A., and S. N. Isobe. 2012. Will genomic selection be a practical method for plant breeding? *Annals of Botany* 110(6):1303–1316, https://academic.oup.com/aob/article-lookup/doi/10.1093/aob/mcs109.

Pardey, P., C. Chan-Kang, S. P. Dehmer, and J. M. Beddow. 2016. Agriculture R&D is on the move. *Nature* 537(7620):301–303.

Phenospex. 2017. How far are we from the 100$ phenome? https://phenospex.com/blog/how-far-are-we-from-the-100-phenome/.

Richards, A. R., G. J. Rebetzke, M. Watt, A. G. Condon, W. Spielmeyer, and R. Dolferus. 2010. Breeding for improved water productivity in temperate cereals: Phenotyping, quantitative trait loci, markers and the selection environment. *Functional Plant Biology* 37(2):85–97.

Shapiro, C. and H. R. Varian. 1999. *Information Rules: A Strategic Guide to the Network Economy*. Harvard Business School Press. Boston, MA.

Singh, B. D., and A. K. Singh. 2015. *Marker-Assisted Plant Breeding: Principles and Practices*. Springer, New Delhi.

Thomas, D. 2010. Gene-environment-wide association studies: Emerging approaches. *Nature Reviews Genetics* 11(4):259–272.

Index

G

H

I

N

O

P

Q

R

S

T

U

V

W